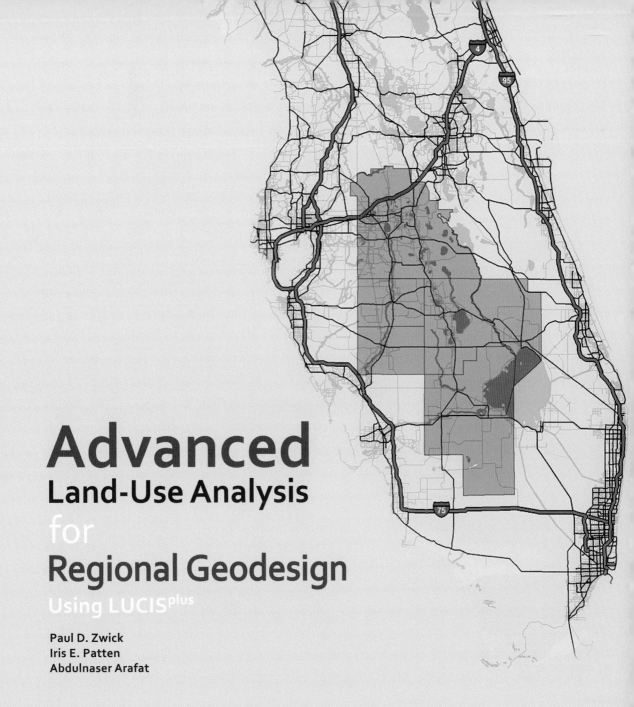

Advanced
Land-Use Analysis
for
Regional Geodesign
Using LUCIS^{plus}

Paul D. Zwick
Iris E. Patten
Abdulnaser Arafat

Esri Press
REDLANDS | CALIFORNIA

Cover image Paul D. Zwick and the Arizona Board of Regents on behalf of the University of Arizona. Water bodies from the US Geological Survey; state boundary from the US Census Bureau-TIGER/Line Files; interstates and highways from the Florida Department of Transportation; conservation areas from the Florida Natural Areas Inventory; railroads from the Federal Railroad Administration.

Esri Press, 380 New York Street, Redlands, California 92373-8100
Copyright © 2016 Esri
All rights reserved. First edition 2016

Printed in the United States of America
20 19 18 17 16   1 2 3 4 5 6 7 8 9 10

Library of Congress Cataloging-in-Publication Data

Zwick, Paul Dean, 1946- Advanced land-use analysis for regional geodesign : using LUCISplus / Paul D. Zwick, Iris E. Patten, and Abdulnaser Arafat. -- First Edition.
    pages cm
  Includes bibliographical references and index.
  ISBN 978-1-58948-389-7 (pbk. : alk. paper) — ISBN 978-1-58948-433-7 (electronic) 1. Land use--Mathematical models. I. Patten, Iris Elaine, 1979- II. Arafat, Abdulnaser Amin, 1965- III. Title.
  HD108.4.Z95 2015
  333.73'13--dc23
  2015019328

The information contained in this document is the exclusive property of Esri unless otherwise noted. This work is protected under United States copyright law and the copyright laws of the given countries of origin and applicable international laws, treaties, and/or conventions. No part of this work may be reproduced or transmitted in any form or by any means, electronic or mechanical, including photocopying or recording, or by any information storage or retrieval system, except as expressly permitted in writing by Esri. All requests should be sent to Attention: Contracts and Legal Services Manager, Esri, 380 New York Street, Redlands, California 92373-8100, USA.
The information contained in this document is subject to change without notice.

US Government Restricted/Limited Rights: Any software, documentation, and/or data delivered hereunder is subject to the terms of the License Agreement. The commercial license rights in the License Agreement strictly govern Licensee's use, reproduction, or disclosure of the software, data, and documentation. In no event shall the US Government acquire greater than RESTRICTED/LIMITED RIGHTS. At a minimum, use, duplication, or disclosure by the US Government is subject to restrictions as set forth in FAR §52.227-14 Alternates I, II, and III (DEC 2007); FAR §52.227-19(b) (DEC 2007) and/or FAR §12.211/12.212 (Commercial Technical Data/Computer Software); and DFARS §252.227-7015 (DEC 2011) (Technical Data – Commercial Items) and/or DFARS §227.7202 (Commercial Computer Software and Commercial Computer Software Documentation), as applicable. Contractor/Manufacturer is Esri, 380 New York Street, Redlands, CA 92373-8100, USA.

@esri.com, 3D Analyst, ACORN, Address Coder, ADF, AML, ArcAtlas, ArcCAD, ArcCatalog, ArcCOGO, ArcData, ArcDoc, ArcEdit, ArcEditor, ArcEurope, ArcExplorer, ArcExpress, ArcGIS, arcgis.com, ArcGlobe, ArcGrid, ArcIMS, ARC/INFO, ArcInfo, ArcInfo Librarian, ArcLessons, ArcLocation, ArcLogistics, ArcMap, ArcNetwork, *ArcNews*, ArcObjects, ArcOpen, ArcPad, ArcPlot, ArcPress, ArcPy, ArcReader, ArcScan, ArcScene, ArcSchool, ArcScripts, ArcSDE, ArcSdl, ArcSketch, ArcStorm, ArcSurvey, ArcTIN, ArcToolbox, ArcTools, ArcUSA, *ArcUser*, ArcView, ArcVoyager, *ArcWatch*, ArcWeb, ArcWorld, ArcXML, Atlas GIS, AtlasWare, Avenue, BAO, Business Analyst, Business Analyst Online, BusinessMAP, CityEngine, CommunityInfo, Database Integrator, DBI Kit, EDN, Esri, esri.com, Esri—Team GIS, Esri—*The GIS Company*, Esri—The GIS People, Esri—The GIS Software Leader, FormEdit, GeoCollector, Geographic Design System, Geography Matters, Geography Network, geographynetwork.com, Geoloqi, Geotrigger, GIS by Esri, gis.com, GISData Server, GIS Day, gisday.com, GIS for Everyone, JTX, MapIt, Maplex, MapObjects, MapStudio, ModelBuilder, MOLE, MPS—Atlas, PLTS, Rent-a-Tech, SDE, SML, Sourcebook•America, SpatiaLABS, Spatial Database Engine, StreetMap, Tapestry, the ARC/INFO logo, the ArcGIS Explorer logo, the ArcGIS logo, the ArcPad logo, the Esri globe logo, the Esri Press logo, The Geographic Advantage, The Geographic Approach, the GIS Day logo, the MapIt logo, The World's Leading Desktop GIS, *Water Writes*, and Your Personal Geographic Information System are trademarks, service marks, or registered marks of Esri in the United States, the European Community, or certain other jurisdictions. CityEngine is a registered trademark of Procedural AG and is distributed under license by Esri. Other companies and products or services mentioned herein may be trademarks, service marks, or registered marks of their respective mark owners.

Ask for Esri Press titles at your local bookstore or order by calling 800-447-9778, or shop online at esri.com/esripress. Outside the United States, contact your local Esri distributor or shop online at eurospanbookstore.com/esri.

Esri Press titles are distributed to the trade by the following:

In North America:
Ingram Publisher Services
Toll-free telephone: 800-648-3104
Toll-free fax: 800-838-1149
E-mail: customerservice@ingrampublisherservices.com

In the United Kingdom, Europe, Middle East and Africa, Asia, and Australia:
Eurospan Group
3 Henrietta Street
London WC2E 8LU
United Kingdom
Telephone: 44(0) 1767 604972
Fax: 44(0) 1767 601640
E-mail: eurospan@turpin-distribution.com

Contents

Foreword — xi
Acknowledgments — xiii

Chapter 1 Introduction — 1
    The paradigm shift — 5
    Where do we go from here? — 6
    What this book covers — 7
    Locations mentioned in this book — 9

Part I Foundations

Chapter 2 Conflict analysis as a decision-making tool — 17
    What this chapter covers — 17
    The five-step LUCIS process: A review — 18
    Reviewing the concept of LUCIS land-use conflict — 25
    LUCIS greenfield conflict and mixed-use opportunity for land-use decisions — 32
    Using the ArcGIS Combine tool to create a multivariable raster for land-use decision-making — 36
    Chapter summary — 44

Chapter 3 Suitability automation tools: A4 and LUCIS^{plus} — 45
    What this chapter covers — 45
    Land-use modeling as automated sequential procedures — 46
    LUCIS^{plus} automation tools — 48
    Chapter summary — 70

Chapter 4 Analyzing and mapping residential land-use futures — 73
    What this chapter covers — 73
    Five simple methods for population projection — 73
    LUCIS^{plus} basic residential analysis of Hillsborough County, Florida — 80

Hillsborough County residential suitability	84
LUCIS^plus residential allocation using the criteria evaluation matrix	88
Summarizing the data	109
Chapter summary	112
Chapter 5 Analyzing and mapping employment land-use futures	113
What this chapter covers	113
Understanding employment	113
Economic base analysis: Getting the basic import and export employment	115
Using economic base location quotients to investigate the shifts and shares within local employment	118
Using shifts and shares by employment sector as a projection technique	119
LUCIS^plus employment land-use allocation for Hillsborough County, Florida	123
Chapter summary	148

Part II Land-use analysis and alternative futures

Chapter 6 Identifying and mapping an alternative urban mixed-use opportunity	151
What this chapter covers	151
Identifying alternative land-use allocations to reduce trend sprawl	152
Exploring increased density as an alternative land-use allocation concept	152
Chapter summary	179
Chapter 7 Analyzing and mapping conservation and agriculture preservation and protection	181
What this chapter covers	181
Criteria to be evaluated	181
Acquisition and land-use options for conservation or agriculture preservation	183
Green infrastructure and land-use planning	184
LUCIS ecological significance goals	187

Using LUCIS to identify conservation opportunities in the landscape	189
Agriculture preservation and land-use planning	193
LUCIS agriculture goals	196
Using LUCIS to identify agriculture lands for the preservation or protection of agricultural productivity and scenic views	198
High-hazard flood areas, connectivity, agriculture, and conservation protection	201
Conservation and agriculture preservation conclusions	203
Chapter summary	206
Chapter 8 Summarizing LUCIS land-use results	209
What this chapter covers	209
Criteria used in selection queries	210
Summarizing LUCIS suitability and conflict	211
Chapter summary	226

Part III Advanced allocation techniques using LUCIS^{plus}

Chapter 9 Analyzing and mapping land use for natural disasters: Hurricane storm surge and sea level rise	229
What this chapter covers	229
A few basics about hurricanes	230
The basic storm surge	232
The Hazus-MH coastal flood model	235
Hazus-MH coastal flood definitions	236
Sea level rise—the problem	241
Level 1 Hazus-MH analysis of a 100-year base storm, with and without SLR impacts	241
2006 property parcel-based level 1 Hazus-MH analysis of a 100-year base storm, with and without SLR impacts	248
Using LUCIS to assess future land-use implications, with and without SLR impacts	251
Chapter summary	257

Chapter 10 Providing LUCIS maps and information to users, clients, and the general public — 259
  What this chapter covers — 259
  Creating ArcGIS for Server services — 260
  Creating the Flex Viewer application — 264
  Using the viewer application — 275
  Chapter summary — 281
Chapter 11 Applications of LUCIS on tribal lands — 283
  What this chapter covers — 283
  The planning process — 284
  Chapter summary — 292
Chapter 12 Analyzing and mapping affordable-housing alternatives — 293
  What this chapter covers — 293
  The housing suitability model — 294
  Physical and neighborhood characteristics preference layer — 296
  Travel cost preference layer — 297
  Demand preference layer — 299
  Transit access preference layer — 301
  Access-driving-demand-transit opportunity layer — 304
  Identifying affordable-housing locations — 305
  Using CEM to allocate affordable housing — 306
  Creating the CEM raster layer for the allocation — 308
  Simple scoring procedures using CEM — 310
  Advanced allocation and scoring procedures using CEM and scenario tables — 311
  Chapter summary — 312
Chapter 13 Advanced A4 tools for accessibility — 315
  What this chapter covers — 315
  Distance estimation methods — 316
  Accessibility measurement — 317
  LUCIS^{plus} tools for estimating accessibility — 321
  Tool validation and estimation accuracy — 330
  Tool limitations — 330
  Chapter summary — 332

Chapter 14 Analyzing and mapping transportation accessibility
and land use 335
  What this chapter covers 335
  Using network distance to replace Euclidean distance 336
  Using the A4 Network Distance tool to estimate accessibility by
    transit system 340
  Proximity of origins to transit stops 341
  Estimating transit access to destinations by frequency/trip length 342
  Using network opportunity to replace point density 344
  Using network gravity access as a suitability layer 345
  The impacts of accessibility on transportation and land-use planning 347
  Chapter summary 350

Index 353

Foreword

Like the authors, I am excited by the emergence of geodesign as a GIS-based process for land planning and design. The potential of geodesign to provide enlightened solutions to many of the problems facing humanity heightens my excitement. I believe that geodesign can completely change our understanding of the design process. It allows us to analyze and simulate scenarios before we implement them, giving us the ability to enhance the quality of our solutions and avoid unintended consequences. This book uses scenarios to explore the Land-Use Conflict Identification Strategy (LUCIS) for regional geodesign.

Like Esri president Jack Dangermond, I have known one of the authors, Paul D. Zwick, for over 20 years. We first met at an Esri User Conference in Palm Springs, California. Zwick was a researcher at the University of Florida's GeoPlan Center. Since those early days, Zwick has taken on leadership in the Department of Urban and Regional Planning and as an associate dean for the College of Design, Construction, and Planning. Throughout his years as an administrator, he has continued his passion for the use of GIS in urban planning, and the GeoPlan Center has become one of the premier spatial analysis centers in the country.

Zwick coauthored his first book, *Smart Land-Use Analysis: The LUCIS Model* (Esri Press, 2007), with Margaret H. Carr, a landscape architect professor at the University of Florida. Through their development of LUCIS, they have brought the power of GIS analysis into the world of land use as a technological modeling process. Their work shows that this process is replicable, incorporates stakeholder preferences, and acknowledges the conflict that often exists among stakeholders in land-use decisions. With this new book, Zwick and coauthors Iris E. Patten and Abdulnaser Arafat dramatically expand the capability of LUCIS for geodesign.

Patten and Arafat, both trained by Zwick and Carr, exemplify the best of what can come from a mentoring relationship. Both Patten and Arafat have interests in planning. Patten uses GIS as a tool in growth management decision-making and collaborative GIS, and Arafat uses GIS as a tool in transportation planning and affordable housing. In this book and through their own work, they demonstrate the flexibility of LUCIS and its wide-ranging applications.

All three authors have developed an interesting approach to land-use allocation that weaves the LUCIS conflict identification strategy into a criteria evaluation matrix (CEM). The use of the LUCIS CEM allows multiple professionals, including GIS analysts, planners, landscape architects, engineers, architects, and environmental scientists, to merge their professional experiences into a flexible land-use allocation process. Using CEM supports both bottom-up and top-down decision-making that is directed by geodesign teams or stakeholders. Professionals can jointly develop scenarios using CEM to test policy and its impacts on the neighborhood, city/town, county, or regional urban form. The expansion of customized LUCIS tools through LUCIS^{plus}, which makes use of planning land-use scenarios (plus), automates the scenario-building process. LUCIS^{plus} enables more efficient suitability and allocation procedures.

This book's concentration on LUCIS^{plus} provides a notable update to LUCIS. I look forward to what lies ahead!

Bill Miller
Director, Geodesign Services, Esri

Acknowledgments

First, we acknowledge our friend and artistic contributor, Wanyi Song, a research specialist at the University of Arizona, for her wonderful development of the graphics in this book. We also thank our undergraduate and graduate students for their dedication and creative use of the Land-Use Conflict Identification Strategy (LUCIS) process. Students at the University of Florida and the University of Arizona use LUCIS and LUCIS^{plus} in research and graduate studies with the authors. Their work helps contribute to the success of the land-use analysis, population, and employment allocations that are the heart and soul of LUCIS^{plus}.

We are grateful for the expertise and dedication of staff at the University of Florida's Geo-Facilities Planning and Information Research Center (also known as the GeoPlan Center) for their continued development of the Florida Geographic Data Library (FGDL). Their efforts on behalf of FGDL make the use of consistent statewide data in Florida a reality. We would like to thank Associate Director Alexis Thomas in particular.

We also appreciate and acknowledge the staff at the Shimberg Center for Housing Studies at the University of Florida for their work in the development of the housing suitability model and support of our research work.

We also acknowledge Professor Margaret H. Carr for her original contribution to LUCIS as coauthor of *Smart Land-Use Analysis: The LUCIS Model* (Esri Press, 2007). And we would like to thank Danny Downing, a researcher in the GeoPlan Center, for his contribution to chapter 10 in our current book. Downing, who has degrees in liberal arts and urban planning, has many years of GIS experience working with data management and web GIS.

The three of us are indebted to our families and loved ones for their patience, support, and thoughtfulness during the many hours of solitary and collaborative work we have put into this book. Without them, we could not have seen it through.

Paul D. Zwick
Iris E. Patten
Abdulnaser Arafat

Chapter 1

Introduction

Iris E. Patten

Jane Jacobs. Ian McHarg. Daniel Burnham. Kevin Lynch. When you consider notable names in planning and environmental design, these names are often at the top of the list. Collectively, these individuals, along with many others, laid the foundation for sound land-use planning through their theories and applications. Their contributions to the discipline have lasted generations. And the expectation is that the discipline will continue to grow, as critical planning theory, applied planning, and technology become more integrated. In turn, new planning and land-use challenges will lead to increased creativity in developing better communities, regions, and environments.

In 2007, Margaret H. Carr and Paul D. Zwick wrote *Smart Land-Use Analysis: The LUCIS Model*. It documents the challenges of "Mrs. Smith" and her community in north central Florida, which undergoes dynamic change in accommodating added development because of significant population increases. The book describes how the Land-Use Conflict Identification Strategy (LUCIS) uses geographic information systems (GIS) to illustrate opportunity for three land-use stakeholders—agriculture, conservation, and urban. In addition to identifying land-use opportunity, the first book also describes conflict in allocating projected population increases. It explains how to use the LUCIS conflict values to prioritize the lands to accommodate expected demand. Allocation procedures discussed in the first book led to various challenges. The processes of constraining population growth to certain physical boundaries, such as specific counties, often proved cumbersome. Without oversimplifying or diminishing the methodology described in *The LUCIS Model*, it is a seminal work on a simple land-use allocation procedure. The book did not address common impacts to suitability or planning, such as affordable housing and transportation accessibility, planning in low-density areas, and variable density allocations. It focused on a theoretical method, yet it encouraged more responsible decision-making through the identification of conflict among land-use stakeholders.

Since the first book, *visioning* and *scenario planning* have become the norm as many communities now review the long-term results of early growth management policy. Regardless of the term, the principle is that citizens and the community define the core of future growth priorities. With a focus on community definition comes a responsibility for jurisdictions to address the wide-ranging needs of their constituents. As a result, communities must assess all the planning impacts, which were not addressed in the first book. An extra challenge for many communities is how to integrate personal or emotional values into the analysis. The process that is now popularly termed *collaborative GIS* is the balance between analyzing the science and integrating community values. The first book addressed this balance by using T. L. Saaty's (2008) analytic hierarchy process (AHP), and then using the resulting weights as the inputs for goals and objectives in the LUCIS model.

This book, *Advanced Land-Use Analysis for Regional Geodesign: Using LUCIS^{plus}*, fills in the gaps from *The LUCIS Model*. It uses new automation tools, new methodologies, and expanded applications of LUCIS, including more complex planning land-use scenarios (the "plus" in LUCIS^{plus}). During the initial development of this book, we, the authors, discussed (and even wrote) an introduction describing the plight of Mrs. Smith since the advent of the first book. It would document how LUCIS changed the planning process in her Florida community and the evolving needs that led to improvements in the LUCIS process. We realized that continuing Mrs. Smith's story would underscore the value of what could be considered an expansion of traditional LUCIS, or LUCIS 2.0. Mrs. Smith's story and location demonstrate what is occurring in many communities. But since the first publication, land-use planning has evolved to include the different topics and scenarios that planners face today. It is important to balance employment, housing options, and transportation in land-use decisions. Additionally, community or regional planning is not a one-size-fits-all format. Variable densities and adaptive design can accommodate varied land-use goals and preserve community character. The concept of threats to land use, stemming from conflicts, has expanded to include the impacts of natural disasters and the elements that only Mother Nature can predict. Further, many underserved communities, especially Native American tribes, are expanding the integration of GIS in their decision-making process. The tribes are using GIS as they strive to achieve the balance between land, water, air, and tradition in the cultural foundation that has sustained them for generations.

Over the next 40 years, almost 129 million new people will live in the United States. Considering the current rate of land consumption, it will take more than 51 million acres to accommodate the projected population (Nelson 2009). In most urban areas, population growth contributes between 97 and 100 percent of sprawl, whereas an increase in per capita land consumption accounts for less than 3 percent (NumbersUSA Action 2009). Why does this matter? Since *The LUCIS Model* was originally published, the United States experienced a notable recession. A significant impact of this recession is a greater focus on balancing inefficiencies related to the costs associated with space and productivity. Since the end of the recession, communities have emphasized creating better-developed connections between housing, economic development, employment, and transportation. Considering the anticipated degree of population growth, the challenges of balancing these connections will remain into the foreseeable future. As a result, more tools that illustrate these connections have emerged, and these tools are now more available to the public as well as to scientists and decision-makers.

The recession in 2008 changed the future of development, probably more so than the introduction of the automobile. It effectively halted the 50-year expansion of suburban development patterns. According to the National Association of Realtors (2007), from 2004 to 2007, the size of the typical home purchased increased by about 100 square feet, and over four-fifths of the homes purchased were detached, single-family homes. More than half of those homes were in the suburbs. Yet the recession helped turn the expansion of the 50 years preceding 2007 into the contraction of the new millennium, according to a June 2009 Urban Land Institute report. New suburban demographics, rising transportation costs, and infrastructure investments played a role. "Shrinking percentages of households with children and a growing market for multiunit housing in the suburbs, an aging population, continued suburban job growth, regional growth patterns that have given leapfrogged suburban areas a new centrality, higher gasoline prices that have made closer-in living more attractive, and local smart-growth policies and transit investments that are limiting sprawl and redirecting growth to existing infrastructure" (Dunham-Jones and Williamson 2009, 42) all contribute to the new paradigm of planning and development. Convenience and connectivity influence neighborhood choice now more than ever before (table 1.1).

Although *sustainability* and *smart growth* are common terms used to justify the new paradigm of more responsible development, it is more an issue of city viability. Market forces and policies now recognize that correcting unhealthy suburban patterns and the performance of streets, blocks, and lots plays a significant role in diversifying people and industry. Directing new development toward existing urban areas reduces the cost of public services and infrastructure more than by developing in open spaces. That is, it is less expensive to develop in vacant (infill) or previously developed (redevelopment) areas than in previously undeveloped (greenfield) open space or agricultural land. According to Arthur C. Nelson, a researcher in metropolitan development patterns, approximately "2.8 million acres (1.1 million ha) of 'greyfields' will become available in the next 15 years. If only one quarter is redeveloped into mixed-use centers, it has the potential to supply half the housing required by 2030" (Dunham-Jones and Williamson 2009, 44). Besides adjusting spatial patterns, mixed uses of development also significantly contribute to quality of life. Sprawled development patterns account for the largest per capita carbon footprints among any type of development (40). Increased residential density; walkability of neighborhoods and communities; and a denser, more cohesive mix of commercial, residential, and recreational uses reduces greenhouse emissions, stimulates economic development, reduces transportation costs, and improves public health (table 1.2).

Table 1.1 Factors influencing neighborhood choice, by location

	BUYERS WHO PURCHASED A HOME IN A ...					
	All Buyers	Suburb/ Subdivision	Small Town	Urban/ Central City	Rural Area	Resort/ Recreation Area
Quality of the neighborhood	63%	69%	57%	64%	45%	51%
Convenient to job	48	49	48	58	37	17
Overall affordability of homes	40	41	41	42	30	26
Convenient to friends/family	38	40	38	36	33	27
Quality of school district	29	35	28	19	21	8
Design of neighborhood	28	31	26	26	21	40
Convenient to shopping	26	29	23	29	15	20
Convenient to schools	22	26	20	20	18	5
Convenient to entertainment/ leisure activities	20	20	14	32	9	30
Convenient to parks/ recreational facilities	19	19	17	27	10	25
Availability of larger lots or acreage	17	13	19	8	50	9
Convenient to health facilities	11	11	12	12	8	22
Home in a planned community	9	10	7	4	5	35
Convenient to public transportation	7	5	2	20	2	2
Convenient to airport	6	6	3	7	5	15
Other	5	4	7	5	7	9

Source: National Association of Realtors 2013.

Table 1.2. Benefits of smart growth

Economic	Social	Environmental
Reduced development costs.	Improved transport options and mobility, particularly for nondrivers.	Green space and habitat preservation.
Reduced public service costs.	Improved housing options.	Reduced air pollution.
Reduced transportation costs.	Community cohesion.	Increased energy efficiency.
Economies of agglomeration.	Preserves unique cultural resources (historic sites, traditional neighborhoods, etc.).	Reduced water pollution.
More efficient transportation.	Increased physical exercise and health.	Reduced "heat island" effect.
Supports industries that depend on high-quality environments (tourism, farming, etc.).		

Source: Victoria Transport Policy Institute 2010.

The paradigm shift

Historic zoning trends encouraged segregated land uses based on the compatibility of their functions: industrial, commercial, residential, institutional, and so forth. In addition to considering specific land uses independently, early development lacked intergovernmental and interregional coordination. History and experience have shown that early trends significantly contributed to the inefficient patterns of land use and methods to transport goods and people. Local governments are now scrambling to reorganize. Smart growth alone will not resolve this problem. Just as the neighborhoods during the Industrial Revolution were built to accommodate automobiles, the neighborhoods of the future must recognize that land use and mass-transit infrastructure are not mutually exclusive. Using transit as a way to enhance or build communities is a new perspective reshaping regional growth. Cities across the country such as Charlotte, North Carolina, and Portland, Oregon, are using transit to deal with the age-old problems of urban decline and traffic congestion. Installing rail lines is not the solution for every community. Yet "transit-oriented development (TOD) principles can serve as a checklist for the development of pedestrian-scale communities that will be suitable for public transportation, either now or in the future. The principles will also be useful for transit agencies and others engaged in new transit projects, to ensure that nearby development will generate sufficient numbers of riders to support transit, and that transit will indeed enhance the community" (Dunphy et al. 2003, v–vi).

Transit is one tool used to support place-making. What is place-making? The Project for Public Spaces (2009) defines *place-making* as "not just the act of building or fixing up a space, but a whole process that fosters the creation of vital public destinations: the kind of places where people feel a strong stake in their communities and a commitment to making things better. Simply put, place-making capitalizes on a local community's assets, inspirations, and potential, ultimately creating good public spaces that promote people's health, happiness, and well-being."

The greatest difficulty in implementing or operating a transit system is the significant amount of public investment needed and the challenge to achieve profitability. For these reasons, such a system's financial justification is rooted in maximizing the benefit to the community and stakeholders. For municipalities, increasing development around transit stations is the most efficient way to achieve financial feasibility and the needed ridership to support transit activities. A transit system and its surrounding TOD lead to opportunities generated by higher densities, mixed product types appealing to a broader spectrum of incomes, and the creation of quality places with greater aesthetic value. Planning for areas around transit is a dynamic exercise in economic development, fiscal feasibility, and place-making.

Where do we go from here?

Urban planning, specifically land-use planning, is a complex institution. Planning is not a one-size-fits-all practice. Development patterns, the efficacy of transit networks, and government policy are all transaction costs that affect the quality of land-use decisions. Although transaction costs are typically considered in terms of the "costs of running the economic system" (Arrow 1969, 48), transaction costs in land-use modeling refer to the aggregation of data that identifies intrinsic suitability. Therefore, "transaction costs are the costs that are made to increase the information available to us and to reduce uncertainty" (Buitelaar 2007, 30). Scenario modeling is a tool that allows the public to measure development outcomes, considering the conditions bounding the model. The LUCIS strategy is a simplified analytical land-use model that captures the essential features of stakeholder groups and identifies conflict between them.

LUCIS is a goal-driven GIS model that produces a spatial representation of where land-use suitabilities are in conflict or where one land-use type is clearly more suitable than another (Carr and Zwick 2007). The foundation of the LUCIS strategy is a set of hierarchical goals and objectives for each suitability group. These goals and objectives ultimately become suitability criteria. The goals and objectives are combined spatially, using weights to indicate preferences for each stakeholder group (agriculture, conservation, and urban), resulting in the final LUCIS conflict layer. LUCIS stops short of representing alternative futures. Instead, it compares the results of three suitability analyses designed to capture the inherent biases in the motivations of three stakeholder groups: conservationists, developers, and farmers and ranchers (Carr and Zwick 2007).

Reallocating resources is essential to better managing future population growth. Regions and communities must balance future growth among infill, redevelopment, and greenfield areas to responsibly achieve regional density goals. In 2003, then-Milwaukee mayor John Norquist suggested that "most of the development in the United States, 90 percent or something like that, is new development on the edge. If we ignore that and just concentrate on infill, the edge city will never repair itself" (Heid 2004, 2). Infill and redevelopment provide significant benefits, including decreased vehicle miles traveled (VMT), lower emissions, saved open land, and decreased infrastructure costs. The problem with relying solely on infill and redevelopment to handle increased population and increase city efficiency is that creating an appropriate environment for infill and redevelopment takes time. In many communities, the lack of available infill sites is the primary barrier to an infill-dependent development plan. Other obstacles include time and costs associated with updating existing infrastructure as well as the possible inability to support affordable housing because of the land prices in close-in locations. The goal of this book is to introduce a comprehensive GIS modeling strategy for making complex land-use, employment, and residential allocations. This book addresses not only the need for analysis to identify more complex land-use relationships, but also the increased demand for tools to aid land-use mapping and analysis.

What this book covers

Understanding the changes and new challenges in land-use planning is as significant as improving the methods used to identify, measure, and evaluate land-use opportunities. Since the first book, we have added, modified, and increased the scope of the LUCIS mission. This book does not teach you the mechanics of suitability or conflict, but how to practically solve real-world land-use problems. It explores the distribution of population and employment, which furthers the new mission of LUCIS, known as LUCIS^{plus}. Fundamentally, the LUCIS process is a sound method to identify conflict between stakeholders. But as mentioned previously, the original method could not efficiently model microclimates of policy and community input. LUCIS has always been described as a what-if land-use scenario model. Now the methods and theory for measuring the what-if are more efficient and widely applicable than ever before. This book introduces LUCIS^{plus}, a planning land-use scenario approach. LUCIS^{plus} allows users to indicate custom policy options outside the standard ArcGIS suitability models that influence population and employment allocation. LUCIS^{plus} achieves this customization through a user-generated table known as a criteria evaluation matrix (CEM), based on a complex raster layer and described in chapter 2. LUCIS^{plus} also introduces three new custom ArcGIS toolsets for automating LUCIS analysis functions and improving the calculation of land-use goals.

Geodesign, although old in practice, is relatively new in terminology. The varied definitions all include basic ideas: collaboration (Steinitz 2012), the integration of science- and value-based design (McElvaney 2012), and improving the quality and efficiency of problems solved at various geographic scales (McElvaney 2012 and Steinitz 2012). Geodesign is a process that integrates design with technologies, scenario development, and simulation across multiple scales to better illustrate the spatial and time-sensitive impacts of design—both intended and unintended. Regarding this book, regional geodesign should integrate scale and provide mobility or opportunity to move easily between scales to aid decision-making. It should guide urban or environmental design locations. Geodesign should simultaneously support design freedom at site-specific locations; aggregation of site-specific development (that is, residential and employment land-use allocations and design alternatives for comparison at multiple scales); and finally, simulation and impact analysis across scales.

The following chapters describe how the improved methodologies of LUCIS^plus address each of these tenets. In addition to identifying site opportunities for particular land uses, LUCIS^plus provides a seamless approach to equating LUCIS-based model outcomes to community feedback that informs design.

Part I: Foundations

Part I is a review of the LUCIS five-step process and the common terms and tools central to LUCIS modeling. Chapter 2 provides an overview of conflict analysis as a decision-making tool. Chapter 3 introduces new customized ArcGIS tools developed to streamline the LUCIS modeling process.

This book is not intended for beginning GIS users. It is meant for users who have been exposed to basic GIS concepts and techniques and are ready to try their hands at land-use analysis. It should also prove interesting to those who are well-versed in land-use analysis but may not have considered multiple suitabilities simultaneously and the conflicts that arise between them. This book expands on the land-use modeling techniques described in The LUCIS Model *(Carr and Zwick 2007). Although reading the first book is not necessary, readers of* Using LUCIS^plus *should familiarize themselves with the techniques and principles described in the first book before attempting to master the advanced techniques described in the following chapters. Additionally, a familiarity with basic statistics will be helpful to readers of this book, but it is not necessary.*

This book is recommended for intermediate to advanced GIS users. Therefore, the review of GIS that takes place in chapter 2 is not for beginners, yet it does review some basic concepts that are important to proper land-use analysis. Chapter 2 also identifies new tools developed specifically to increase LUCIS model efficiency and automate several common LUCIS processes. Chapter 2 does not teach you how to use the ArcGIS tools to create LUCIS models, but it describes their role in each step of the LUCIS process.

Until now, the real value of LUCIS has been its ability to use land-use conflict to determine opportunities for future growth (hence, the name, Land-Use Conflict Identification Strategy). This book realizes the full potential of LUCIS by integrating flexible tools that identify conflict and allocate population and employment. Chapter 3, perhaps one of the most important chapters in the book, introduces tools that are new to the traditional ArcGIS environment. These tools complete the intended goal of the LUCIS and "plus" strategy. Before now, these tools were unavailable to GIS users outside the University of Florida. Chapters in this book discuss the importance of each tool, the data formats necessary for tool execution, and the flexibility of the tools in allocating population or employment. In chapters 4 and 5, the LUCIS strategy is applied as a continued trend of existing residential and employment policy. As a unit, part I can be considered the lecture portion or knowledge base of the book, because it reviews the basic (and familiar) LUCIS structure. It then lays the foundation for the explanation of the advanced methods and applications of LUCIS, found in parts II and III.

Part II: Land-use analysis and alternative futures

Part II focuses on the nuts and bolts of the five steps of LUCIS. Chapters 6, 7, and 8 describe new methods that provide more flexibility in allocation options. Allocations are done in an urban mixed-use future and in a conservation and agriculture future. The results of the allocation process are then summarized. These chapters evaluate each of the alternative futures using environmental and development indicators. Although CEM is applied throughout this book, these chapters illustrate the true flexibility and ease of scenario development using the revised LUCIS methodologies of LUCIS^{plus}.

Part III: Advanced allocation techniques using LUCIS^{plus}

The value of LUCIS is three-tiered. The first tier, described in parts I and II, determines land-use conflict. The process includes (1) determining land-use suitability based on predetermined goals and objectives, (2) determining land-use preference, and (3) identifying conflict. The second tier illustrates alternative futures through the allocation of population, employment, or both. The third tier is described in this part of the book. The LUCIS^{plus} tools automate spatial modeling processes in all three tiers. This book comprehensively details for the first time the wide range of applications using LUCIS. Chapter 9 describes how the identification of land-use opportunities can guide better decision-making and policy development in the event of storm surge and sea level rise. Chapter 10 uses ArcGIS Online tools to publish and explore LUCIS opportunity raster layers for web users. In the face of limited data resources and an emphasis on respecting culture and tradition in location decisions, chapter 11 expands the application of LUCIS to a new type of community: Native American tribes. Chapters 12 through 14 expand applications of customized LUCIS tools to affordable housing and transportation accessibility.

Locations mentioned in this book

Hillsborough County, Florida. Hillsborough County, Florida (figure 1.1), is the fourth-most populous county in the state. Tampa, the county seat, has seen a steady increase in wealth since 2006. Its economy includes a diversity of businesses, from financial services, bioscience, and technology to international trade and education. Although the county has long roots in agriculture, 96 percent of its population is located in urban areas. In the 2010 Census, almost half the population was between 18 and 49 years old. The greatest percentage change in residents since the 2000 Census was in the 50–64 age group. Hillsborough, on the Tampa Bay peninsula, is a county rich in natural features, ranging from natural springs to hardwood swamp forests. Hillsborough County has used GIS-based decision support tools for land-use analysis since 2002. Considering the rapid pace of development in the mid-2000s, the use of LUCIS tools has proved beneficial.

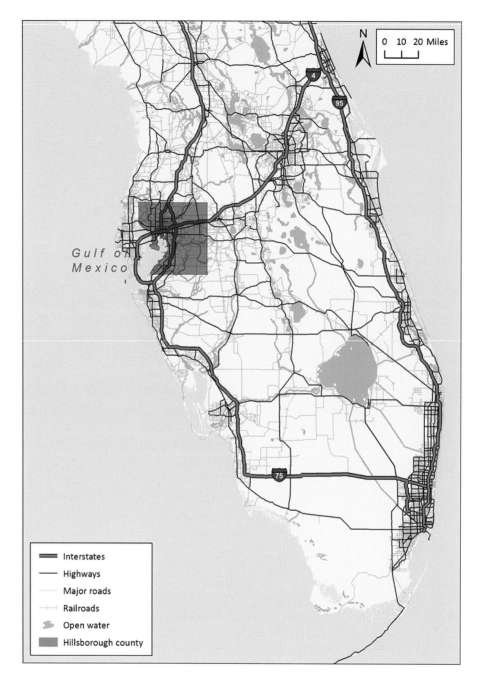

Figure 1.1. Hillsborough County, Florida, is on the west central coast of the state. It is rich in natural features that provide a balanced quality of life for its urban and rural residents.

Figure from the Arizona Board of Regents on behalf of the University of Arizona. Water bodies from the US Geological Survey; state boundary from the US Census Bureau-TIGER/Line Files; interstates and highways from the Florida Department of Transportation; railroads from the Federal Railroad Administration.

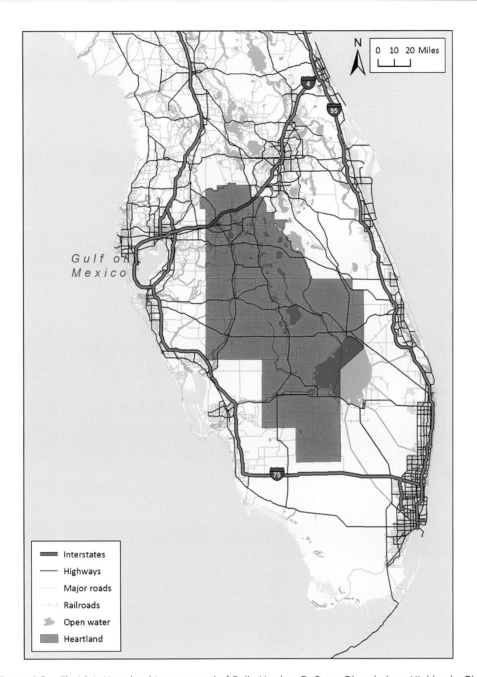

Figure 1.2. Florida's Heartland is composed of Polk, Hardee, DeSoto, Okeechobee, Highlands, Glades, and Hendry Counties. Physical and economic features make this south central Florida region unlike any other region in the state.

Figure from the Arizona Board of Regents on behalf of the University of Arizona. Water bodies from the US Geological Survey; state boundary from the US Census Bureau-TIGER/Line Files; interstates and highways from the Florida Department of Transportation; railroads from the Federal Railroad Administration.

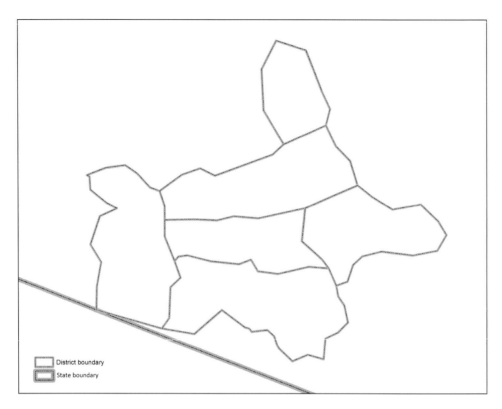

Figure 1.3. The Aldrich Nation is a fictitious tribe in Kentucky. This tribe is composed of multiple political districts totaling almost two million acres.

Figure from the Arizona Board of Regents on behalf of the University of Arizona.

Heartland region, Florida. The Heartland (figure 1.2) is composed of seven counties: Polk, Hardee, DeSoto, Okeechobee, Highlands, Glades, and Hendry. This south central Florida region "encompasses 5,000 square miles and is home to an estimated 300,000 people" (Florida's Heartland REDI 2014). "The region supports a large concentration of threatened and endangered species, high-quality habitats, endemic natural communities, and significant hydrological resources" (The Nature Conservancy 2010, 4). Large ranches, agriculture, and the largest freshwater lake in Florida create physical and economic considerations unlike other parts of the state.

Aldrich Nation. The Aldrich Nation (figure 1.3) and Ellis University are a fictitious Native American tribe and university, respectively. They were created for the sole purpose of demonstrating applications of LUCIS in this book. The methods used in chapter 11 were applied to a real tribe for a land-use planning exercise that employed LUCIS. The geography of the Aldrich Nation in Kentucky is also fictitious.

References

Arrow, K. J. 1969. "The Organization of Economic Activity: Issues Pertinent to the Choice of Market versus Nonmarket Allocation." In *The Analysis and Evaluation of Public Expenditure: The PPB System*, 1:47–63, US Joint Economic Committee, 91st Congress, First Session. Washington, DC: US Government Printing Office.

Buitelaar, E. 2007. *The Cost of Land-Use Decisions: Applying Transaction Cost Economics to Planning and Development*. Malden, MA: Blackwell Publishing.

Carr, Margaret H., and Paul D. Zwick. 2007. *Smart Land-Use Analysis: The LUCIS Model*. Redlands, CA: Esri Press.

Dunham-Jones, E., and J. Williamson. 2009. *Retrofitting Suburbia: Urban Land*. Washington, DC: Urban Land Institute. http://www.uli.org/wp-content/uploads/2009/10/Sustainable-Suburbs-Retrofitting-Suburbia.pdf.

Dunphy, Robert, Deborah Myerson, and Michael Pawlukiewicz. 2003. *Ten Principles for Successful Development around Transit*. Washington, DC: Urban Land Institute.

Florida's Heartland REDI. 2014. Florida's Heartland: The Region. Accessed September 1, 2014. http://flaheartland.com/the-region.

Heid, J. 2004. "Greenfield Development without Sprawl: The Role of Planned Communities." Washington, DC: Urban Land Institute.

McElvaney, Shannon. 2012. *Geodesign: Case Studies in Regional and Urban Planning*. Redlands, CA: Esri Press.

National Association of Realtors. 2007. *2007 Profile of Buyers' Home Features Preferences*. Washington, DC: National Association of Realtors.

The Nature Conservancy. 2010. *Heartland Ecological Assessment Report*. Babson Park, FL: The Nature Conservancy.

Nelson, A. 2009. "American Mega Trends." Lecture, College of Architecture and Landscape Architecture at the University of Arizona, Fall 2009 Lecture Series, Tucson, AZ, October 28.

NumbersUSA Action. 2009. "Population Growth Is Half the Problem in Sprawl." Retrieved December 2009. http://www.sprawlcity.org/cgpg/index.html#graph.

Project for Public Spaces. 2009. "What Is Place-Making?" Retrieved September 2009. http://www.pps.org/what_is_placemaking.

Saaty, T. L. 2008. "Decision Making with the Analytic Hierarchy Process." *International Journal of Services Sciences* 1 (1): 83.

Steinitz, Carl. 2012. *A Framework for Geodesign: Changing Geography by Design*. Redlands, CA: Esri Press.

Victoria Transport Policy Institute. 2010. "Smart Growth: More Efficient Land-Use Management." Retrieved April 2010. http://www.vtpi.org/tdm/tdm38.htm.

Part I

FOUNDATIONS

Chapter 2
Conflict analysis as a decision-making tool

Paul D. Zwick

What this chapter covers

Chapter 1 sets the stage for land-use conflict analysis as a decision-making tool. This chapter describes the basic technical concepts used to accomplish that analysis and decision-making. It begins with a review of the LUCIS (Land-Use Conflict Identification Strategy) five-step process. It includes a short discussion of three relevant terms: *suitability*, *community preference*, and *conflict*, as presented in *Smart Land-Use Analysis: The LUCIS Model* (Carr and Zwick 2007). The original LUCIS concept provides 27 categories of major land-use conflict between agriculture, conservation, and urban use. This chapter expands LUCIS to accommodate urban mixed-use opportunity. The LUCIS urban mixed-use opportunity matrix, which builds on the 27 conflict categories, identifies combinations of overlapping land-use opportunity for mixed-use redevelopment in urban and suburban areas. Using the matrix, you learn how to create multiple combinations of three urban uses—multifamily, retail or service, and commercial—to arrive at 27 opportunities for improved mixed-use design.

In this chapter, you will learn how to create and use a criteria evaluation matrix (CEM) for more complex land-use analysis. Two basic ArcGIS tools, the Combine and Add Field tools, are used to create a single multivariable raster. The raster makes complex land-use decisions and analysis easier. CEM uses conflict, mixed-use opportunity, and other variables to analyze complex land-use decisions. These decisions include the allocation of employment (commercial, retail, service, institutional, and industrial); residential population (single-family and multifamily); and urban and suburban mixed-use development (employment and population). Adding variables representing development policy or development incentives to CEM creates brand-new allocation opportunities. Planning and design can meld these allocation opportunities through *geodesign*.

The five-step LUCIS process: A review

In *The LUCIS Model,* Margaret H. Carr and Paul D. Zwick (2007) describe a five-step process to identify, and develop a better understanding of, land-use conflict. The process involves (1) defining goals and objectives, (2) data inventory and preparation, (3) defining and mapping land-use suitability, (4) integrating community values to determine land-use preference, and (5) identifying potential land-use conflict.

Step 1: Defining goals and objectives

Land-use planning and the design process should never result from a plan-as-you-go mentality (Carr and Zwick 2007). By its fundamental nature, the land-use planning process requires a statement of intent. It also requires goals and objectives that provide direction for the sound, efficient, and effective allocation of future lands. Without a statement of intent and an accompanying set of goals and objectives that incorporate a community's values, any recommendations made in the future land-use plan or implementation of the plan can be considered suspect. To paraphrase an adage, any road will suffice if it does not matter where you are going. The land-use planning process must result in the allocation of lands for urban, agriculture, and conservation uses that represent the community's values. The resulting product must adhere to sound land-use goals and objectives.

LUCIS is the antithesis of the plan-as-you-go mentality. The first step of the LUCIS process is to use community input to develop a *statement of intent* and the requisite goals and objectives to support that statement of intent. The LUCIS statement of intent concisely describes the task at hand. For example, "The intent of this project is to develop a set of land-use alternatives that help community leaders and citizens visualize future allocations of a proposed population increase of three million new people in the east central Florida region."

Next, a set of goals and supporting objectives are developed to support the statement of intent. The goals and objectives hierarchically define what to attain (the *goals*) and how to accomplish each goal (*the objectives*). For example, the LUCIS urban suitability models are based on a set of goals to identify the most suitable areas for residential (both multifamily and single family), commercial, retail, industrial, service, and institutional land use. Each goal has an accompanying set of objectives and subobjectives. The single-family residential goal has four suitability objectives: (1) physical characteristics of the area, (2) proximal characteristics of neighboring uses, (3) the historical residential growth pattern, and (4) existing local residential density. The single-family residential physical objective has five suitability subobjectives: (1) noise, (2) soil construction characteristics, (3) air quality, (4) surface drainage, and (5) hazardous materials.

Step 2: Data inventory and preparation

Step 2 in LUCIS modeling is the traditional step in a GIS activity: collect and prepare the data required for land-use modeling. (See chapter 7 in *The LUCIS Model* [Carr and Zwick 2007] for a detailed discussion of the subject.)

Step 3: Defining and mapping land-use suitability

Before reviewing the mapping of land-use suitability, we must define the terms *suitability*, *preference*, and *conflict*.

Suitability categorizes the usefulness of a particular spatial unit of area (designated by a cell within a raster) for a specific land use. Levels of utility, in most cases, range from 1 to 9, low to high utility. There are two types of raster suitability. The first is single-utility assignment (SUA). An SUA assigns a level of utility to a single land-use category by asking a single, specific question through a selection query. For example, you can assign a physical residential utility for soil drainage based on the drainage characteristic of the soil in a particular cell. You can also assign a utility to suitability for the proximity of a location (an individual cell) to a community service, such as the nearest hospital. The second type of raster suitability is a multiple-utility assignment (MUA). Combine multiple SUAs to form an aggregate MUA for a particular category. For example, combine the following three SUAs: (1) for proximity to interstate highway intersections, (2) for proximity to state and county roads, and (3) for proximity to local roads. This weighted combination creates an MUA for proximity to automobile transit opportunities. The MUA represents the combined utility for access to transit from any location (that is, individual cell) within a suitability raster. The weighted combination of multiple MUAs or multiple SUAs and MUAs can also create an MUA. A weighted combination of multiple SUAs and MUAs is considered a *complex MUA*. For example, using the Weighted Sum tool, you can develop a combination of factors, some physical and some proximal, to identify the suitability of locations for residential development.

Preference is the weighted combination of goal suitability raster data. The individual weights represent community values. The Weighted Sum tool helps establish preferences. The individual community weights represent a value that community stakeholders place on a specific goal or objective data when combined with other goal or objective data. You can combine community values to create an urban preference raster. For example, combine transit opportunities, land values, proximity to community and social services, proximity to employment opportunities, proximity to amenities, and areas subject to physical flooding and hazardous areas. The goal is to support the identification of areas that are of value to the community for a major land use (in this case, urban development).

Conflict is the spatial combination of community land-use preferences to identify areas in which the community's values are preferred or not preferred for individual land-use categories. The LUCIS model identifies conflict between three major land-use categories: agriculture, conservation, and urban. A detailed review of conflict follows later in this chapter.

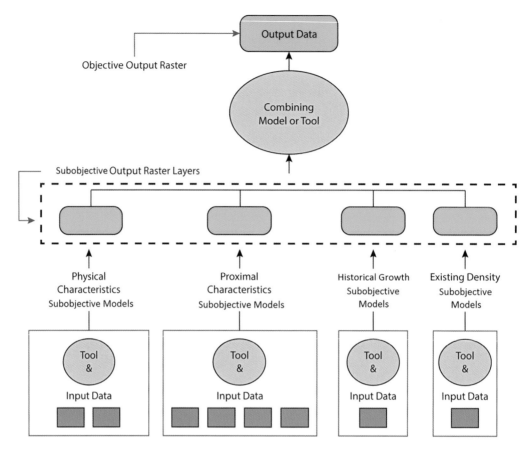

Figure 2.1. A conceptual land-use model for an objective with four subobjectives: physical, proximal, historical growth, and existing density.

Figure from the Arizona Board of Regents on behalf of the University of Arizona.

The LUCIS modeling process begins with the development of individual subobjective models that create a unique suitability raster for a specific land use. The subobjectives include physical characteristics, proximal characteristics, historical growth pattern, and existing density. Figure 2.1 is a conceptual diagram of the objective modeling process. The model components include four subobjective models, which contain eight input datasets (the blue boxes), four tools (the yellow circles), and output raster data (the green boxes). The proximal characteristic model depicted in this figure processes input data using distance or proximity to create a specific subobjective suitability layer (for example, proximity to interstate highway intersection features). Other examples of proximity include proximity to railroads, racetracks, and county and local roadways. The aggregate of the subobjective suitability raster data creates a suitability raster for the objective of residential proximity to transit access.

Other subobjective models could model historical growth and use the residential growth patterns to determine the region's suitability for residential development. Another subobjective model could identify the suitability for residential density. Combining the subobjectives for physical, proximal, historical growth, and existing density suitability with land values then identifies the region's residential suitability.

Figure 2.2 shows the subobjective model for single-family residential soil suitability. This ArcGIS ModelBuilder graphic uses the same color scheme as figure 2.1: the blue rectangles are input data, the yellow circles are model tools, the green boxes are output or intermediate data, and the red boxes are layer weights. The process and diagrammatic flow of the model is from bottom to top, with the input data being concrete corrosion and drainage. The final subobjective output for soil suitability, named ugsf1o13so132, is a green box centered at the top in the graphic. Figure 2.3 is the raster output data produced by the model subobjective for soil suitability. A red to green color ramp shades the suitability values. Green values indicate the most suitable location for single-family residential development based on soil corrosion and drainage characteristics. The brown areas are existing conservation lands; the gray lines are interstate highways; the thin gray lines are major roadways; and the blue areas are open water, including the Gulf of Mexico.

You can then combine multiple-objective suitability layers to produce a single goal layer (in this case, for single-family residential suitability). These layers can include noise level; flood potential; soil drainage; land values; proximity to shopping opportunities, schools, employment centers, and entertainment; residential density; and the county's historical residential growth pattern. Next, you can combine raster suitability layers for multiple urban goals using community values and other criteria to create a single weighted urban preference layer. The goals can include single-family residential, multifamily residential, commercial, service, retail, industrial, and institutional. The weighted urban preference layer represents the community's spatial preference for urban land use.

Step 4: Integrating community values to determine land-use preference

Integrating community values into any land-use visioning process is important. It is perhaps the most important component of LUCIS modeling efforts. Sound regional and local land-use analysis supported by a functional visioning component is intrinsic to good land-use planning. After all, the best environmental planning is sound urban planning. Visualizing future land-use change is important for:

- a community's transportation plan,
- creation of a future sustainable development pattern,
- development of functional strategies that decrease urban sprawl,
- protection of the region's biodiversity,
- local and regional agriculture preservation,
- creation of a vibrant urban redevelopment strategy, and
- better understanding of the land-use relationships between the built and natural environments.

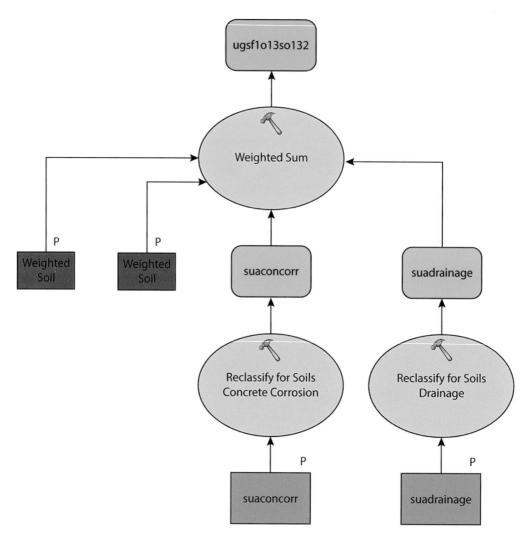

Figure 2.2. The ModelBuilder model for the LUCIS single-family residential soil suitability. The two input datasets, suaconcorr and suadrainage, are reclassified into the suitability scale. Then the outputs are combined. A weight is applied to each output dataset, reflecting how much they contribute to measuring residential soil suitability. The final output, ugsf1o13so132, represents the final soil suitability for single-family residential.

Figure from the Arizona Board of Regents on behalf of the University of Arizona.

Without community input, the regional visualization process is most often suspect at the local level and, therefore, ineffective. In fact, the reason for visualizing land-use alternatives is to aid informed decision-making. This strategy leads to a more realistic long-term plan for the region that its varied communities and local interest groups can validate. Community leaders, citizens, and support organizations must feel they have fully participated in the visioning/planning process. However, you do not need to include exercises in community

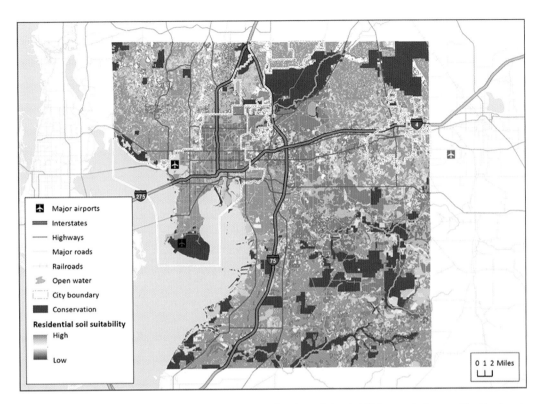

Figure 2.3. Raster output for single-family residential soil suitability in Hillsborough County, Florida. A red (low suitability) to green (high suitability) color ramp depicts suitability.

Figure from Paul D. Zwick and the Arizona Board of Regents on behalf of the University of Arizona. Water bodies from the US Geological Survey; state boundary from US Census Bureau TIGER/Line Files; interstates and highways from the Florida Department of Transportation; conservation areas from the Florida Natural Areas Inventory.

visioning to determine the area's trend (or existing) development pattern. The best method for identifying trends is data analysis of existing and historical land-use patterns and conditions. In fact, capturing the actual spatial statistics through property analysis of, say, the historical density patterns of single-family development is important when analyzing trend land-use patterns. LUCIS^{plus} relies on trend analysis to set the baseline for scenario comparisons.

It is vital to understand the difference between community land-use visioning and land-use prediction. LUCIS is a process of land-use planning and visioning, *not* prediction. Land-use prediction is not land-use visioning or planning. It is an extrapolation of the components deemed significant to an existing land-use pattern. It is not a tool for community leaders, planners, or other professionals to envision the land-use possibilities resulting from an implementation of a specific set or combination of community values. The LUCIS modeling of community values produces various land-use visioning alternatives. You can then compare these alternatives with the trend or pattern extrapolation to assess alternative planning options and land-use futures.

LUCIS uses a pairwise comparison process called the *analytic hierarchy process* (AHP) (Saaty 1980) to incorporate community values. LUCIS implements AHP when combining goal data to create the major land-use preference layers. You can incorporate pairwise comparisons in a visioning session in a number of ways. Although it is fundamentally simple to comprehend the process for incorporating community values, it is complex to organize and difficult to complete without the use of GIS tools. What results is a new tool for LUCIS modeling, the A4 LUCIS Community Values Calculator (figure 2.4). Chapter 3 provides an in-depth discussion of the A4 LUCIS Community Values Calculator and other A4 LUCIS tools.

Planning meetings and town hall/community visioning meetings can use the A4 LUCIS Community Values Calculator to develop the multiple weights for combining land-use goals. The community value weights are used to create the major LUCIS land-use preference layers (that is, agriculture, conservation, and urban). The A4 LUCIS Community Values Calculator can organize the pairwise comparison of multiple raster layers, either by multiple users or on the fly during a stakeholder meeting. Using the tool, you can now collect community values from gatherings of community leaders, special-interest groups, and the public to help identify preferences. The A4 LUCIS Community Values Calculator is flexible. It can collect and save the results of multiple meetings or visioning sessions. This tool makes it easier to rapidly and correctly produce alternative land-use preference layers reflecting various community factions. The A4 LUCIS Community Values Calculator has greatly improved the capabilities of LUCIS to collect valuable community input for land-use visioning.

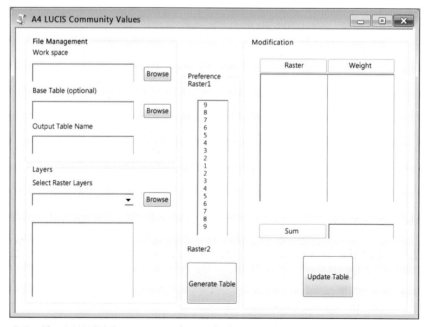

Figure 2.4. The A4 LUCIS Community Values Calculator can be used as a stand-alone tool or within the ArcGIS ModelBuilder environment. This tool is used to develop the multiple weights for combining land-use criteria.

Figure from the Arizona Board of Regents on behalf of the University of Arizona.

Step 5: Identifying potential land-use conflict

Step 5 in the LUCIS process is creating a conflict layer to spatially identify the conflict between major land-use categories. The following sections of this chapter review LUCIS conflict and expand the use of LUCIS conflict for urban infill, redevelopment, and mixed-use analysis within LUCIS^{plus}.

Reviewing the concept of LUCIS land-use conflict

In LUCIS land-use conflict methodology, community preference layers are each reclassified into three levels: low, moderate, and high. These values are combined to generate a conflict matrix containing 27 land-use conflict or preference values.

To start the conflict identification process, aggregate the final community weighted preference layers for each of the major land-use categories into low, moderate, and high preferences. You can use any of the ArcGIS methods available to reclassify the preferences, but make sure to apply the same method to each of the major land-use categories. Table 2.1 is a selection of ArcGIS methods for data reclassification. The following section, "LUCIS collapsed preference: Reclassification of community values," explains and compares three specific methods of reclassification for land-use preference: (1) equal interval, (2) standard deviation, and (3) geometric interval.

LUCIS collapsed preference: Reclassification of community values

The *equal-interval reclassification* method does exactly what its name implies. It reclassifies the collection of individual preference values in a raster into equal intervals. Again, the three categories of collapsed LUCIS preference are low, moderate, and high. The upside of an equal-interval reclassification is that almost everyone has an intuitive understanding of equal intervals. The downside is its inability to adjust skewed data.

The second reclassification used is the *standard deviation reclassification*. The standard deviation reclassification is more complicated. Standard deviation reclassification of the three preference intervals determines the deviation of values in the preference layer from the mean value of all values in the layer. That is, the standardized numerical difference is between an individual cell preference value and the average of all cell preference values for the layer. The standard deviation method of reclassification is excellent for normally distributed data. Normally distributed data is symmetrically distributed above and below the mean value. That is, 50 percent of the data is above the mean value, and 50 percent is below the mean. As with the equal-interval distribution, the standard deviation cannot always accommodate skewed data. However, it does a better job of reclassification for skewed data.

The third method presented is the *geometric interval reclassification*. It is even more obtuse than the other methods but can produce intervals that are reasonable for skewed data. The advantage of a geometric interval distribution is that it aggregates preference data into a better distribution in cases in which suitability is not normally distributed. This distribution offers a spatial advantage when analyzing conflict because it produces a better mix of conflict within the region of interest.

Table 2.1. Standard data classification methods available within ArcGIS

Manual	Allows you to set the class breaks manually. Use this choice if, for example, you want to emphasize particular patterns by placing breaks at important threshold values, or if you need to comply with a particular standard that demands certain class breaks. The Classes drop-down list is disabled when you choose this method. You specify the classes by working with the histogram in this dialog box: • To insert a class break, right-click in this histogram and click Insert Break. • To remove a class break, select it by clicking in the histogram or in the break values list to the right of the histogram (it will turn red when selected), and then right-click it in the histogram and click Delete Break. • To move a class break, either click on it in the histogram and drag it, or edit its value in the break values list to the right of the histogram.
Equal Interval	This method divides the attribute range into equally sized classes and is best applied to familiar data ranges such as percentages and temperatures. Use this method to emphasize the relative amount for attribute values compared with other values.
Defined Interval	This method is similar to the Equal Interval method, except here you define the size of the interval. The Classes drop-down list is disabled when you choose this method because it adjusts automatically to reflect the number of classes needed for the entire interval size you defined once you pressed OK in the Classification dialog box.
Quantiles	Each class will contain an equal number of features. This method is well suited to linearly distributed data.
Natural Breaks (Jenks)	Classes are based on natural groupings of data values. In this method, data values are arranged in order. The class breaks are determined statistically by finding adjacent feature pairs, between which there is a relatively large difference in data value. This is the default classification method.
Geometric Interval	This method creates class ranges based on intervals that have a geometric sequence based on a multiplier (and its inverse). It creates these intervals by minimizing the square sum of elements per class; this ensures that each interval has an appropriate number of values within it and the intervals are fairly similar. This algorithm was specifically designed to accommodate continuous data. It produces a result that is visually appealing and cartographically comprehensive. The Geometric Interval method minimizes the variance between classes, and even works on data that is not normally distributed. This classification method is called Smart Quantiles in the ArcGIS Geostatistical Analyst extension.
Standard Deviation	Use this method to emphasize how much feature values vary from the mean. It is best used on normally distributed data.

Source: Esri 2011.

Table 2.2 compares the three described methods of reclassification for the urban preference layer. Each reclassification has a range of 1 to 9 and the same mean value, approximately 5.135. However, each reclassification has three distinctly different classification intervals. Range 1 for each reclassification shows the values for low preference, range 2 for moderate preference, and range 3 for high preference. Clearly, the geometric interval reclassification has the largest low and high preference ranges. This characteristic is expected because most preference values in this layer are distributed closer to the mean than at lower and higher values. The equal-interval reclassification generates the largest moderate range. The standard deviation and geometric interval reclassifications create variable low, moderate, and high preference ranges. As expected, the equal interval does not. Figure 2.5 shows the final urban preference layer for equal-interval reclassification.

Table 2.2. A comparison of the reclassification methods for the urban preference layer

Type	Minimum	Maximum	Mean	Range 1	Range 2	Range 3
Equal Interval	1	9	5.135532	1–3.6667	3.6668–6.6667	6.6668–9.0000
Standard Deviation	1	9	5.135532	1–2.9048	2.90489–5.6790	5.6791–9.0000
Geometric Interval	1	9	5.135532	1–4.5406	4.5407–5.4593	5.4594–9.0000

Figure 2.6 shows the histograms of the aggregated preference layers for the three reclassifications: (a) the equal-interval reclassification, (b) the standard deviation reclassification, and (c) the geometric interval reclassification. Both the equal-interval and standard deviation reclassifications generate far fewer low and high preference cells than the geometric interval reclassification. Clearly, the geometric interval reclassification produces a more spatially uniform distribution of aggregate preference areas.

Once land-use preference has been reclassified into low, moderate, and high ranges, you can aggregate the three layers containing land-use preferences to create the LUCIS conflict layer. Figure 2.7 illustrates the value attribute table (VAT) of the base LUCIS greenfield conflict raster layer. Also, to review, map algebra equation (2.1) combines the three reclassified preference layers.

$$\text{LUCIS conflict raster layer} = ((\text{agriculture preference} * 100) + (\text{conservation preference} * 10) + (\text{urban preference})). \quad (2.1)$$

For example, a conflict value of **313** represents a high preference for agriculture (**3**13), a low preference for conservation (3**1**3), and a high preference for urban development (31**3**). When two of the three digits in the conflict value are equal, there is a *moderate conflict*. Therefore, the conflict value **313** can be labeled as "Moderate Conflict with High Agriculture-Urban Preference." The conflict value **333** indicates a high preference for all

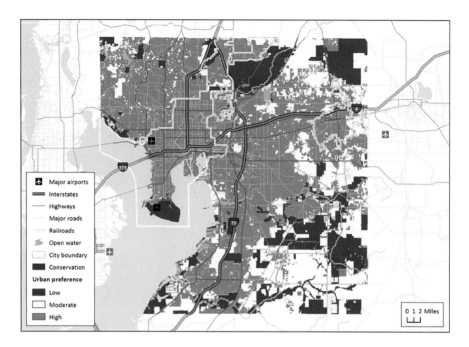

Figure 2.5. An equal-interval classification scheme transforms LUCIS suitability to preference. This raster illustrates urban preference, in which red is low preference, yellow is moderate preference, and green is high preference.

Figure from Paul D. Zwick and the Arizona Board of Regents on behalf of the University of Arizona. Water bodies from the US Geological Survey; state boundary from US Census Bureau TIGER/Line Files; interstates and highways from the Florida Department of Transportation; conservation areas from the Florida Natural Areas Inventory.

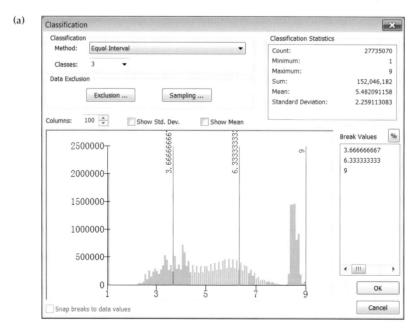

(a)

(b)

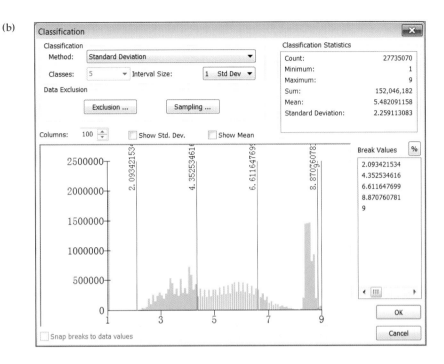

(c)

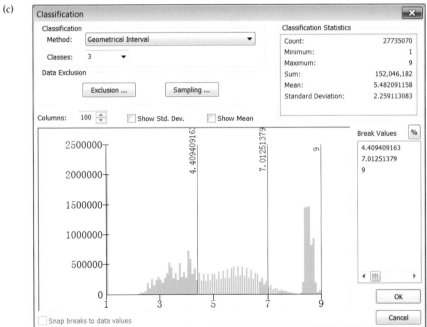

Figure 2.6. Reclassification histograms of the aggregated urban preference layer: (a) equal interval, preceding page; (b) standard deviation; and (c) geometric interval.

Figure from the Arizona Board of Regents on behalf of the University of Arizona.

three major land-use categories. Because all three digits in the conflict value are equal, it is called a major conflict. Therefore, 111, 222, and 333 represent major conflicts among the three (agriculture, conservation, and urban) land-use categories. The conflict value **111** indicates "Major Conflict with Low Preference." Alternatively, the conflict value **222** indicates "Major Conflict with Moderate Preference." Conflict can be categorized as "major," even if the three preference values indicate a low or moderate preference. *Major conflict* is defined as equal preference between the three land-use categories. Understanding the conflict values and preference levels allows users of the LUCIS model to make simple land allocations quickly and within a standardized selection process. Some conflict values represent no conflict. For example, conflict value 113 represents "Urban High Preference." By definition, it is a no-conflict value. Therefore, it is a prime area for urban land use or future urban development. Table 2.3 presents the 27 greenfield land-use conflict categories and the three-digit value associated with each category.

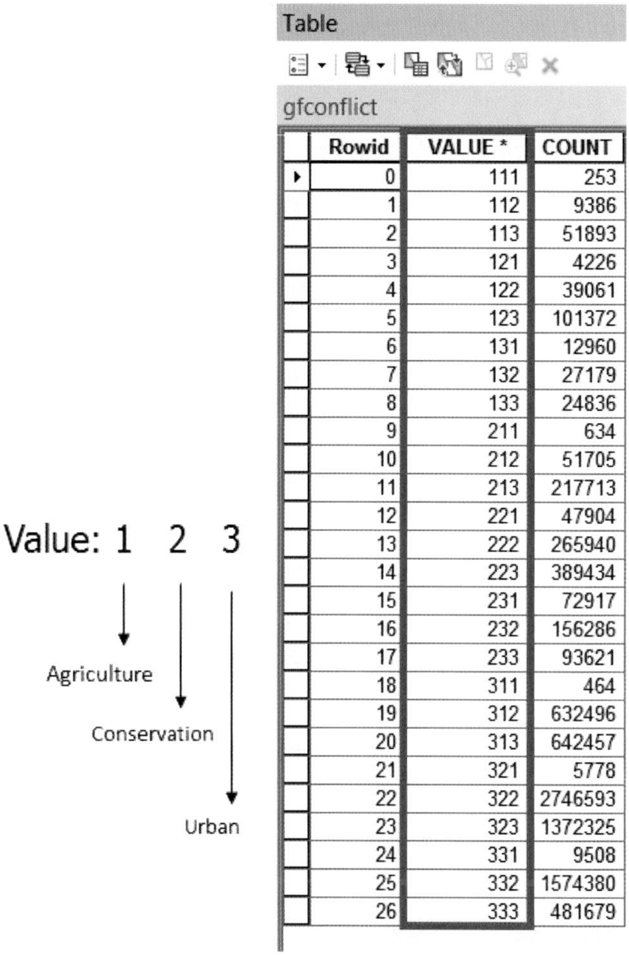

Rowid	VALUE	COUNT
0	111	253
1	112	9386
2	113	51893
3	121	4226
4	122	39061
5	123	101372
6	131	12960
7	132	27179
8	133	24836
9	211	634
10	212	51705
11	213	217713
12	221	47904
13	222	265940
14	223	389434
15	231	72917
16	232	156286
17	233	93621
18	311	464
19	312	632496
20	313	642457
21	321	5778
22	322	2746593
23	323	1372325
24	331	9508
25	332	1574380
26	333	481679

Value: 1 2 3 → Agriculture, Conservation, Urban

Figure 2.7. VAT of the LUCIS greenfield conflict (gfconflict) layer. The second column in the VAT represents the LUCIS conflict value. In a gfconflict layer, the first digit of the conflict value represents agriculture preference; the second digit, conservation preference; and the third digit, urban preference. The third column represents the number of cells assigned that conflict value.

Figure from the Arizona Board of Regents on behalf of the University of Arizona.

Table 2.3. The 27 LUCIS greenfield conflict matrix values and their individual conflict identification descriptions

\multicolumn{2}{c	}{Areas of Conflict}	\multicolumn{2}{c}{Areas of No Conflict}
Code	Description	Code	Description
111	Major Conflict with Low Preferences	112	Urban Moderate Preference
122	Minor Conflict Conservation and Urban with Moderate Preferences	113	Urban High Preference
133	Minor Conflict Conservation and Urban with High Preferences	121	Conservation Moderate Preference
212	Minor Conflict Agriculture and Urban with Moderate Preferences	123	Urban High Preference
221	Minor Conflict Agriculture and Conservation with Moderate Preferences	131	Conservation High Preference
222	Major Conflict with Moderate Preferences	132	Conservation High Preference
233	Minor Conflict Conservation and Urban with High Preferences	211	Agriculture Moderate Preference
313	Minor Conflict Agriculture and Urban with High Preferences	213	Urban High Preference
323	Minor Conflict Agriculture and Urban with High Preferences	223	Urban High Preference
331	Minor Conflict Agriculture and Conservation with High Preferences	231	Conservation High Preference
332	Minor Conflict Agriculture and Conservation with High Preferences	232	Conservation High Preference
333	Major Conflict with High Preferences	311	Agriculture High Preference
		312	Agriculture High Preference
		321	Agriculture High Preference
		322	Agriculture High Preference

Note: In conflict values, 1 = low preference, 2 = moderate preference, and 3 = high preference. The combination of preferences leads to conflict.

LUCIS greenfield conflict and mixed-use opportunity for land-use decisions

The LUCIS greenfield conflict matrix clearly identifies the spatial conflicts among the three major land-use categories. The LUCIS process uses these conflict categories in the allocation process. You can make selections in the conflict raster to (1) identify areas for new urban development with little or no conflict for competing use, (2) identify areas for potential conservation, (3) help decision-makers identify locations preferred for agricultural preservation, and (4) identify areas in which agriculture land use should remain as the dominant use. Figure 2.8 illustrates the LUCIS greenfield conflict layer for Hillsborough County, Florida. The gray areas in the map are existing urban or suburban areas within the county.

Figure 2.9 shows a simple example of conflict selection using the LUCIS greenfield conflict matrix. This selection is for a community that wants to identify preferred areas that

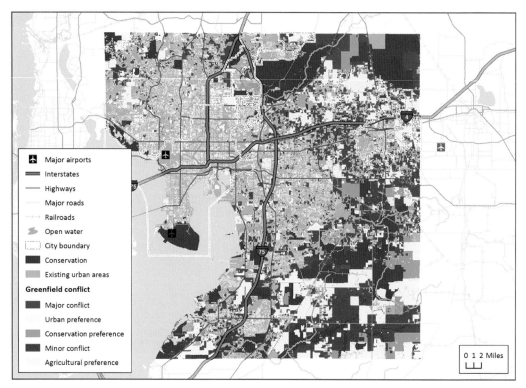

Figure 2.8. A simplified LUCIS greenfield conflict raster layer for Hillsborough County, Florida, illustrates three individual preference types—agriculture, conservation, and urban—and major and minor land-use conflict.

Figure from Paul D. Zwick and the Arizona Board of Regents on behalf of the University of Arizona. Water bodies from the US Geological Survey; state boundary from US Census Bureau TIGER/Line Files; interstates and highways from the Florida Department of Transportation; conservation areas from the Florida Natural Areas Inventory; parcels from the Florida Department of Revenue.

are suitable for agriculture preservation. The selection extracts cells from the conflict matrix that share an equal preference for agriculture and conservation as potential agriculture preservation areas. To select areas preferred for agriculture preservation, choose the cells in the LUCIS greenfield conflict matrix with conflict values of **333** (major conflict high preference), **222** (major conflict moderate preference), **331** or **332** (minor agriculture-conservation conflict high preference), and **221** (minor agriculture-conservation conflict moderate preference). A text attribute `"CONFLICT"` added to the LUCIS conflict layer contains a text description of the category of LUCIS conflict. The ArcGIS Select By Attributes tool is used to select cells from the LUCIS conflict layer. The selected cells satisfy the SQL query (`"CONFLICT" = 'Major Conflict High Preference' OR "CONFLICT" = 'Major Conflict Moderate Preference' OR "CONFLICT" = 'Minor Conflict Ag-Con High Preference' OR "CONFLICT" = 'Minor Conflict Ag-Con Moderate Preference'`). These cells are candidates for potential agriculture preservation. In figure 2.10, the dark-green cells satisfy the LUCIS conflict selection for potential agriculture preservation. The brown areas are the region's existing conservation lands.

Although the selection of cells from the LUCIS conflict layer can effectively identify particular areas within the 27 base LUCIS greenfield conflict categories, land-use

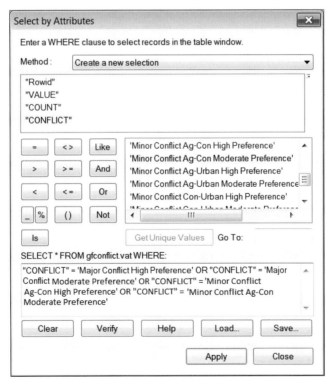

Figure 2.9. Using the ArcGIS Select By Attributes tool, a SQL query extracts possible agriculture preservation areas from the LUCIS greenfield conflict raster layer.

Figure from the Arizona Board of Regents on behalf of the University of Arizona.

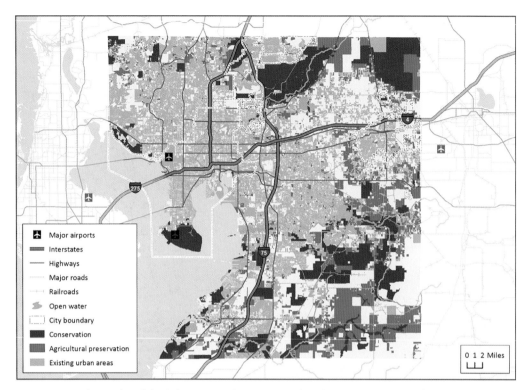

Figure 2.10. The results of the Select By Attributes query identify areas appropriate for agriculture preservation in Hillsborough County, Florida.

Figure from Paul D. Zwick and the Arizona Board of Regents on behalf of the University of Arizona. Water bodies from the US Geological Survey; state boundary from US Census Bureau TIGER/Line Files; interstates and highways from the Florida Department of Transportation; conservation areas from the Florida Natural Areas Inventory; parcels from the Florida Department of Revenue.

selections can be more complex. Selections can often be directed more toward a search for opportunity than the identification of basic conflict. Many planners, community activists, and planning commissioners want to increase existing urban density as a means of reducing the sprawl created by rapid incremental greenfield development. The desire to increase urban density is yet another example of a more complex land-use problem. To increase density within an urban area, identify areas appropriate for new infill development and redevelopment (replacement of older structures with an accompanying change in the use of existing occupied lands).

LUCIS urban mixed-use opportunity matrix

The LUCIS urban mixed-use opportunity matrix lists multiple preference criteria in a single matrix for a particular land-use type. This amalgamated matrix aids the analysis of more complex land-use issues. Not surprisingly, the LUCIS greenfield conflict matrix can be

Table 2.4. LUCIS urban mixed-use opportunity matrix values and their mixed-use preference definitions

Mixed-Use Value	Description
111	Mixed Use with Low Preference
112	Retail Moderate Preference
113	Retail High Preference
121	Multifamily Moderate Preference
122	Multifamily and Retail with Moderate Preferences
123	Retail High Preference
131	Multifamily High Preference
132	Multifamily High Preference
133	Multifamily and Retail with High Preferences
211	Commercial Moderate Preference
212	Commercial and Multifamily with Moderate Preferences
213	Retail High Preference
221	Commercial and Multifamily with Moderate Preferences
222	All with Moderate Preferences
223	Retail High Preference
231	Multifamily High Preference
232	Multifamily High Preference
233	Multifamily and Retail with High Preferences
311	Commercial High Preference
312	Commercial High Preference
313	Commercial and Retail with High Preferences
321	Commercial High Preference
322	Commercial High Preference
323	Commercial and Retail with High Preferences
331	Commercial and Multifamily with High Preferences
332	Commercial and Multifamily with High Preferences
333	All with High Preferences

adapted to identify opportunity instead of conflict. You can easily modify the LUCIS greenfield conflict matrix to identify potential opportunities for mixed-use redevelopment within an existing urban core. This opportunity matrix will remain true to the concept of the original conflict matrix. The newly developed mixed-use opportunity matrix is a tool to tackle the more complex issue of mixed-use redevelopment. To remain true to the LUCIS concept, the matrix must provide a mechanism to identify opportunities between commercial, retail or service, and multifamily residential uses. The concept of weighting community values to convert suitability to preference is the same method for developing weighted mixed-use preference. The evolution occurs in combining the three reclassified preference layers (commercial, multifamily residential, and retail or service) to identify mixed-use opportunity instead of conflict. Again, the values representing low, moderate, and high preference are 1 = low, 2 = moderate, and 3 = high. Three urban-use preference layers for commercial, multifamily residential, and retail are combined in the same process as the agriculture, conservation, and urban major land-use preference layers. Map algebra equation (2.2) combines the reclassified preference layers.

$$\text{LUCIS mixed-use raster layer} = ((\text{commercial preference} * 100) + (\text{multifamily preference} * 10) + (\text{retail or service preference})). \quad (2.2)$$

Table 2.4 lists 27 urban mixed-use opportunities and associated mixed-use preferences. The LUCIS urban mixed-use opportunity matrix is used to analyze mixed-use redevelopment in later chapters.

Using the ArcGIS Combine tool to create a multivariable raster for land-use decision-making

Another solution for modeling land-use complexity is using the ArcGIS Combine tool to develop a single complex multivariable raster layer. The *A4* Allocation tools can then be used to process the complex multivariable raster layer. This tool is described in greater detail in chapter 3. The two solutions, the urban mixed-use opportunity matrix and land-use complex multivariable raster layer, are not mutually exclusive. They can be used together for advanced land-use analysis.

The Combine tool does what its name implies. It combines multiple input raster layers, one cell at a time, to create a single multivariable raster layer. This layer contains all the information present from the input raster layers. The advantage of the Combine tool is that it can create a single raster layer that contains incredible amounts of spatial information. Once the newly combined raster layer is completed, you can add more data to the raster layer's VAT to make it more useful for land-use decision-making. Figure 2.11 illustrates the general process of creating a multivariable raster layer. Figure 2.12 shows the input raster layers. Many of the fields in the raster layer VAT, which is the top layer in figure 2.11, were added to the table using the Add Field tool after the table was created (figure 2.13). (You can see the VAT fields in detail in table 2.5.)

Allocation scenario 1

Making land-use and population allocations is complex. A reasonable example of this complexity follows.

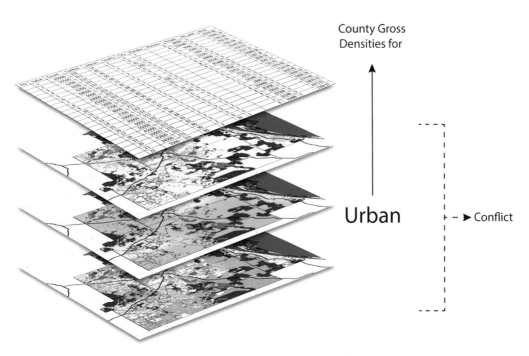

Figure 2.11. Creating a multivariable complex raster layer. The complex raster layer, representing county gross densities for employment and residential development and urban conflict, are combined into a single table. More fields can later be added and calculated to help in employment and population allocations.

Figure from Paul D. Zwick and the Arizona Board of Regents on behalf of the University of Arizona.

A developer in one of the region's counties just purchased four thousand acres of pasture and timberland. The area is northeast of the major city in the region and next to a large existing conservation area. The area is not identified as a development hub in the existing regional land-use plan because there is no existing urban development in the area. The only exception is an existing county roadway. A LUCIS greenfield conflict layer for the county has cells in the developer's four thousand acres categorized as mostly "minor agriculture-conservation conflict with high or moderate preference" (that is, 332, 331, or 221). It has some cells with "urban preferred in the moderate range" (112) and other cells categorized as "major conflict with moderate preference" (222). The county planning department has granted a land-use change for the property, and the developer is intent on constructing a new urbanism development. The development will contain mixed-use development, with single-family residential, commercial, and retail/sales opportunities. The region has been experiencing rapid growth, and land development corporations have made 10 large-area purchases in the past five years. The regional planning Council of Governments (COG) staff wants to know the combined effect of the newly proposed planned developments on the proposed future land-use allocations within the region. Moreover, COG has proposed greater redevelopment within the existing urbanized areas of the region. Yet many of the new planned developments are not close enough to the existing urbanized areas to offer the desired density increase proposed in the region's future land-use plan. Because the planned developments are located in separate though adjacent counties, they are subject to different developmental guidelines, which include different density requirements and different commercial and retail space allocations. Also, many of the planned developments have varied conservation suitability as identified by the COG LUCIS visioning exercise.

LUCIS adopts a sequential process to allocate population and employment, allowing the use of automated allocation models. In the allocation process, Carr and Zwick (2007, 167) identify six general steps to visualize future land use:

1. Start allocating in the area that does not include conflict and in which urban preference is highest.
2. Continue allocating, if needed, in moderate-conflict and major-conflict areas, if necessary, in which the values for urban are highest.
3. Create a "remaining lands" mask to account for the cells allocated in steps 1 and 2.
4. Allocate remaining cells for future agricultural land that are not in conflict and in which the preference is greater than for conservation or urban.
5. Allocate remaining cells for future conservation land that are not in conflict and in which the preference is greater than for agriculture or urban.
6. Allocate remaining conflicting cells between agriculture and conservation, according to the greater preference.

The layer VAT shown in table 2.5 can help explain the use of a complex multivariable raster layer. You can use the table to answer a simple question using a hypothetical LUCIS allocation. Using a complex multivariable raster VAT for employment or residential population land-use allocation can be as complicated or as easy as you want to make it. This book uses a scenario guide for land-use allocations. A scenario guide outlines the hierarchical process (sequence of SQL queries) for identifying cell characteristics that fulfill the allocation criteria. For example, in the land-use problem described, the regional COG has identified a desire to increase existing urban densities within areas of the county. The question arises, how will the new planned developments affect that policy? If the planned developments siphon off residential population and employment, will the urban densification plan have the desired effect and reduce regional sprawl? Conversely, if the regional COG incentivized the policy to increase urban density with tax reductions for infill and redevelopment in the region's urban areas, how would it affect the marketability of proposed planned developments?

The fields in the raster layer VAT are as follows:

OID: the object identification number created by ArcGIS software.

Value: unique values assigned when the raster was generated (integer). It is generated during the raster combination process.

Table 2.5. VAT for a complex multivariable raster layer for the hypothetical LUCIS allocation

OID	Value	Count	Conflict	Urbsuit	County	City	Plandev	Sfresden	People	Res_year	Acres
1	1	10000	333	57	1	0	1	8	18992	0	2374.68
2	2	55500	313	85	1	0	1	8	104480	0	13060.74
3	3	6700	112	76	3	10	0	6	9546	0	1591.04
4	4	3200	113	90	1	100	0	6	4560	0	759.90
5	5	6700	122	66	2	0	0	2	3182	0	1591.04
6	6	68000	311	23	2	0	0	2	32296	0	16147.82
7	7	9900	223	87	2	0	1	8	18808	0	2350.93
8	8	10600	213	88	1	0	1	8	20136	0	2517.16
9	9	45000	313	88	1	100	0	8	85488	0	10686.06
10	10	5000	113	89	3	10	0	6	7122	0	1187.34
11	11	7700	113	90	1	0	0	2	3658	0	1828.50
12	12	8600	213	84	3	10	0	6	12252	0	2042.22
13	13	2200	123	77	3	100	0	6	3132	0	522.43
14	14	10000	112	67	1	0	0	2	4750	0	2374.68
15	15	30000	111	10	1	0	0	2	14248	0	7124.04
16	16	6000	222	44	1	0	0	2	2848	0	1424.81
17	17	55000	333	88	2	0	2	8	104488	0	13060.74
18	18	45000	313	83	2	0	2	8	85488	0	10686.06
19	19	10000	133	82	3	0	2	8	19024	0	2374.68
20	20	45800	323	80	3	0	2	8	87008	0	10876.03
21	21	34000	233	90	2	0	0	2	16148	0	8073.91
22	22	5500	113	90	3	10	1	8	10488	0	1306.07

Note: This raster layer VAT is an example CEM.

Count: the cell count for the row or the number of cells in the unique zone with the attribute characteristics as defined by the remaining items in the row. It is generated during the raster combination process.

Conflict: the LUCIS conflict value (integer), which comes from an existing raster during the combination process.

Urbsuit: the LUCIS urban suitability value (integer), which comes from an existing raster during the combination process. It is created from a decimal field (ranging from 1 to 9) and multiplied by 10 during the raster combination process.

County: the county identifier (integer), which comes from an existing raster during the combination process.

City: the city identifier. There are two cities (10, 100) and the rural county (0).

Plandev: planned development identifier (integer), which comes from an existing raster during the combination process.

Sfresden: single-family residential density for development, which is the number of people in a 31 × 31 meter (approximately one-quarter acre) cell size. It is a calculated field added to the raster.

People: single-family potential or the number of single-family residential people that can possibly be allocated to that raster zone. It is a calculated field (count × Sfresden).

Res_year: a placeholder for the allocation year (integer). It will be calculated as the row cells are assigned in the allocation process.

Acres: the number of acres occupied by the cells identified within each row.

The ArcGIS Geoprocessing tools used to create figure 2.11 are the Combine tool shown in figure 2.12 and the Add Field tool shown in figure 2.13.

For example, follow the creation of a simple scenario guide for the allocation of residential development and single-family residential population within the planned unit developments. The conceptual question is, How many acres and what accompanying single-family residential population are in the proposed planned development areas with high or moderate urban preference?

Allocation scenario 1 seeks to identify in county 1 what the planned unit development (PUD) allocations will produce in land-use allocations. The scenario does not look at the four thousand acres for the newly proposed PUD in the county. Instead, it is analyzing the existing PUD number one (PUD1) in county 1 that was approved within the past five years.

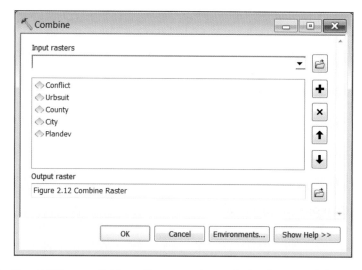

Figure 2.12. The LUCIS greenfield conflict raster layer, final urban suitability raster layer, county boundaries, city boundaries, and plan development locations are added to the Combine tool to create a multivariable raster layer.

Figure from the Arizona Board of Regents on behalf of the University of Arizona.

Figure 2.13. Extra fields are added to the multivariable raster using the Add Field tool.

Figure from the Arizona Board of Regents on behalf of the University of Arizona.

Allocation scenario 1: allocate proposed single-family growth in the planned developments using LUCIS conflict with urban preference (and using the variables Conflict, Plandev, and Urbsuit).

The following selections are for CEM cells that fulfill this criteria: (1) LUCIS Conflict values identified as holding urban preference, (2) the urban suitability layer has Urbsuit values above the mean suitability, and (3) the cell is located in a proposed plan unit development (Plandev 1, Plandev 2, or Plandev 3). To be part of the selection, all three criteria must be true.

1. Conflict = 113, or Conflict = 213, or Conflict = 123, or Conflict = 223, or Conflict = 112. (All these cells occupy acres with high or moderate urban preference.)
2. Urbsuit >= 50. (All these cells have urban suitability higher than the average for the existing urban preference layer.)
3. Plandev > 0. (All the cells are in proposed planned unit developments.)

Before proceeding with an examination of the allocation scenario, examine some of the following basic descriptive statistics. These statistics are included within the data presented in the multivariable raster VAT, or CEM. This information should provide an idea of the possibilities for land-use and population allocation.

- There is a total of 113,960 acres.
- The mix is 15,435 acres in major conflict (high preference) for all three major categories, 1,424 acres in major conflict (moderate preference), and 7,124 acres in major conflict (low preference).
- There are 16,147 acres in agriculture preference and 16,480 acres in urban preference.
- The average acreage for the LUCIS conflict or preference categories is 5,180 acres, with a standard deviation of 4,984 acres.
- The minor conflict (in which two LUCIS categories are in conflict) occupies 57,348 acres, with the largest portion of minor conflict between agriculture and urban preference.
- The total acres available in the two planned unit developments represented in the table are 58,607 acres, which are distributed in three counties.

Table 2.6. Resulting selection and population allocation for allocation scenario 1

OID	Value	Count	Conflict	Urbsuit	County	City	Plandev	Sfresden	People	Res_year	Acres
1	1	10000	333	57	1	0	1	8	18992	0	2374.68
2	2	55500	313	85	1	0	1	8	104480	0	13060.74
3	3	6700	112	76	3	10	0	6	9546	0	1591.04
4	4	3200	113	90	1	100	0	6	4560	0	759.90
5	5	6700	122	66	2	0	0	2	3182	0	1591.04
6	6	68000	311	23	2	0	0	2	32296	0	16147.82
7	7	9900	223	87	2	0	1	8	18808	0	2350.93
8	8	10600	213	88	1	0	1	8	20136	0	2517.16
9	9	45000	313	88	1	100	0	8	85488	0	10686.06
10	10	5000	113	89	3	10	0	6	7122	0	1187.34
11	11	7700	113	90	1	0	0	2	3658	0	1828.50
12	12	8600	213	84	3	10	0	6	12252	0	2042.22
13	13	2200	123	77	3	100	0	6	3132	0	522.43
14	14	10000	112	67	1	0	0	2	4750	0	2374.68
15	15	30000	111	10	1	0	0	2	14248	0	7124.04
16	16	6000	222	44	1	0	0	2	2848	0	1424.81
17	17	55000	333	88	2	0	2	8	104488	0	13060.74
18	18	45000	313	83	2	0	2	8	85488	0	10686.06
19	19	10000	133	82	3	0	2	8	19024	0	2374.68
20	20	45800	323	80	3	0	2	8	87008	0	10876.03
21	21	34000	233	90	2	0	0	2	16148	0	8073.91
22	22	5500	113	90	3	10	1	8	10488	0	1306.07

Table 2.6 provides the information for allocation scenario 1 in OID rows 1, 2, and 8. With a total acreage available in all planned unit developments in County 1 of 17,786 acres, only 2,517 acres, or 14 percent, are in the urban preferred category. Further, all 2,517 acres are in planned unit development number 1 (Plandev 1). The remainder of acres for Plandev 1 in County 1 are in the LUCIS minor- or major-conflict category. The LUCIS preference value is a good indication that these areas, if developed, will contribute to the county's urban sprawl. Using an average household size of two people and four units per acre, the 2,517 acres will support 20,136 new people in single-family residential units. There will be no new commercial or retail space in the development. Such new greenfield development will require more residential infrastructure, such as roadway maintenance, sewer and waterlines, emergency services, fire and police services, and new classroom space. Because LUCIS urban suitability includes proximity to existing urban infrastructure, these lands will most likely have reduced infrastructure costs compared with other areas in the planned unit development. However, there will be new increased costs to the county for services and schools, which research shows are never completely covered from the property tax base generated by new development.

Even more interesting is that Plandev 1 is located in all three counties, making it a megadevelopment. As stated, the counties are interested in urban densification to reduce services and support decreased transportation costs. The development of single-family homes in Plandev 1 will surely decrease the three counties' average density (units per acre or people per acre). And it will have an adverse impact on COG's desire to increase density within the existing urban boundaries. The total new single-family residential population in Plandev 1 that is in the urban preferred category supports 49,432 people in the three counties. With an average household size of two people per unit, that is 24,716 new single-family units if the developer is restricted from development in agriculture or conservation preference areas.

This simple example using LUCIS conflict and a multivariable raster layer provides some insight into LUCIS land-use analysis and population allocation. The remaining chapters in this book expand this concept to analyze complex land-use issues, including transit-oriented development, increasing density within existing urban areas, urban mixed-use development opportunities, and natural disasters such as flooding and hurricane damage.

Chapter summary

The following points are discussed in this chapter:

- LUCIS relies on suitability to create preference and collapsed preference to define land-use conflict, using 27 greenfield categories.
- LUCIS conflict can be modified to create urban and suburban mixed-use opportunity, again creating 27 categories, for opportunity.
- The best method to capture trend development is to identify the existing conditions within the study area.
- The Combine tool can combine the LUCIS conflict or mixed-use opportunity raster data with other raster data to develop a multivariable raster VAT or CEM.
- The Add Field tool can add extra data to the LUCIS CEM. This data can include important information regarding urban development policy, transit opportunity, and urban development incentives to enhance the land-use allocation opportunities presented in *The LUCIS Model* (Carr and Zwick 2007).
- The LUCIS CEM is a complex raster layer that is the spatial representation of multiple combined raster layers, including the greenfield conflict or mixed-use opportunity raster layers. CEM is used to develop multiple future land-use scenarios within a visioning process.

ArcGIS tools referenced in this chapter

Tool Name	Version 10.2 Toolbox/Toolset
A4 Allocation by Table	A4 LUCIS Tools/A4 Allocation
A4 LUCIS Community Values Calculator	A4 LUCIS Tools/A4 Suitability Overlay
Add Field	Data Management Tools/Fields
Combine	Spatial Analyst Tools/Local
Select (By Attributes)	Analysis Tools/Extract
Raster Calculator	Spatial Analyst Tools/Map Algebra
Weighted Overlay	Spatial Analyst Tools/Overlay
Weighted Sum	Spatial Analyst Tools/Overlay
Zonal Statistics as Table	Spatial Analyst Tools/Zonal

References

Carr, Margaret H., and Paul D. Zwick. 2007. *Smart Land-Use Analysis: The LUCIS Model*. Redlands, CA: Esri Press.

Esri. 2011. *ArcGIS 10 Help Documentation*. Redlands, CA: Esri.

Saaty, T. L. 1980. *The Analytic Hierarchy Process*. New York: McGraw-Hill.

Chapter 3
Suitability automation tools: A4 and LUCIS^{plus}

Abdulnaser Arafat

What this chapter covers

Land-use planning has become more complex with the increased focus on achieving sustainable development goals. The modern planning process involves conflicting and contradicting interests among conservation, economic development, and the accommodation of changing lifestyle preferences and burgeoning populations. The evolution of suitability analysis requires current methods to be "more accurate, legally defensible, technically valid, ecologically sound, and open to scrutiny by the public" (Ndubisi 2002, 102). This evolving new modeling approach adds to the complexity of how planners handle suitability analysis and the new role of geographic information systems (GIS) in land-use decision-making. However, the evolution of GIS analysis and the development of tools that automate suitability modeling have reduced this complexity. The new tools also help estimate urban form metrics and develop alternative land-use scenarios.

This chapter describes how LUCIS uses programming methods to enhance ArcGIS tools. Programming creates customized automated tools that can run complex procedures that are difficult to run using standard tool technology. This chapter also introduces A4 LUCIS Tools. This package of tools automates Land-Use Conflict Identification Strategy (LUCIS) processes used in suitability modeling, and also LUCIS^{plus}, a strategy for creating planning land-use scenarios. You do not need a prior knowledge of programming or a desire (or a lack thereof) to learn programming to understand the purpose, applicability, or usefulness of the A4 LUCIS Tools.

Land-use modeling as automated sequential procedures

GIS layer overlay is the core of suitability analysis. Even suitability analysis done in the time of hand-drawn maps depended on map overlay (Collins et al. 2001; McHarg 1969). The overlay procedure in GIS raster analysis depends on four logical spatial overlay rules: enumeration, dominance, contribution, and interaction. According to Margaret H. Carr and Paul D. Zwick (2007, 50–57):

> *Enumeration* preserves all attribute values from multiple input layers. Enumeration creates an output layer that combines all attributes from the spatial input layers to provide a clear and distinct set of unique attribute combinations from the input. … The *dominance* rule depends on the selection of a single value that is preferred over all other values found at the same spatial location. The selection is defined or governed by external rules, not simply the combination of values. … The *contributory* rule is applied by performing a group of operations [which are] values from one input contributing to the results without regard for the values from other inputs. … Lastly, the *interaction* rule, unlike the contributory rule, considers the interaction between factors. However, to consider interactions between factors, the factors must be translated into the same standard intervals.

These rules represent logical operations that translate into equivalent functions in land-use modeling. The land-use functions include layer weighting and combining different utility layers into a suitability layer.

The dynamic relationship between land characteristics and land use illustrates the complexity of land-use suitability analysis (Driessen and Konijn 1992). Through interaction, utilities—in the form of single-utility assignments (SUAs) and multiple-utility assignments (MUAs)—are combined to create suitability. However, utility is a measurement of human satisfaction. If utility is applied to land use, it could represent the appeal of land characteristics to a segment of the population.

The GIS overlay techniques for the multicriteria decision-making (MCDM) method, a land-use modeling scheme using a hierarchal structure, can be divided into two main methods. These methods are the multiobjective method and multiattribute method. The multiobjective method uses a set of constraints to combine two or more objectives. Standard linear programming methods can perform this combination. The problem is that adding constraints will help the planner in decision-making but add computational complexity. This complexity makes it difficult to apply in a GIS environment. The multiattribute method uses GIS map algebra techniques. It uses weighted linear combination (WLC) and the Boolean operators AND and OR in the overlay process. However, this process gives the same weight regardless of the geographic location. WLC is based on the concept of a weighted average. In this method, relatively more importance is given to the attributes because of the assumption that location is taken into account in generating each layer combined by the Boolean operator.

So the multiobjective method is a programming procedure, and the multiattribute method is similar to the LUCIS process.

The MCDM method also uses ordered weighted averaging (OWA) to overcome the disadvantages of WLC. The OWA method involves two sets of weights. The criterion importance weight is constant for the criterion at all locations. The order weight is associated with the criterion on a location-by-location basis (geographic or spatial weights). MCDM also uses the analytic hierarchy process (AHP) to generate the linear combination weights. AHP is also used as a consensus-building tool in situations involving group decision-making (Malczewski 2004).

Various ranking and ordering procedures are used to assess the importance of weights. They include programmed and automated procedures, as well as community participation using Delphi or pairwise comparison methods (Carr and Zwick 2007; Malczewski 1999 and 2004). The pairwise comparison technique, developed by T. L. Saaty in the 1970s and 1980s in the context of AHP multiple-criteria evaluation methods, outlines the relative importance of criteria. According to Timothy L. Nyerges and Piotr Jankowski (2009, 140–41):

> Weights are not assigned directly but represent a "best fit" set of weights derived from the eigenvector of the square reciprocal matrix used to compare all possible pairs of criteria. The advantage of this technique is that information can be used from handbooks, regression output, or decision modelers/experts can be asked to rank order individual factors.

John Malczewski (1999) defines *weight* as a value assigned to the output of criterion evaluation. The weight represents the relative importance of the output. The criterion is more important if the weight is higher and less important if the weight is lower.

Both MCDM and LUCIS integrate weights into their methods. MCDM provides four methods for assessing criterion weights: ranking, rating, pairwise comparison, and trade-off analysis. Malczewski (1999) recognized that the choice of method depends on the trade-offs the user is willing to perform, the availability of software, and the method of incorporating GIS-based criteria evaluation.

The value of LUCIS is twofold. The first stage determines land-use conflict. It includes (1) determining land-use suitability based on predetermined goals and objectives, (2) determining land-use preference, and (3) identifying conflict. The second stage illustrates alternative futures through the allocation of population and employment. As stated in chapter 2, the conflict raster layer is a preference matrix using the cumulative suitability of the goals within each land use. Early applications of LUCIS allocated people and employment according to a general "urban" category. The development of a conflict layer does not manipulate the original preference values. A conflict layer can also be generated between goals for a more detailed analysis of land-use preference. Thus, allocation of urban uses has evolved from areas classified solely as urban to allocating projected residential populations to areas with high multifamily and single-family preference.

LUCIS^{plus} is an improved strategy of land-use decision-making. It carries the traditional LUCIS allocation procedure to a new level with the addition of automated planning land-use scenarios (plus). The criteria evaluation matrix (CEM) and scenario and policy tables (SPT) help automate this method. (CEM is described in chapter 2.) SPT contains the conditions and controls over CEM that form the scenario allocation instructions.

The MCDM scenario-building approach evaluates the criteria for different model alternatives. The process uses a selection of the most appropriate scenarios, running the process for each alternative. However, in the LUCIS structure and LUCIS allocation procedure, scenario building is performed on multiple levels. The first level occurs as a result of changing the weights when combining suitability raster layers for each hierarchical level. MCDM uses the same analysis. The second level occurs in the flexible allocation scenario. The conflict and suitability assignments are used in the CEM raster layer. The population allocation is performed according to priorities specified in different scenarios in SPT. CEM joins the conflict and suitability layers. Then it preserves the attributes for these layers in the overlay.

LUCIS^{plus} automation tools

Automation toolsets support the automatic allocation procedures in the A4 LUCIS Tools toolbox and in LUCIS^{plus}. This chapter explains three types of automation toolsets. They include automation tools for suitability assignment, suitability overlay, and population allocation based on CEM and SPT. This book uses automation tools that go beyond the three mentioned categories. However, this chapter discusses only the tools developed for the LUCIS land-use models. The first toolset, A4 Suitability, contains only one tool, the suitability assignment tool. A4 LUCIS Suitability is a raster reclassification tool that reduces the manual calculation used outside the GIS environment. The second toolset is A4 Suitability Overlay, which contains two tools. The first tool calculates suitability overlay weights according to community values (A4 LUCIS Community Values Calculator). The second tool performs the overlay between suitability assignments (A4 LUCIS Weights). Finally, the third toolset, A4 Allocation, contains three tools: A4 Standard Trend Allocation, A4 Allocation by Table, and A4 Detailed Allocation.

Automation tools for suitability assignment: A4 Suitability tools

The suitability index includes values that represent the relative usefulness of a land use. The utility values are classified in a range of 1 to 9. Different methods of reclassification are described in chapter 2. Reclassification depends on the nature of the criteria being evaluated. It also depends on the utility being classified as a suitability layer. This classification procedure can be performed using interval, ratio, nominal, or ordinal data. Some of the procedures are simple, and some are more complex.

Proximity-based indicators of change are probably the most important in land-use analysis. They integrate transaction costs in determining land-use opportunity. Before the introduction of the A4 LUCIS Suitability tool, the planner would use the ArcGIS Spatial

Analyst tool Zonal Statistics as Table to generate the mean (MEAN), standard deviation (STD), and minimum (MIN) or maximum (MAX) values. Then the planner would manually calculate the suitability intervals for nonbinary classifications. After determining the values for each interval, the planner would manually input these values into the Reclassify tool. This method proved to be time consuming, cumbersome, and prone to error.

The A4 LUCIS Suitability tool performs these calculations for the planner. The tool creates nine intervals between the MEAN and MIN or MAX values. The resultant suitability indexes depend on whether the suitability is decreasing (figure 3.1) or increasing (figure 3.2) in the input distance layer. The tool (figure 3.3) can work as a stand-alone tool in the GIS environment or a customized tool within the ArcGIS ModelBuilder application. Zonal Statistics as Table and Euclidean Distance are typical ArcGIS tools used in LUCIS suitability models. The A4 LUCIS Suitability tool is added to the ModelBuilder model to automate the reclassification process (figure 3.4).

You can easily insert the output of the Zonal Statistics as Table tool into the A4 LUCIS Suitability tool. The tool uses the values from that table. The tool does all the required calculations to generate the suitability table ranges between the minimum and maximum values. The tool will also generate the reclassification table for the raster. The reclassification table resulting from the A4 LUCIS Suitability tool is a listing of the LUCIS suitability index assignments. It also lists the range of values it assigns to the specific utility value. To determine this utility value, the average of the mean values of all zones listed acts as the baseline for suitability and one-quarter standard deviation ranges. Figure 3.5 shows an example zonal statistics table for distance to shopping. Figure 3.6 shows the output reclassification table internally generated by the tool. This table is automatically used to reclassify the input raster. The tool can also accept a user-generated reclassification table in any database file (DBF) format. Use the same DBF format for the table containing the reclassification ranges and suitability score as the output table generated by the tool. You can also use Microsoft Excel or Access software to generate the reclassification table. You can manually modify the remap table produced by the A4 LUCIS Suitability tool and use the modified table for subsequent model analysis.

LUCIS employs two possible suitability index classification value ranges. Increasing suitability ranges from 1 to 9, and decreasing suitability ranges from 9 to 1. Increasing suitability means that the farther away a feature, such as a noise source, is from its objective (for example, residential development), the more suitable the land (suitability increases from 1 to 9 with increasing distance). Decreasing suitability means that the closer a feature, such as roads, is to its objective (again, residential development), the more suitable the land (suitability decreases from 9 to 1 with increasing distance). Because proximity to roads is desired for accessibility in residential development, the farther from roads, the less suitable the accessibility. Therefore, suitability decreases as the distance increases away from roads. The A4 LUCIS Suitability tool allows you to indicate the suitability index as decreasing or increasing within the A4 LUCIS Suitability tool interface. If you choose the decreasing suitability option, the tool will use the mean and a one-quarter standard deviation to compose ranges that correspond to the suitability index values from 9 to 1. It starts with a suitability index of 9 for all values up to the MEAN value. It decreases by one-quarter standard deviation increments for eight intervals between the MEAN distance and MAX distance values (see figure 3.1). Because suitability index 1 is the last value calculated, this value range may be larger or smaller than the other eight suitability index ranges. If the one-quarter standard deviation value is less than the cell size, the suitability index values are divided into equal intervals between the MEAN distance and MAX distance values.

Increasing suitability is calculated in a similar manner. If you choose the increasing suitability option, the tool will prepare suitability index values from 1 to 9. It starts with a suitability index of 9 for all values above the MEAN. It decreases by one-quarter standard deviation increments for eight intervals between the MEAN and MIN (see figure 3.2). Because a suitability index of 1 is the last value calculated, this value range may be larger or smaller than the other eight suitability index ranges. If the one-quarter standard deviation value is less than the cell size, the suitability index values will be divided into equal intervals between the MEAN and MIN.

The accuracy of the A4 LUCIS Suitability tool has been verified many times. The tool results have been compared with the results of reclassifying the same raster doing careful and time-consuming manual calculations. The results have also been tested using the ArcGIS raster reclassification tools. The reclassified output raster layers are always identical, but the A4 LUCIS Suitability tool automates the process in ArcGIS. The A4 LUCIS Suitability tool reduces the chance of errors caused by manual calculations and significantly reduces the time involved in processing.

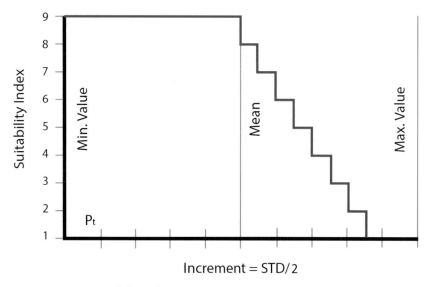

Figure 3.1. Decreasing suitability indexing is calculated using the mean and one-quarter standard deviation to compose ranges that correspond to the suitability index values from 9 to 1. If the one-quarter standard deviation value is less than the cell size, the suitability index values are divided into equal intervals between the MEAN and MAX values.

Figure from Abdulnaser Arafat and the Arizona Board of Regents on behalf of the University of Arizona.

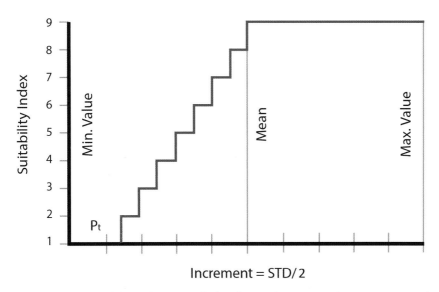

Figure 3.2. Increasing suitability indexing is calculated using the mean and one-quarter standard deviation to compose ranges that correspond to the suitability index values from 1 to 9. If the one-quarter standard deviation value is less than the cell size, the suitability index values are divided into equal intervals between the MEAN and MIN values.

Figure from Abdulnaser Arafat and the Arizona Board of Regents on behalf of the University of Arizona.

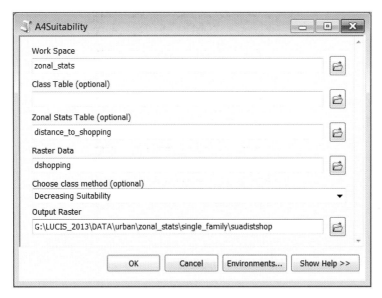

Figure 3.3. The A4 LUCIS Suitability tool interface requires raster inputs and creates an output raster that illustrates suitable distance ranges according to the zonal statistics table or a user-created table. The user selects whether the output suitability ranges are based on an increasing or decreasing suitability method.

Figure from Abdulnaser Arafat and the Arizona Board of Regents on behalf of the University of Arizona.

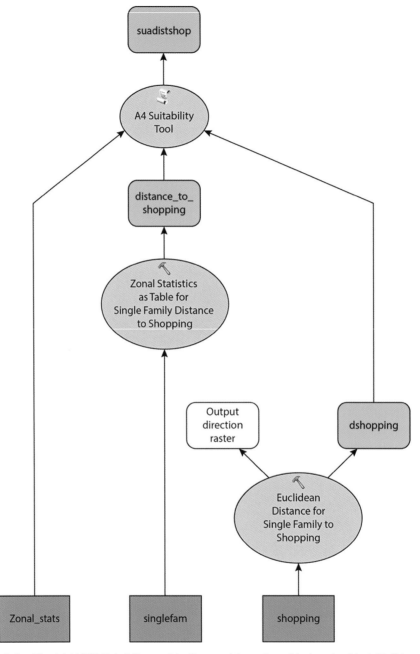

Figure 3.4. The A4 LUCIS Suitability tool (yellow ovals) can be added to the ModelBuilder model, just as any ArcGIS tool. The A4 LUCIS Suitability tool can substitute for the Reclassify tool used in the original LUCIS method.

Figure from the Arizona Board of Regents on behalf of the University of Arizona.

Table: distance_service_bus_tran

Rowid	DESCRIPT2	ZONE-CODE	COUNT	AREA	MIN	MAX	RANGE	MEAN	STD	SUM
1	CONDOMINIUM	1	3996	399600	3.506871	1482.755	1479.248	418.0105	382.5372	1670370
2	TOWNHOUSE/VILLA	2	23524	2352400	2.42691	5734.66	5732.233	399.9405	598.4563	9408200
3	MULTI RES DWELLINGS	3	5401	540100	12.01107	1533.837	1521.826	189.1579	172.7457	1021642
4	OFF MULT-STY CONDO	4	8	800	12.98778	40.92805	27.94026	18.94187	8.785759	151.535
5	CONDO APARTMENT	5	6	600	63.89187	105.3076	41.4157	78.9265	15.17257	473.559
6	MIXED USE MULTI FAM	6	400	40000	3.947716	156.161	152.2132	63.92339	41.79302	25569.36
7	MH CONDOMINIUM	7	1031	103100	38.16571	269.9223	231.7566	178.6725	62.27672	184211.3

Figure 3.5. The output table created by the Zonal Statistics as Table tool includes a breakdown of MIN, MAX, MEAN, and STD, all fields used to calculate increasing or decreasing suitability in the A4 LUCIS Suitability tool.

Figure from the Arizona Board of Regents on behalf of the University of Arizona.

A4 Suitability

Rowid	FROM1	TO	OUT	MAPPING
1	0	484334.61328	9	ValueToValue
2	48434.613281	71225.847656	8	ValueToValue
3	71225.847656	94017.082031	7	ValueToValue
4	94017.082031	116808.31640	6	ValueToValue
5	116808.31640	139599.55078	5	ValueToValue
6	139599.55078	162390.78515	4	ValueToValue
7	162390.78515	185182.01953	3	ValueToValue
8	185182.01953	207973.25390	2	ValueToValue
9	207973.25390	1000000	1	ValueToValue

Figure 3.6. The output table of the A4 LUCIS Suitability tool lists the distance ranges for each suitability classification. The values created by the A4 Suitability tool were compared with values calculated by hand, and the results were identical.

Figure from the Arizona Board of Regents on behalf of the University of Arizona.

Automation tools for suitability overlay

The LUCIS suitability workflow is illustrated in figure 3.7. Once the suitability of each objective and subobjective is determined, they are combined according to their hierarchical level. The utility weights add up to 1.00 (100 percent). The weights at the objective and subobjective levels are citizen driven. The weights obtained at this level reflect localized knowledge of community values. These weights are obtained from existing plans, community meetings, and focus groups. Surveys are also often used to gauge community values.

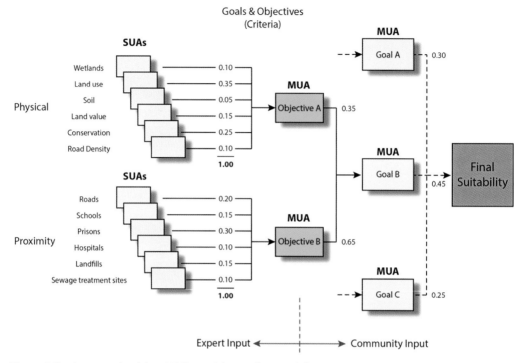

Figure 3.7. An example of the LUCIS suitability analysis workflow consists of the identification of geospatial data reflecting physical and proximity-based characteristics to measure what to accomplish (goal) and how to achieve that goal (objective). Suitability for each subobjective is measured, and weights that reflect how much each criterion influences the objective are applied. Weights are then applied to each objective reflecting community values. The objective MUAs are combined to illustrate the final goal suitability. Goals are combined to create the final suitability layer for the stakeholder.

Figure from Paul D. Zwick and the Arizona Board of Regents on behalf of the University of Arizona.

A4 LUCIS Community Values Calculator

The A4 LUCIS Community Values Calculator is used to determine the numeric weight, particularly between goals. Based on pairwise comparison methods, this program blurs the line between planner and land-use modeler. A planner with minimal experience in modeling can easily use this program within a GIS environment to complete a values survey among stakeholders. When evaluating the importance between objectives or between alternatives, the A4 LUCIS Community Values Calculator integrates any number of objective and subobjective raster suitability layers as inputs. The A4 LUCIS Community Values Calculator is a program that can be used within the ArcGIS ArcMap application as an icon on the toolbar. Its interface prompts the user to specify the usefulness of each pair of raster layers and dynamically compares the raster pair. As you indicate the values for each pair, the A4 LUCIS

Community Values Calculator automatically populates a pairwise comparison matrix. The calculator then outputs a parameter table of the raster names and their corresponding relative weights.

A reasonable example of this tool follows in which the A4 LUCIS Community Values Calculator weights subobjectives for multifamily residential suitability in a regional visioning planning meeting.

During a two-hour regional visioning exercise in east central Florida, the 12 participants engaged in a facilitated values assessment and mapping exercise. They placed stickers on a paper map, reflecting spatial patterns of development and generalized land uses and density. Throughout the evening, participants voiced concerns that the process did not take into account what was best for the land. There did not seem to be a clear connection between how these random yet creative exercises informed decision-making. Participants were concerned that the process ignored future land-use decision-making that would include science or a more thorough analysis of land suitability. The meeting moderator, who was the local planning director, agreed that there should be a connection between community values and typical factors of development. At the least, the factors considered during the subdivision platting process should be observed. The planning director then listed the three generalized factors used to determine land suitability for multifamily residential use. The factors were physical characteristics, such as noise and infrastructure; density of multifamily; and proximity. The planning director then asked each participant to compare two criteria at a time, such as physical characteristics and density or physical characteristics and proximity. Participants were told to indicate which one they believed was more important and to what degree, using the AHP scale of importance (table 3.1). As each participant provided their feedback, the GIS technician in the room recorded their responses directly into the A4 LUCIS Community Values Calculator (figure 3.8). When the participants completed their preferences, the technician ran the tool and a table was generated (figure 3.9). The table reflects the weights for each raster based on community input.

Table 3.1. **AHP importance categories**

Category	Value
Extremely more important	9
Very strongly to extremely more important	8
Very strongly more important	7
Strongly to very strongly more important	6
Strongly more important	5
Moderately to strongly more important	4
Moderately more important	3
Equally to moderately more important	2
Equally important	1

Source: Saaty (1980).

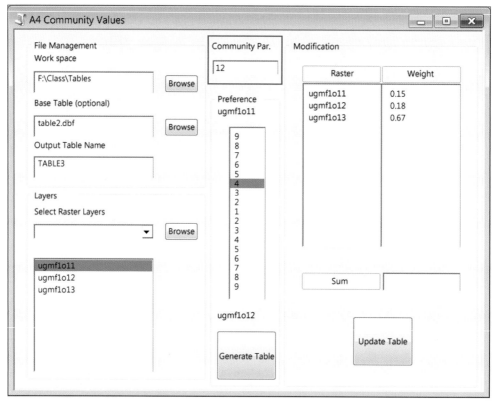

Figure 3.8. The A4 Community Values Calculator calculates SUA weights from user, expert, or community comparison of input criteria. The tool can be used in ArcMap or as a stand-alone tool.

Figure from Abdulnaser Arafat and the Arizona Board of Regents on behalf of the University of Arizona.

To gauge community participation, the tool also uses an algorithm to update the weights based on the different pairwise comparison assignments for a group of people or a panel meeting. The result is a table of weights reflecting group values. These weighted values are then used as an input for the A4 LUCIS Weights tool. This tool is used to create complex MUAs. Although many multicriteria decision support tools are available, having this tool available within the GIS unit saves time. It eliminates the expense of purchasing a third-party software package and reduces error when inputting values from a stand-alone software package.

The A4 Community Values Calculator integrates any number of objective and subobjective raster suitability layers as inputs to evaluate the importance between goals, objectives, or alternatives. The A4 LUCIS Community Values Calculator interface prompts the user to specify the usefulness of each pair of raster layers and dynamically compares the raster pair. As you indicate values for each pair, the calculator automatically populates an internal pairwise comparison matrix. When you enter the values for all pairwise comparisons, the calculator then outputs a parameter table of the raster names and their corresponding relative weights.

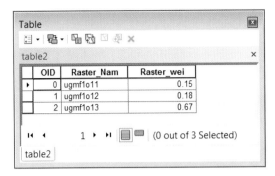

Figure 3.9. The DBF output table of the A4 Community Values Calculator. The output table illustrates the weights generated by the tool. Weights are generated after all participants provide their comparison scores, and a pairwise comparison method calculates the weights shown.

Figure from the Arizona Board of Regents on behalf of the University of Arizona.

More features of the tool include the ability to accept feedback from more than one participant, thus integrating community participation into the tool. You can use the community participation field (Community Par) and participants' votes to generate weights. The tool uses the geometric mean to calculate the pairwise comparison scores generated from the votes of all participants.

Other methods update existing weight tables. If, for example, you are using the LUCIS method in a Delphi process, you can customize the table of weights outside the ArcMap environment. Then navigate to the table in the Base Table field of the tool to accommodate input on the weight values. Run the pairwise comparisons again and update the table using the Update Table button instead of the Generate Table button. You can also manually enter the new weights in the table provided by the tool interface and then click the Update Table button.

A4 LUCIS Weights

To determine the final suitability of each land use, the weights generated by the A4 LUCIS Community Values Calculator measure the degree of interaction between each goal MUA. The A4 LUCIS Weights tool is similar to the Weighted Sum tool in the ArcGIS Spatial Analyst toolbox. Both tools can multiply multiple raster layers by a specified weight and then sum the layers together. Instead of the weights being entered manually for each goal layer, the A4 LUCIS Weights tool uses the parameter table generated from the A4 LUCIS Community Values Calculator as an input. The A4 LUCIS Weights tool can also use a table of similar structure generated outside the A4 Community Values Calculator as an input.

The A4 LUCIS Weights tool relies on the DBF output table generated by the A4 Community Values Calculator. Updating this table updates the associated weights in the suitability model, which allows you to perform weighting scenarios and use the new weights to run the models. This feature is useful within a community meeting or other public participation activity. Figure 3.10 shows the weights tool in ModelBuilder.

The input layers to the A4 LUCIS Weights tool are raster datasets. Figure 3.11 illustrates raster datasets for physical characteristics, proximity to services, and density for east central Florida. The A4 LUCIS Weights tool uses map algebra to combine the input raster layers. It generates the output raster (figure 3.12) according to the weights in the input table.

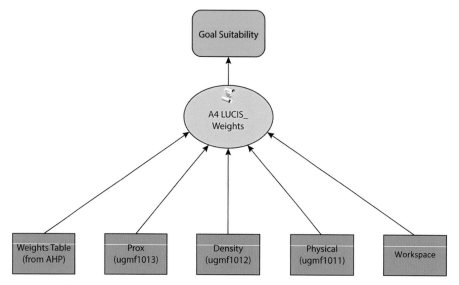

Figure 3.10. The A4 LUCIS Weights tool is added to a model and applies the weights generated from the A4 LUCIS Community Values Calculator to create an MUA of the goal suitability.

Figure from the Arizona Board of Regents on behalf of the University of Arizona.

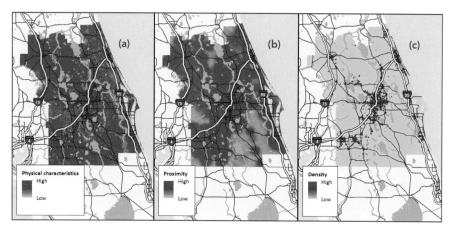

Figure 3.11. The suitability layers for (a) physical characteristics, (b) proximity, and (c) density are objectives of the residential goal. These suitability layers will serve as input layers for the A4 LUCIS Weights tool.

Figure from Abdulnaser Arafat and the Arizona Board of Regents on behalf of the University of Arizona. Water bodies from the US Geological Survey; state boundary from US Census Bureau TIGER/Line Files.

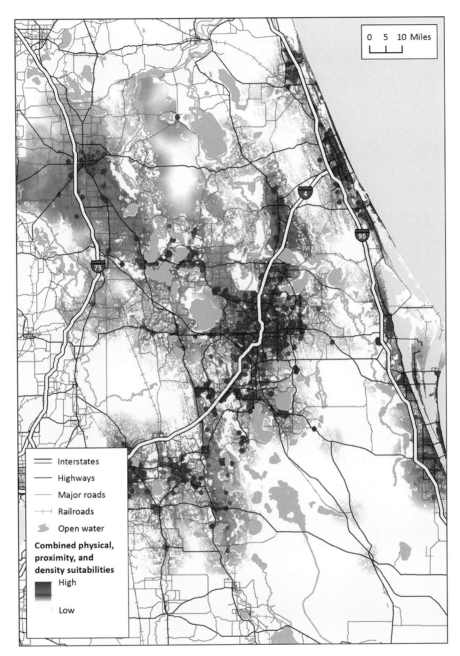

Figure 3.12. The A4 LUCIS Weights tool output raster after combining the individual suitability layers from figure 3.11. The layers are combined using the resultant weighted values of the A4 Community Values Calculator. The weights tool can also use a table that is populated with user-determined weights based on policy or expert knowledge.

Figure from Abdulnaser Arafat and the Arizona Board of Regents on behalf of the University of Arizona. Water bodies from the US Geological Survey; state boundary from US Census Bureau TIGER/Line Files.

The output raster is added to the workspace directory specified in the A4 LUCIS Weights tool dialog box. The model used in figure 3.10 shows the input weights table as a variable added in ModelBuilder as Weights Table (from AHP). The tool automates models in ModelBuilder. However, the tool can also be used on its own as a stand-alone tool. It has a user interface, so you can run the tool inside or outside the ArcGIS user interface. The output of the tool has been verified by manually generating the weights using the AHP procedure and combining the layer using the Weighted Sum tool in ArcGIS. Each time, the results were identical.

Automation tools for population allocation

The goal of LUCIS is to identify land-use opportunities and potential land-use conflicts. Conflict identification is useful in various applications, including visioning efforts and future land-use allocation scenarios (Carr and Zwick 2007). The allocation process based on LUCIS outputs identifies growth opportunities and helps planners distribute residential uses and employment based on density, policy, and other factors.

You can automate the allocation procedures in the base LUCIS model by integrating a set of GIS tools for future allocation, scenario building, and testing policies. For example, the automatic allocation of urban land using the LUCIS procedure works according to development type: infill, redevelopment, or greenfield. To set up the scenario guide, establish the priority of each development type. *Infill* "refers to development in urban areas with existing streets, infrastructure, and development. *Greenfield* refers to development on previously undeveloped ('green') parcels in suburban or nonurban locations with limited existing infrastructure and development" (EPA 1999, 2). *Redevelopment* refers to the reuse and development of underused sites.

The use of GIS with LUCIS facilitates scenario building and the allocation process. The identification of land-use opportunity and allocation of residential uses and employment are executed based on the process outlined in the scenario guide. These priorities may depend on growth patterns, current or proposed densities, transportation policy, and so forth. Considering the demand for accuracy and the time spent on this complex procedure, the allocation process calls out for automation that can be done more feasibly and flexibly.

The A4 Allocation tools are used to allocate future population in land-use models such as LUCIS. This toolset contains three tools: the A4 Standard Trend Allocation tool, A4 Detailed Allocation tool, and A4 Allocation by Table tool. The allocation tools create an iterative environment that runs and displays queries to find the most suitable lands and populate them by proposed population values. The tools run on the CEM complex raster layer. A CEM acts like an enumeration container made up of values of many participating raster layers that compose the combined raster layer. The tools read the raster attribute table and use that table as a database for applying the selection queries. However, because the combined raster layer is typically a large file that can contain millions of records, some allocation processes can take time to execute. Thus, programming methods are used to reduce the size of the combined raster layer and increase the query processing speed.

A4 Standard Trend Allocation tool

The A4 Standard Trend Allocation tool (figure 3.13a) works on a CEM that contains a conflict layer as well as other masks and constraints to prioritize the allocation process, using the power of combined raster layers. The tool uses an enumeration rule on CEM that combines many raster layers and retains their attribute values. This tool works in an iterative procedure to allocate all the available spaces specified by the conditions or constraints. These conditions could be the suitability values and conflict scores.

The A4 Standard Trend Allocation tool works on two conditions and up to six masks. The conditions mainly consist of a query that can hold any value, while the mask query can hold only the values of 1 or 0. The overall query typically consists of using the contributory

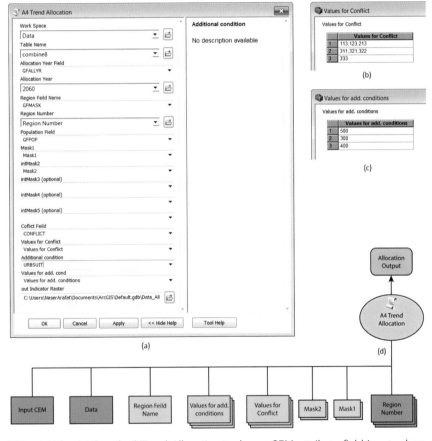

Figure 3.13. (a) The A4 Standard Trend Allocation tool uses CEM attribute field input values and automates allocation based on (b) the rule for allocation based on the conflict numbers and (c) the rule provided for allocation based on the additional condition, which is urban suitability in this figure. The A4 Standard Trend Allocation tool automates the procedure spelled out in the scenario guide discussed in chapter 2. (d) The A4 Standard Trend Allocation tool can be added to the ModelBuilder model.

Figure from the Arizona Board of Regents on behalf of the University of Arizona.

relationship between the two conditions and the first three masks and then a dominancy relationship between the next three masks. Carr and Zwick (2007, 51, 55) explain the dominance and contributory rule as follows: "The dominance rule depends on the selection of a single value (the dominant value) that is preferred over all other values found at the same spatial location. The contributory rule indicates that the values from one input contribute to the results without regard for the values from other inputs." If the user leaves a field empty, it will be considered optional and thus removed from the query.

Understanding the concept of generating the query is the first step to using the tool. However, the next important step is understanding the internal iteration of the A4 Standard Trend Allocation tool. The input of the conditions can be multivariate and separated by semicolons (figure 3.13b). The tool takes the value of the first number of the first condition and generates queries that iterate through all the numbers in the condition. It generates iteration query statements that are used to allocate population. The tool will also loop to all the values of the first condition, which acts like an external iteration loop (figure 3.13c). The tool output comes in two forms. First, the tool generates an output population raster; second, the tool populates year and population fields in CEM (figure 3.13d). If you are extracting the output from CEM, CEM can be used in different reclassifications or summary methods that allow the display of allocation results.

A4 Detailed Allocation tool

The A4 Detailed Allocation tool has the same function as the A4 Standard Trend Allocation tool but with more conditions. The A4 Standard Trend Allocation tool works on up to six masks and two conditions for the allocation. The Detailed Allocation tool can work on six different masks and six extra conditions for an allocation procedure. These added conditions act as inner loops for the iteration. Therefore, careful use of this tool can lead to allocating the population in a region in one step, using the power of iteration. Even though the detailed tool has a user-friendly interface, it is complex. The user should understand exactly how the iteration is performed before using the tool. The simplicity of the A4 Standard Trend Allocation tool and the details of the A4 Detailed Allocation tool were combined to create a crossover tool, the A4 Allocation by Table tool. This tool allows you to perform simpler iterations inside the tool and other external iterations in ModelBuilder.

A4 Allocation by Table tool

The A4 Allocation by Table tool is more sophisticated than the A4 Standard Trend Allocation tool and simpler than the detailed tool. The Allocation by Table tool has two conditions and six masks. The tool can work with up to eight conditions, two of which are iterative inside the tool. The last four mask conditions have a dominancy relationship between them. Their results have a contributory relationship with the first mask and the other conditions. The tool performs larger external iterations in ModelBuilder. The SPT table directs these iterations, which adds more flexibility to the process. The conditions the tool uses vary in complexity. The conditions depend on what you want to do in your scenario. A list performs the model iteration in ModelBuilder (figure 3.14). These iteration lists are taken from an SPT

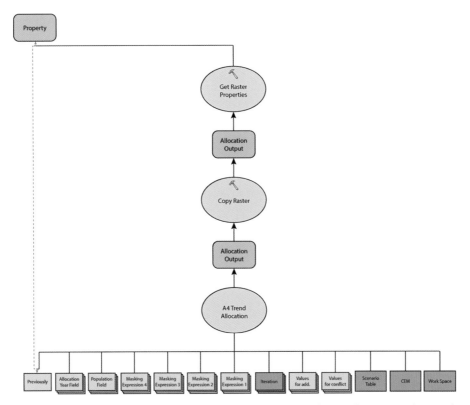

Figure 3.14. The Allocation by Table tool uses up to eight CEM rules to allocate population. The value of the Allocation by Table tool is that in addition to the CEM rules, the tool also iterates population distribution using an iteration list provided in an SPT.

Figure from the Arizona Board of Regents on behalf of the University of Arizona.

(table 3.2). The tool uses the values of the iteration table, row by row, to allocate the population. It uses the conditions specified by the fields of the table. The tool populates the POP field in SPT, which summarizes the population allocated for each iteration.

Allocation scenario 2

Allocation scenario 2 is a growth scenario for sustainable development that encourages transit-oriented development (TOD), mixed use, and redevelopment.

Orange County, Florida, anticipates the largest share of population growth within the east central Florida region over the next 20 years. To encourage more sustainable land-use practices and satisfy goals from a recent regional visioning exercise, Orange County planning officials believe that the community "has spoken." They believe the community desires a future that includes more redevelopment and mixed-use TOD options. As described in chapter 2, planning officials must first develop a scenario guide to outline the order of population allocation. The Orange County planning officials choose to test the A4 automated allocation procedures to model transit-oriented allocation instead of creating queries and using the Select By Attributes tool. They complete the allocation using the Allocation by Table tool (table 3.3) and the following steps:

1. The multifamily transit redevelopment allocation starts by constructing query 3.1 in CEM, using the region condition in addition to Mask1 and Mask2 conditions.

   Query 3.1.
   `Region = 2 AND "CITY" > 0 AND "TRANACCESS" > 50.`

   The query says that the tool will select all places inside cities in the region that have a transit access score of more than 50. The city and transit access raster layers are represented by fields in CEM. The region number is input directly into the tool. However, it can also be added as a field to the SPT table.

2. The results of the query are further refined by adding the query generated from Mask3 to Mask5. Any empty or repeated fields are optional and therefore above Mask4 in the SPT table. Mask5 is not included, and the further refinement is executed by the query `"LUMIX" >= 3`. The query says that the selection must satisfy a land-use mix score of 3 and above. The land-use mix raster layer is also represented by a field in the CEM complex raster layer.

The conflict and additional fields are iterative fields that work together. The semicolon in the fields signifies moving from one variable to another in the allocation. Therefore, for the first row in SPT, the tool will take the results of steps 1 and 2 and refine them by selecting the conflict number `CONF = 464`. CONF 464 is a no-conflict area suitable for multifamily. The tool will perform the population allocation inside this category using the additional condition field (ACOND) for land value. Here, the allocation will start with `ACOND = 9` and then `ACOND = 8`, sequentially. When all the cells in the 464 category are allocated, the loop will move to `CONF = 465` in a similar fashion.

The SPT rows are external iterations performed in ModelBuilder. The internal iterations are performed according to the values of conditions, separated by semicolons, as explained for the A4 Standard Trend Allocation tool.

The most important development in the A4 Allocation by Table tool is the use of conditions instead of masks. Masks are a 1 or 0 value raster, whereas the condition can be any query, as illustrated in the mask fields (MASK1, MASK2, MASK3, and MASK4). The A4 Allocation by Table tool is a planning table or scenario builder. The planner enters the conditions for an allocation depending on each conflict score or multiple sets of scores. The planner can use this tool to perform the allocation for a specified year or for multiyear increments at the same time.

The automation tools for population allocation are central to the LUCIS[plus] methodology. As described in detail in this section, the tools are used collectively to automate the allocation procedure to illustrate various planning scenarios. The output for allocation scenario 2 for Orange County, Florida, is shown in figure 3.15. The allocation is concentrated around transit lines and mixed-used areas (figure 3.16). It is an expected outcome, considering the scenario.

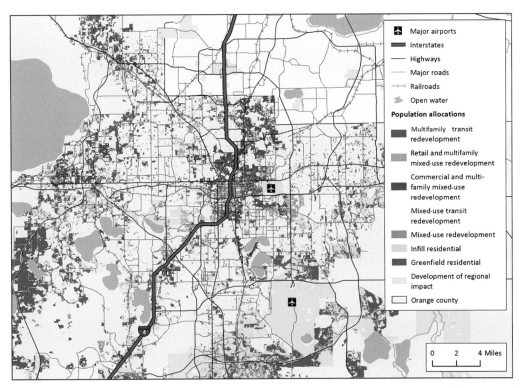

Figure 3.15. The A4 allocation tools support LUCIS^plus population allocation, creating an illustration of planning scenarios. The Allocation by Table tool is used to allocate future population in a transit scenario for Orange County, Florida. The scenario also includes a transit accessibility raster layer, land-use mix, and other proximity conditions in the allocation rules.

Figure from Abdulnaser Arafat and the Arizona Board of Regents on behalf of the University of Arizona. Water bodies from the US Geological Survey; state boundary from US Census Bureau TIGER/Line Files; interstates and highways from the Florida Department of Transportation.

The tool was checked and verified using the query statements generated by the Allocation by Table tool against manually generated queries based on the same allocation conditions. Both automatically and manually generated queries are used to allocate population. The validation procedure produced identical results.

CEM raster layers facilitate the automation procedure of the allocation process. In turn, the tools produce a raster output file and update the input CEM raster layer. The CEM raster layer is also used to present output results through different reclassifications or summaries. Additional fields that are not used as allocation conditions are added to the combined raster layer. Examples of these raster layers are traffic analysis zones (TAZs), census tracts, and census block groups. These raster layers are used in the scenario to summarize the output of the allocation to their zones. Figure 3.17 summarizes the scenario population to the corresponding block groups, while figure 3.18 summarizes the output to different land uses.

Table 3.2. Example of the SPT for Orange County, Florida

Type	CONF	MASK1	MASK2	MASK3	MASK4
Multifamily Transit Redev	464;465;564;565	"CITY" > 0	"TRANACCESS" > 50	"LUMIX" >= 3	"LUMIX" >= 3
Mixed-Use Transit Redev Commercial Multifamily	664;665;554	"CITY" > 0	"TRANACCESS" > 50	"LUMIX" >= 5	"LUMIX" >= 5
Mixed-Use Transit Redev Retail Multifamily	466;566;455	"CITY" > 0	"TRANACCESS" > 50	"LUMIX" >= 5	"LUMIX" >= 5
Mixed-Use Transit Redev	666;555	"CITY" > 0	"TRANACCESS" > 50	"LUMIX" >= 5	"LUMIX" >= 5
Multifamily Redev	464;465;565;564	"CITY" > 0	"TRANACCESS" >= 0	"LUMIX" >= 3	"LUMIX" >= 3
Mixed-Use Redev Commercial Multifamily	664;665;554	"CITY" > 0	"TRANACCESS" >= 0	"LUMIX" >= 5	"LUMIX" >= 5
Mixed-Use Redev Retail Multifamily	466;566;455	"CITY" > 0	"TRANACCESS" >= 0	"LUMIX" >= 5	"LUMIX" >= 5
Mixed-Use Redev	464;465;565;564	"CITY" > 0	"TRANACCESS" >= 0	"LUMIX" >= 5	"LUMIX" >= 5
Infill Residential	113;213;123;223;112	("IMASK" = 1 AND "CITY" >= 0)	"DRI" > 0	"LUMIX" <= 9	"LUMIX" <= 9
Infill Residential	113;213;123;223;112	("IMASK" = 1 AND "CITY" >= 0)	"DRI" >= 0	"LUMIX" <= 9	"LUMIX" <= 9
Greenfield Residential	113;213;123;223;112;313;323;311	"CITY" = 0	"DRI" >= 0	"LUMIX" <= 9	"LUMIX" <= 9

Note: SPT controls the allocation procedure as it contains the instructions for the allocation. The allocation tool works to satisfy the conditions in each row. The Type field describes the allocation.

Table 3.2. (*continued*)

ITER	POP	PREVPOP	ACOND	Pop Field	YearField
1	0	0	9;8	REDPOP	REDALLYR
2	0	0	9;8	REDPOP	REDALLYR
3	0	0	9;8	REDPOP	REDALLYR
4	0	0	9;8	REDPOP	REDALLYR
5	0	0	7;5	REDPOP	REDALLYR
6	0	0	7;5	REDPOP	REDALLYR
7	0	0	7;5	REDPOP	REDALLYR
8	0	0	7;5	REDPOP	REDALLYR
9	0	0	7;5	GFPOP	GFALLYR
10	0	0	2;1	GFPOP	GFALLYR
11	0	0	2;1	GFPOP	GFALLYR

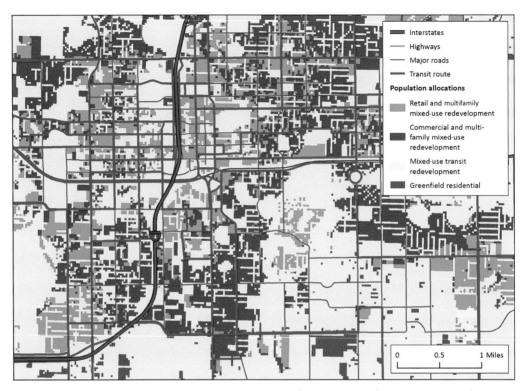

Figure 3.16. A transit-oriented scenario concentrates development around transit centers and encourages mixed-use amenities within a walkable distance.

Figure from Abdulnaser Arafat and the Arizona Board of Regents on behalf of the University of Arizona. Water bodies from the US Geological Survey; state boundary from US Census Bureau TIGER/Line Files; interstates and highways from the Florida Department of Transportation.

OID	BLKGRP	TYPE	FREQUENCY	SUM_GFPOP	SUM_REDPOP	SUM_ACRES	totpop
0	436	Green Fields Residential	17	138	0	11.393616	138
1	464	Green Fields Residential	58	269	0	53.882309	289
2	465	Green Fields Residential	29	107	0	14.716754	107
3	465	Infill Residential	1	2	0	0.237367	2
4	466	Green Fields Residential	30	688	0	71.447467	688
5	467	Green Fields Residential	71	3346	0	303.355026	3346
6	468	Green Fields Residential	84	2320	0	173.040543	2320
7	469	Green Fields Residential	82	2257	0	136.011291	2248

Figure 3.17. Allocating population into CEM using the A4 allocation tools (as shown here), illustrating allocation results on a map, and summarizing the population allocation all help create a complete picture of planning opportunities and implications. Fields within CEM, such as census block groups, can be used to define the allocation rules or summarize the final allocation results.

Figure from the Arizona Board of Regents on behalf of the University of Arizona.

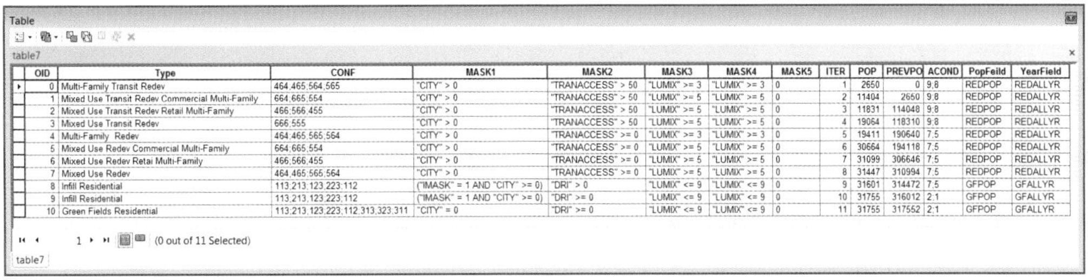

Table

table7

OID	Type	CONF	MASK1	MASK2	MASK3	MASK4	MASK5	ITER	POP	PREVPO	ACOND	PopFeild	YearField
0	Multi-Family Transit Redev	464;465;564;565	"CITY" > 0	"TRANACCESS" > 50	"LUMIX" >= 3	"LUMIX" >= 3	0	1	2650	0	9;8	REDPOP	REDALLYR
1	Mixed Use Transit Redev Commercial Multi-Family	664;665;554	"CITY" > 0	"TRANACCESS" > 50	"LUMIX" >= 5	"LUMIX" >= 5	0	2	11404	2650	9;8	REDPOP	REDALLYR
2	Mixed Use Transit Redev Retail Multi-Family	466;566;455	"CITY" > 0	"TRANACCESS" > 50	"LUMIX" >= 5	"LUMIX" >= 5	0	3	11831	114048	9;8	REDPOP	REDALLYR
3	Mixed Use Transit Redev	666;555	"CITY" > 0	"TRANACCESS" > 50	"LUMIX" >= 5	"LUMIX" >= 5	0	4	19064	118310	9;8	REDPOP	REDALLYR
4	Multi-Family Redev	464;465;565;564	"CITY" > 0	"TRANACCESS" >= 0	"LUMIX" >= 3	"LUMIX" >= 3	0	5	19411	190640	7;5	REDPOP	REDALLYR
5	Mixed Use Redev Commercial Multi-Family	664;665;554	"CITY" > 0	"TRANACCESS" >= 0	"LUMIX" >= 5	"LUMIX" >= 5	0	6	30664	194118	7;5	REDPOP	REDALLYR
6	Mixed Use Redev Retai Multi-Family	466;566;455	"CITY" > 0	"TRANACCESS" >= 0	"LUMIX" >= 5	"LUMIX" >= 5	0	7	31099	306646	7;5	REDPOP	REDALLYR
7	Mixed Use Redev	464;465;565;564	"CITY" > 0	"TRANACCESS" >= 0	"LUMIX" >= 5	"LUMIX" >= 5	0	8	31447	310994	7;5	REDPOP	REDALLYR
8	Infill Residential	113;213;123;223;112	("IMASK" = 1 AND "CITY" >= 0)	"DRI" > 0	"LUMIX" <= 9	"LUMIX" <= 9	0	9	31601	314472	7;5	GFPOP	GFALLYR
9	Infill Residential	113;213;123;223;112	("IMASK" = 1 AND "CITY" >= 0)	"DRI" >= 0	"LUMIX" <= 9	"LUMIX" <= 9	0	10	31755	316012	2;1	GFPOP	GFALLYR
10	Green Fields Residential	113;213;123;223;112;313;323;311	"CITY" = 0	"DRI" >= 0	"LUMIX" <= 9	"LUMIX" <= 9	0	11	31755	317552	2;1	GFPOP	GFALLYR

1 (0 out of 11 Selected)

table7

Figure 3.18. Categories used to summarize allocation results are derived either from fields within CEM, as shown in figure 3.17, or land-use types identified in SPT.

Figure from the Arizona Board of Regents on behalf of the University of Arizona.

Table 3.3. A4 allocation tools description

Tool Name	Description	Output
A4 Standard Trend Allocation	Works on a CEM that contains a conflict surface as well as other masks and constraints to prioritize the allocation process using the power of combined grids. Works on an iterative procedure to allocate all the available spaces specified by the conditions or constraints, which could be the suitability values and conflict scores.	The tool generates an output population raster and populates year and population fields in CEM. If the output is to be extracted from CEM, CEM can be used in different reclassifications or summary methods that allow the display of allocation results.
A4 Detailed Allocation	Same function as the A4 Standard Trend Allocation tool but with more conditions added. Works on six different masks and six extra conditions for an allocation procedure.	The tool generates an output population raster and populates year and population fields in CEM. If the output is to be extracted from CEM, CEM can be used in different reclassifications or summary methods that allow the display of allocation results.
A4 Allocation by Table	Works with up to eight conditions, two of which are iterative inside the tool. Larger external iterations are performed in the ModelBuilder application and directed by the SPT table, which adds more flexibility to the process.	The tool uses the values of the iteration table row by row to allocate the population using the conditions specified by the fields of the table. The tool also populates a field in the scenario table containing the summary of population allocated in each of the iterations (rows).

Chapter summary

The following points are discussed in this chapter:

- LUCIS^{plus} improves decision-making by automating allocation procedures for land-use planning scenarios.
- A4 LUCIS Tools enhance standard ArcGIS tools in facilitating suitability determination and population allocation procedures. They encourage participatory GIS in the creation of community weights. The LUCIS^{plus} tools automate complex spatial procedures typically seen in land-use planning.
- The scenario policy table adds flexibility to the allocation procedure described in chapter 2 that relies on CEM alone. SPT incorporates iteration in allocation. Combined with the attribute fields in CEM, it automates the manual allocation procedure of the scenario guide and the querying statements used in the Select By Attributes tool.

ArcGIS tools referenced in this chapter

Tool Name	Version 10.2 Toolbox/Toolset
A4 Allocation by Table	A4 LUCIS Tools/A4 Allocation
A4 Detailed Allocation	A4 LUCIS Tools/A4 Allocation
A4 LUCIS Community Values Calculator	A4 LUCIS Tools/A4 Suitability Overlay
A4 LUCIS Weights	A4 LUCIS Tools/A4 Suitability Overlay
A4 Standard Trend Allocation	A4 LUCIS Tools/A4 Allocation
A4 LUCIS Suitability	A4 LUCIS Tools/A4 Suitability
Add Field	Data Management Tools/Fields
Combine	Spatial Analyst Tools/Local
Euclidean Distance	Spatial Analyst Tools/Distance
Reclassify	Spatial Analyst Tools/Reclass
Select (By Attributes)	Analysis Tools/Extract
Raster Calculator	Spatial Analyst Tools/Map Algebra
Weighted Overlay	Spatial Analyst Tools/Overlay
Weighted Sum	Spatial Analyst Tools/Overlay
Zonal Statistics as Table	Spatial Analyst Tools/Zonal

References

Carr, Margaret H., and Paul D. Zwick. 2007. *Smart Land-Use Analysis: The LUCIS Model*. Redlands, CA: Esri Press.

Collins, M., Frederick Steiner, and Michael Rushman. 2001. "Land-Use Suitability Analysis in the United States: Historical Development and Promising Technological Achievements." *Environmental Management* 28 (5): 611–21.

Driessen, P. M., and N. T. Konijn. 1992. *Land-Use Systems Analysis*. Wageningen Agricultural University: Department of Soil Science and Geology.

EPA (Environmental Protection Agency). 1999. *The Transportation and Environmental Impacts of Infill versus Greenfield Development: A Comparative Case Study Analysis*. EPA publication number 231-R-99–005.

Malczewski, John. 1999. *GIS and Multicriteria Decision Analysis*. New York: John Wiley and Sons.

Malczewski, John. 2004. "GIS-Based Land-Use Suitability Analysis: A Critical Overview." *Progress in Planning* 62:3–65.

McHarg, I. L. 1969. *Design with Nature*. Garden City, NY: Natural History Press.

Ndubisi, F. 2002. *Ecological Planning: A Historical and Comparative Synthesis*. Baltimore: Johns Hopkins University Press.

Nyerges, Timothy L., and Piotr Jankowski. 2009. *Regional and Urban GIS: A Decision Support Approach*. New York: Guilford Press.

Saaty, T. L. 1980. *The Analytic Hierarchy Process*. New York: McGraw-Hill.

Chapter 4

Analyzing and mapping residential land-use futures

Paul D. Zwick and Iris E. Patten

What this chapter covers

A quick Internet search on land-use models will yield numerous results, many demonstrating a unique perspective on the use of ArcGIS tools and programming methods to determine opportunities or predict future land-use patterns. But the tools and methods are only half the product in land-use modeling. The real value of Land-Use Conflict Identification Strategy planning land-use scenarios (LUCIS^{plus}) is the rationale behind identifying the land-use opportunity to allocate population that most closely resembles reality. Decisions to develop land include the constraints and feasibility provided in site planning. But growth management and land-use practice must also respond to policy and the public perception of how land should be used and managed in the future. This chapter explains the LUCIS^{plus} population allocation process and how it responds to policy constraints and sustainability priorities.

Five simple methods for population projection

Any land-use analysis, planning process, or visioning process must start by understanding the projected growth or decline in the area's population. Many models can be used to project population. These models include basic linear, exponential, carrying capacity, proportion, cohort projection, multivariate regression, and curve fitting. In Florida, the University of Florida Bureau of Economic and Business Research (BEBR) develops county and state population projections. It provides an annual high-, medium-, and low-population projection for all planning agencies in the state. BEBR uses multiple projection methods and then averages the results to produce its population projections. However, population projections are not always readily available in all areas of the United States or international locations. The following section reviews five population projection methods that include time as a variable: (1) basic linear, (2) basic exponential, (3) the modified exponential often called *carrying capacity*, (4) proportional projection, and (5) multivariate regression. Cohort population projection is not reviewed here because it is most useful for large regions or areas such as a state, multiple states, or a nation. However, a cohort population projection is often used as a data input for the proportional projection method. Therefore, the cohort projection method is discussed in the section of this chapter dedicated to the "proportional population projection method." It should also be noted that projections spanning many years into the future tend to be less accurate. The uncertainty within the projection methodology increases as the range of the projection increases.

The basic linear projection

You can easily calculate basic linear projection within a spreadsheet or by using simple linear regression. Microsoft Excel provides a trend feature in its spreadsheet that works for this type of projection. Figure 4.1a depicts this basic process. First, define the basic time intervals for the projection in a spreadsheet. In this example, the time intervals are in 10-year increments from 2000 to 2040, located in cells A3 to E3. Next, fill in the increments for which there is observed population data. In figure 4.1a, this includes the years

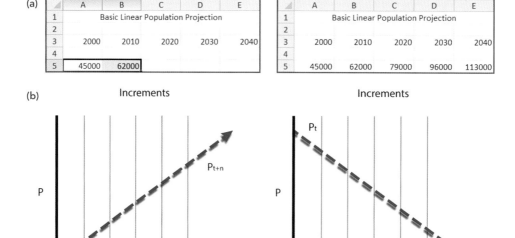

Figure 4.1. Projecting population from 2000 to 2040 (a) in a spreadsheet using basic linear population projection and (b) graphically plotting population growth versus time, which indicates linearity over time.

Figure from Paul D. Zwick and the Arizona Board of Regents on behalf of the University of Arizona.

2000 and 2010—cells A3 and B3. The known population in 2000 was 45,000 people. In 2010, for the same spatial area, it was 62,000 people. Next, select the two cells containing the known population values for the years 2000 and 2010. Notice that the selected cells have a small black square on the lower-right corner of cell B3. Click on the black square, and drag it to the right. Stop on the right side of cell E3. Excel does a linear interpolation for the selected cells, producing a basic linear population projection for the years 2020, 2030, and 2040. Using this projection method, the population is projected to grow from 45,000 people in 2000 to 113,000 people in 2040. Although linear projection is easy to accomplish, it is often not a true representation of actual growth. The best method for determining whether the growth in a specific area is linear is to plot the area's population growth versus time (figure 4.1b) for a period of years long enough to determine what the curve looks like. Equation (4.1) is an example of the trend model in the Microsoft Excel spreadsheet.

$$P_{t+n} = P_{t-1} + B_1(P_t)$$
$$B_1 = Slope. \quad (4.1)$$

The basic exponential projection

You can also easily calculate the basic exponential projection in a spreadsheet using the appropriate equation. Figure 4.2a shows the basic exponential projection for the same hypothetical region. Population increases annually at a rate of approximately 37.8 percent. Ultimately, the population grows much faster exponentially, reaching a 2040 population of 162,155 persons (figure 4.2b). Exponential population grows (or declines) at a constant rate, increasing the growth dramatically from year to year. A constant rate grows not only from the existing population, but from the 37.8 percent annual addition to the new population. If you have more than one time increment with known population, average the individual incremental rates by dividing the sum of the rates by the number of increments, as in equation (4.2).

$$P_{t+n} = P_t(1+r)^n$$
$$r = (1/n)\sum_{t=2}^{n}(P_t - P_{t-1})/P_{t-1}, \quad (4.2)$$

where:
$P_t = Base\_Population\_Time = t$
$P_{t+n} = Future\_Population\_Time = t+n$
$r = Rate\_of\_Population\_Change$
$n = Length\_Population\_Projection.$

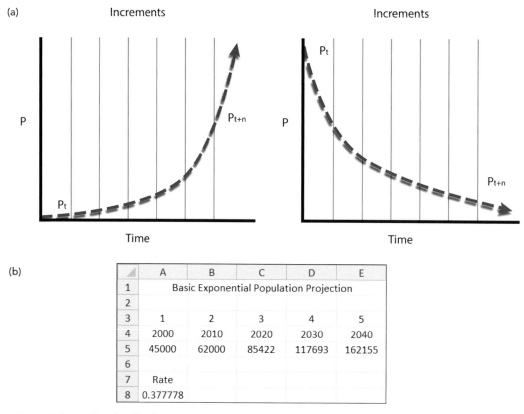

Figure 4.2. (a) Graphically plotting population growth versus time indicates a dramatic growth (or decline) because people are added from the existing population as well as from the annual rate of increase. (b) An example of the spreadsheet methodology results for a population projected from the year 2000 to 2040 using an annual population increase of approximately 37.8 percent.

Figure from Paul D. Zwick and the Arizona Board of Regents on behalf of the University of Arizona.

The modified exponential method

The modified exponential method uses a rate identified as v in equation (4.3). This method is often described as the *carrying capacity* because population approaches an ultimate population asymptotically. Therefore, it can never actually reach the ultimate population (figure 4.3a). With each new year, the population approaches the ultimate population. But each year's population increase is smaller than the previous year's increase. Equation (4.3) calculates population projections using the carrying capacity methodology. Notice that a prevalent variable in the model, K, is the carrying capacity, or the maximum population that can be sustained within the region. So P_t is the population at time t, and P_{t+n} is the future population at time $t + n$, where n is equal to the number of time increments into the future. As in the basic exponential projection, a simple spreadsheet example can illustrate the process. The modified exponential projection increases rapidly in the first years from 2010 through 2030, but each new increment is drastically reduced as it approaches the carrying capacity K equal to 94,000 people. Although it is difficult to estimate K in land-use applications, it could be estimated if you knew the average number of people per acre in the study area and the number of undeveloped acres. The resulting population prediction would be a trend that approached the built-out population.

$$P_{t+n} = K - [(K - P_t)(v)^n]$$

$$v = (1/n)\sum_{t=2}^{n}(K - P_t)/(K - P_{t-1}). \quad (4.3)$$

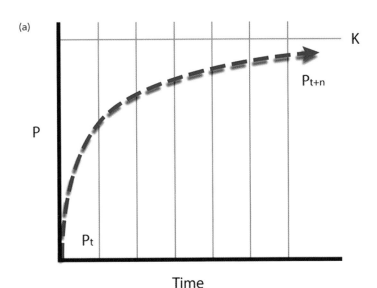

Figure 4.3. (a) Graphically plotting population growth versus time illustrates that as population approaches a threshold population asymptotically, it never actually reaches the ultimate population. (b) An example of the spreadsheet methodology with results showing a rapid population increase in the first three projection increments but a declining rate of growth in the period between 2030 and 2040 as the population reaches the carrying capacity.

Figure from Paul D. Zwick and the Arizona Board of Regents on behalf of the University of Arizona.

The proportional population projection method

The proportional population projection method is useful for estimating a small area population based on its historical proportion of a regional population (figure 4.4). For example, a state population might be projected based on the cohort population projection. Again, cohort projection is good for large regional projections. If the local area population has historically been stable at about 4 percent of the state population, it could be argued that a future state population would increase from a year 2000 population of 1,250,000 to a 2040 population of 1,560,000. It would be an increase in population of 310,000 people. Next, the local area population could be projected to increase by 4 percent of the state population. It would be 0.04 times 310,000, for an increase of 12,400 people by 2040. Clearly, the region's local proportion of the state's population increase must rely on a stable historical regional proportion of the state population. The method can still be used if the proportion is variable and relatively stable by using the county's mean proportion of state population within a selected historical time frame. For example, if the county's annual proportion of state population or of a region has the following annual population proportion during a five-year period of 3 percent, 3.5 percent, 3.35 percent, 3.9 percent, and 3.1 percent, clearly it is relatively stable but variable. The proportional population projection method might employ the five-year mean of 3.37 percent for the county's proportion of the state's

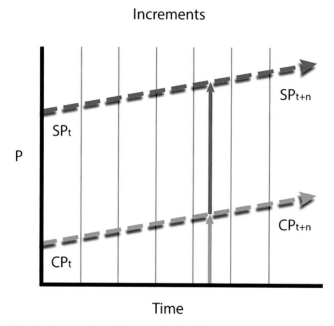

Figure 4.4. The historical proportion of regional to local population growth is maintained in the proportional population projection method. SP_t is the state population at time t, SP_{t+n} is the state's population at time $t + n$, and CP is the county's population using the same subscripts.

Figure from Paul D. Zwick and the Arizona Board of Regents on behalf of the University of Arizona.

population projection. Equation (4.4) calculates population increase as a proportion of a larger regional population.

$$CP_t / SP_t = CP_{t+n} / SP_{t+n}$$
$$CP_{t+n} = (CP_t / SP_t) * SP_{t+n}. \quad (4.4)$$

The multivariate regression method

The multivariate regression method is useful for many scales of population projection. It works for linear or nonlinear (curve-fitting) projections (figure 4.5). Typically, this method requires selection of statistical software, such as R, SPSS, or SAS, to develop the model for projection. Equation (4.5) uses five hypothetical variables to project population at time P_{t+n}. These variables are time (time), the incremental lag of population (lagP), new housing starts (HS), median income (MI), and retirement investments (RI). The nonlinear model in the figure uses the square of the number of housing starts (HS^2) and the square of retirement income (RI^2) instead of the simple numbers. Some software packages offer nonlinear modeling that can fit a curve to the existing data and provide a measure of the closeness of fit.

$$P_{t+n} = P_t + B_1(P_{t-1\cdots4}) + B_2(HS) + B_3(RI) + B_4(MI). \quad (4.5)$$

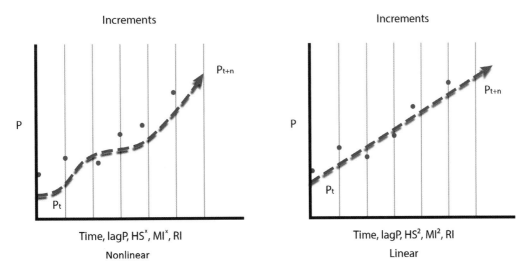

Figure 4.5. The multivariate regression method can be graphically illustrated either as nonlinear (curve fitting) or linear. Considering the complexity of the multivariate regression method equation, projections for this method are typically calculated in a statistical software package.

Figure from Paul D. Zwick and the Arizona Board of Regents on behalf of the University of Arizona.

LUCIS^plus basic residential analysis of Hillsborough County, Florida

To begin the discussion of residential allocation, start with a summary review of the suitability for multifamily and single-family residential development. Residential suitability is composed of five primary characteristics: (1) physical, (2) proximal or location, (3) historical growth, (4) land use, and (5) land values. A weighted sum comprising noise quality, soils, flooding, air quality, and hazardous sites is used to model residential physical characteristic suitability. The noise model includes inputs for airports, industrial noise generators, interstates, major roadways, utilities, racetracks, and railways and rail crossings.

The concept of baseline residential analysis is to use physical data for quantifying and identifying to assign suitability for multifamily and single-family residential location. Use central tendency and dispersion statistics whenever possible. When physical data does not support the collection of dispersion statistics, use expert opinion to assign suitability. For example, you can map soil drainage using the following categories: excessively well drained, well drained, moderately well drained, poorly drained, very poorly drained, and water. Then assign suitability. Assign the highest suitability to excessively well-drained soils, decreasing progressively until very poorly drained. Assign NoData to the water category, reflecting the unsuitable use of water for residential housing. Alternatively, consider the count of residential structures within the individual soil drainage categories. Assign suitability based on the proportion of units per individual category (table 4.1). Next, take the proportions and assign suitability. You can also use a linear transformation to convert these proportions into a scale of 1 to 9.

You can use equation (4.6) to transform residential parcel proportions in table 4.1 to suitability values for residential soil drainage. In the example, there are 45,000 existing residential units in "Well Drained" soils. Equation (4.6) transforms the well-drained soils from a 0.2206 proportion of total residential units to a suitability value of 5.446056. Within a range of 1 to 9, a suitability value of 5.446056 is above the average suitability but not the highest

Table 4.1. Residential suitability for soils drainage as a portion of the parcels by soils category

Category	Residential Parcels	Parcel Proportion	Suitability
Excessively Well Drained	50,000	0.2451	6.334542
Well Drained	45,000	0.2206	5.446056
Moderately Well Drained	65,000	0.3186	9.000000
Poorly Drained	24,000	0.1176	1.710789
Very Poorly Drained	20,000	0.0980	1.000000
Water	0	0.0000	NoData
Totals	**204,000**	**1.0000**	**4.698300**

suitability. The highest suitability is for "Moderately Well Drained" soils, which has a 0.3186 proportion of the total residential units in the region. Figure 4.6 is a map of the residential soil suitability for Hillsborough County, Florida. Suitability is illustrated using a red to green color ramp, from red (low suitability) to green (high suitability). The soils with the highest residential suitability are the green areas in the center to north center part of the county.

$$TV = ((SV - OMinSV)*(NMaxV - NMinV)/(OMaxSV - OMinSV)) + 1$$
$$TV = ((0.2206 - 0.098)*(9 - 1))/(0.3186 - 0.098)) + 1$$
$$TV = 5.446056, \qquad (4.6)$$

where:

TV = transformed value (that is, 5.446056),

SV = sample value (that is, 0.2206),

$NMinV$ = new minimum value for transformation (that is, 1),

$NMaxV$ = new maximum value for transformation (that is, 9),

$OMaxSV$ = old maximum value of sample (that is, 0.3186), and

$OMinSV$ = old minimum value of sample (that is, 0.098).

Next, the LUCIS^{plus} residential modeling process analyzes or identifies proximal or location suitability. As with physical suitability, the concept is to identify or analyze residential location for scenario development or current location statistics for trend development. As with the description of physical residential suitability, an example should clarify the concept and analysis. The example for proximal suitability is residential proximity to schools. Schools are categorized into primary, middle, and high schools. Proximal suitability is used to determine residential suitability for primary schools in the study area. The mean distance (1 mile) and standard deviation of the distances (0.4 miles) are used to calculate raster suitability. You can use the Zonal Statistics as Table tool to calculate the mean and standard deviation for residential distance to schools.

Existing locations for primary schools and residential parcels can be used to create location suitability for the historical trend describing residential proximity to primary schools (table 4.2). The suitability index is based on statistical data produced from the area's existing urban form. The index sets suitability in areas in which roughly 50 percent of the residential structures are highly suitable. The proximal suitability decreases in one-quarter (0.1-mile) standard deviations until approximately 95 percent of all residential structures are identified. The range of decreasing suitability is from 9 to 2 within an increasing range of distance from 0 to 1.7 miles. By developing a location index using the mean and standard deviation for distance, the raster layer for residential proximity to primary schools has a defendable and known statistical basis for the index of suitability. Finally, the same analysis can be performed for middle schools and high schools.

You can then use the Weighted Overlay tool to combine the three suitability raster layers for primary, middle, and high school proximity. The weights used to combine the three

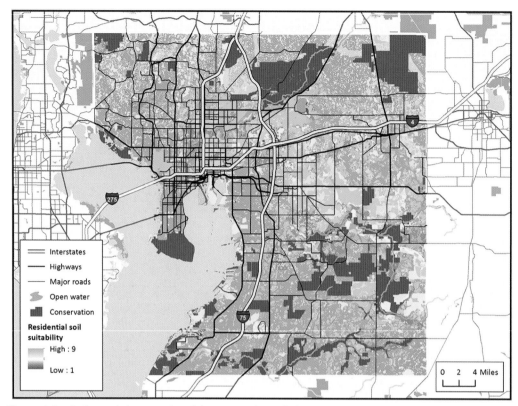

Figure 4.6. Soils suitability map. Suitability can traditionally be generated by assigning values for physical categories or through statistical analysis using proportions and linear transformation.

Figure from Paul D. Zwick and the Arizona Board of Regents on behalf of the University of Arizona. Water bodies from the US Geological Survey; state boundary from US Census Bureau TIGER/Line Files; interstates and highways from the Florida Department of Transportation; conservation areas from the Florida Natural Areas Inventory.

Table 4.2. Suitability for proximal distance for residential access to primary schools

Distance to Primary Schools	Suitability Value
0 to 1 mile	9
1 to 1.1 miles	8
1.1 to 1.2 miles	7
1.2 to 1.3 miles	6
1.3 to 1.4 miles	5
1.4 to 1.5 miles	4
1.5 to 1.6 miles	3
1.6 to 1.7 miles	2
Greater than 1.7 miles	1

residential-to-school proximity raster layers are based on (1) the proportion of students in the three types of schools, (2) the proportion of schools in the category, and (3) the acreage proportion of each category of school. However these weights are calculated, using the index reflects the historical or trend distance for residential structures to schools in the region.

Interestingly, but not unexpectedly, the mean and standard deviation of distances for primary schools in Hillsborough County, Florida, is smaller than for middle schools. Also, the area of high suitability is larger because the number of primary schools is larger than middle and high schools. That places Hillsborough County, Florida, in line with other counties in Florida, and probably with other counties in the United States. At least spatially, families seem to be more concerned with the distance younger children travel to primary school than to middle school. The region also has fewer high schools, which have a larger population of students per school than primary or middle schools. That fact may account for the longer distances, on average, to high schools from residential units in the county. Figure 4.7 shows the weighted residential suitability for proximity to all three types of schools.

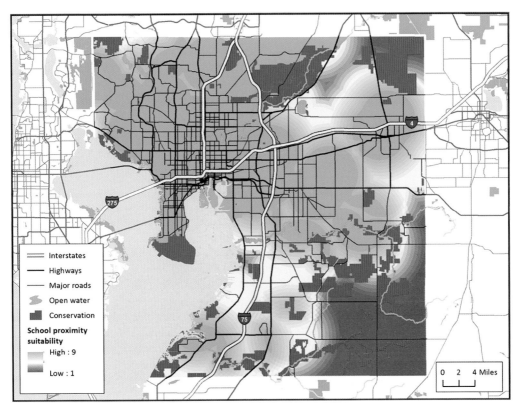

Figure 4.7. Proximal weighted suitability for primary, middle, and high schools in Hillsborough County, Florida.

Figure from Paul D. Zwick and the Arizona Board of Regents on behalf of the University of Arizona. Water bodies from the US Geological Survey; state boundary from US Census Bureau TIGER/Line Files; interstates and highways from the Florida Department of Transportation; conservation areas from the Florida Natural Areas Inventory.

Population projections for Hillsborough County

In Florida, BEBR provides population projections for state agencies. The methodology is explained in *Florida Population Studies* Volume 44, Bulletin 159 (June 2011). Hillsborough County's population projection from 2010 to 2040 is provided in table 4.3.

Table 4.3. Baseline BEBR Hillsborough County population projection (from 2010 to 2040)

Projection Level	2010	2015	2020	2025	2030	2035	2040
Low	1,229,226	1,277,500	1,328,400	1,368,000	1,395,300	1,410,100	1,414,100
Moderate		1,325,300	1,439,000	1,549,200	1,652,700	1,747,900	1,836,700
High		1,384,000	1,559,400	1,741,100	1,926,800	2,115,200	2,307,100

The projected moderate population during the period from 2010 to 2040 increases the county's population from 1,229,226 to 1,836,700, or a net gain of 607,474 people. In this example, the projected period should be 2010 to 2045. Because the projection is linear, the Excel spreadsheet trend method is used. The new projected population from 2010 to 2045 is 1,229,226 to 1,940,526 people (table 4.4), or a total increase in population of 711,300 people.

Table 4.4. Extended trend BEBR moderate population projection for Hillsborough County (from 2010 to 2045)

2010	2015	2020	2025	2030	2035	2040	2045
1,229,226	1,325,300	1,439,000	1,549,200	1,652,700	1,747,900	1,836,700	1,940,526

Hillsborough County residential suitability

Allocation of residential population and its accompanying land-use categories relies heavily on two types of residential suitability, single family and multifamily. Single-family residential development is often called low-density development. Unit densities are as low as two to four units per acre. Multifamily residential development is much denser. It can reach unit densities of more than one hundred units per acre in urbanized areas. In Hillsborough County, redevelopment unit density ranges from a minimum of five units per acre to a maximum of one hundred units per acre. LUCIS^{plus} uses greenfield conflict and mixed-use redevelopment opportunity to determine the best locations for new greenfield single-family residential development. LUCIS^{plus} also identifies the best locations for mixed-use or multi-family residential redevelopment.

LUCIS^{plus} modeling for single-family residential suitability (figure 4.8) follows the general LUCIS suitability analysis. It uses A4 LUCIS Tools, including transit tools (see chapters 13 and 14), to model five characteristics: (1) proximity (30 percent weight); (2) physical (16 percent weight); (3) land-use change (17 percent weight); (4) density (18.5 percent weight); and

(5) historical growth (18.5 percent weight). The weights are used to combine the output raster layers for the final single-family residential suitability raster layer.

Single-family proximity suitability includes models for (1) service amenities, (2) transit roads, (3) open water, (4) shopping, (5) existing single-family residential, (6) infrastructure, (7) entertainment, (8) prisons, and (9) land values. Single-family physical suitability includes models that capture local characteristics for (1) noise, (2) soil quality, (3) drainage, (4) air quality, and (5) hazardous sites or materials. Single-family modeling captures the suitability of existing land-use types other than existing residential for new residential development. Next, single-family density is modeled, and it is just what it sounds like—a raster layer with the density of existing single-family residential units per square mile. Finally, historical growth suitability is developed by modeling three decades of single-family residential development within the study area. In Hillsborough County, the historical growth pattern was mapped for new single-family residential housing added during the decades of 1976–85, 1986–95, and 1996–2005. The three decades were weighted, from the lowest weight for the oldest decade to the largest weight for the most recent decade. The suitability raster layers for proximity, physical, land-use change, density, and historical growth were combined using the weighting method described in chapter 2. (For a detailed discussion of suitability weighting, see *Smart Land-Use Analysis: The LUCIS Model* [Carr and Zwick 2007].) The final suitability raster layer illustrates single-family suitability for Hillsborough County.

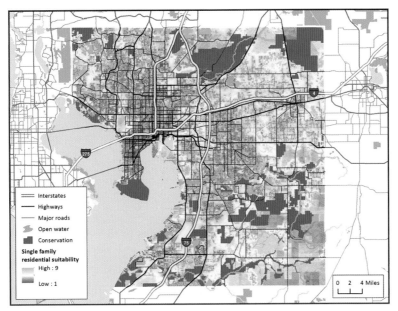

Figure 4.8. The multiple-utility assignment suitability raster layers for proximity, physical, land-use change, density, and historical growth are combined to create the final suitability raster layer for single-family residential in Hillsborough County, Florida.

Figure from Paul D. Zwick and the Arizona Board of Regents on behalf of the University of Arizona. Water bodies from the US Geological Survey; state boundary from US Census Bureau TIGER/Line Files; interstates and highways from the Florida Department of Transportation; conservation areas from the Florida Natural Areas Inventory.

Residential preference

Single-family residential preference is developed by collapsing single-family residential suitability into three preference categories: low, moderate, and high (figure 4.9). The method we prefer for collapsing residential suitability into residential preference is equal-interval reclassification.

An equal-interval reclassification of the single-family residential suitability is preferred because it is easy for the public to understand and inherently fair. Using natural breaks, standard deviation, quantiles, or other methods of reclassification for land-use preference often results in long, contentious public meetings about the fairness of the reclassification. Clearly, contention is not the optimum outcome sought from a visualization process. Argument about the fairness of the GIS reclassification method is not productive. It does not help to achieve the ultimate goal, which is to formulate a comprehensive future land-use vision for the region.

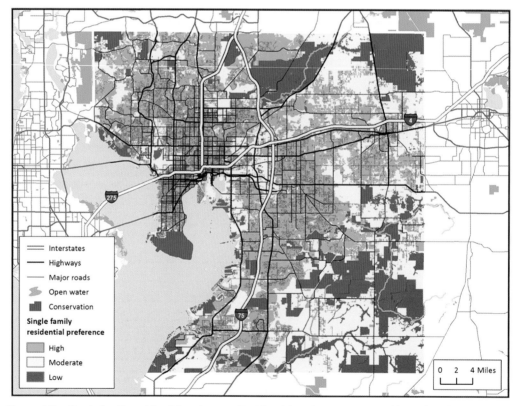

Figure 4.9. Single-family residential preference. Residential preference is created by reclassifying suitability into three categories (low, moderate, and high). The equal-interval reclassification methodology is used.

Figure from Paul D. Zwick and the Arizona Board of Regents on behalf of the University of Arizona. Water bodies from the US Geological Survey; state boundary from US Census Bureau TIGER/Line Files; interstates and highways from the Florida Department of Transportation; conservation areas from the Florida Natural Areas Inventory.

Conflict

Once suitability is reclassified to preference, the single-family residential preference can be combined with other urban, agricultural, and ecologically significant land-use preferences to identify opportunity or conflict (Carr and Zwick 2007). Table 4.5 shows the LUCIS

Table 4.5. LUCIS greenfield preference/conflict matrix categories with descriptions

Unique Numeric Identifier	Preference/Conflict Description
111	Low Preference Agriculture, Ecological, and Urban
112	Low Preference Agriculture and Ecological; Moderate Urban
113	Low Preference Agriculture and Ecological; High Urban
121	Low Preference Agriculture and Urban; Moderate Ecological
122	Low Preference Agriculture; Moderate Ecological and Urban
123	Low Preference Agriculture, Moderate Ecological, and High Urban
131	Low Preference Agriculture and Urban; High Ecological
132	Low Preference Agriculture, High Ecological, and Moderate Urban
133	Low Preference Agriculture; High Ecological and Urban
211	Moderate Preference Agriculture; Low Ecological and Urban
212	Moderate Preference Agriculture and Urban; Low Ecological
213	Moderate Preference Agriculture, Low Ecological, and High Urban
221	Moderate Preference Agriculture and Ecological; Low Urban
222	Moderate Preference Agriculture, Ecological, and Urban
223	Moderate Preference Agriculture and Ecological; High Urban
231	Moderate Preference Agriculture, High Ecological, and Low Urban
232	Moderate Preference Agriculture and Urban; High Ecological
233	Moderate Preference Agriculture; High Ecological and Urban
311	High Preference Agriculture; Low Ecological and Urban
312	High Preference Agriculture, Low Ecological, and Moderate Urban
313	High Preference Agriculture and Urban; Low Ecological
321	High Preference Agriculture, Moderate Ecological, and Low Urban
322	High Preference Agriculture; Moderate Ecological and Urban
323	High Preference Agriculture and Urban; Moderate Ecological
331	High Preference Agriculture and Ecological; Low Urban
332	High Preference Agriculture and Ecological; Moderate Urban
333	High Preference Agriculture, Ecological, and Urban

greenfield preference/conflict matrix. The matrix has 27 categories of preference and conflict that define the interaction of the collapsed preference. For example, the conflict identifier 223 identifies the unique greenfield areas that have moderate preference for agriculture (2) and ecological significance (2) and high preference for urban use (3).

LUCIS mixed-use opportunity

The process to identify multifamily land-use suitability is the same as the preceding discussion for single-family residential. Again, the conversion from suitability to preference uses equal-interval reclassification. The most interesting aspect of multifamily residential use is its major role in mixed-use development. Greenfield mixed-use development can be for single-family as well as multifamily residential combined with retail, service, or commercial uses. In LUCIS^{plus} for urban redevelopment, multifamily residential preference is combined with commercial and retail preference to identify mixed-use redevelopment opportunities. Following the basic LUCIS conflict numbering schema shown in table 4.5, the LUCIS^{plus} mixed-use redevelopment opportunity produces 27 categories of mixed-use preference or opportunity (table 4.6).

Under a mixed-use urban redevelopment strategy, the combination of the three uses (commercial, multifamily, and retail) provides a *unique numeric identifier*. The identifier describes opportunity for mixed-use redevelopment. Here, the mixed-use opportunity is clearly not a conflict among the three uses. Instead, it is an opportunity for higher-density urban development. In LUCIS^{plus}, residential mixed-use opportunity is defined as a high preference for unique combinations of commercial, multifamily residential, and retail uses. The LUCIS^{plus} mixed-use redevelopment categories follow the same numeric identification strategy. The three categories are low, moderate, and high. The first digit represents the three preferences for commercial; the second, multifamily; and the third, retail. Therefore, the mixed-use redevelopment identifier 223 indicates moderate preference (2) for commercial and multifamily and high preference (3) for retail land use. The planner can then conclude that opportunity exists for mixed use but with a greater preference for retail shopping. Figure 4.10 shows the mixed-use redevelopment opportunity raster layer for the City of Tampa, the county seat of Hillsborough County. The map shows multiple combinations for commercial, multifamily, and retail mixed-use opportunity.

LUCIS^{plus} residential allocation using the criteria evaluation matrix

The process for creating a LUCIS^{plus} criteria evaluation matrix (CEM) is straightforward. Allocating land use requires three types of CEM: greenfield, infill, and redevelopment. The following sections of this chapter discuss the pertinent variables in the three CEMs used for the LUCIS^{plus} residential land-use and population allocations in Hillsborough County, Florida.

Table 4.6. LUCIS mixed-use redevelopment opportunity categories

Unique Numeric Identifier	Preference/Conflict Description
111	Low Opportunity Commercial, Multifamily, and Retail
112	Low Opportunity Commercial and Multifamily; Moderate Retail
113	Low Opportunity Commercial and Multifamily; High Retail
121	Low Opportunity Commercial and Retail; Moderate Multifamily
122	Low Opportunity Commercial; Moderate Multifamily and Retail
123	Low Opportunity Commercial, Moderate Multifamily, High Retail
131	Low Opportunity Commercial and Retail; High Multifamily
132	Low Opportunity Commercial, High Multifamily, Moderate Retail
133	Low Opportunity Commercial; High Multifamily and Retail
211	Moderate Opportunity Commercial; Low Multifamily and Retail
212	Moderate Opportunity Commercial and Retail; Low Multifamily
213	Moderate Opportunity Commercial, Low Multifamily, High Retail
221	Moderate Opportunity Commercial and Multifamily; Low Retail
222	Moderate Opportunity Commercial, Multifamily, and Retail
223	Moderate Opportunity Commercial and Multifamily; High Retail
231	Moderate Opportunity Commercial, High Multifamily, Low Retail
232	Moderate Opportunity Commercial and Retail; High Multifamily
233	Moderate Opportunity Commercial; High Multifamily and Retail
311	High Opportunity Commercial; Low Multifamily and Retail
312	High Opportunity Commercial, Low Multifamily, Moderate Retail
313	High Opportunity Commercial and Retail; Low Multifamily
321	High Opportunity Commercial, Moderate Multifamily, Low Retail
322	High Opportunity Commercial; Moderate Multifamily and Retail
323	High Opportunity Commercial and Retail; Moderate Multifamily
331	High Opportunity Commercial and Multifamily; Low Retail
332	High Opportunity Commercial and Multifamily; Moderate Retail
333	High Opportunity Commercial, Multifamily, and Retail

The LUCIS process progresses from identifying land-use suitability to converting that land-use suitability to collapsed preference to ultimately combining multiple collapsed preferences to identify land-use conflict or opportunity. The "plus" in LUCIS^{plus} uses all three CEMs to combine conflict/opportunity, individual land-use suitability, and other data. It uses the resulting opportunities and other data to spatially allocate residential population and regional and local employment. Finally, to allocate population and employment, each

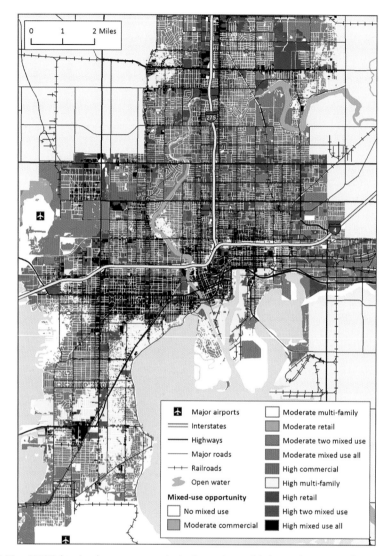

Figure 4.10. LUCIS^{plus} mixed-use opportunity is shown for multiple combinations of commercial, multifamily, and retail development. The areas in black indicate LUCIS high-preference opportunity for a mix of all three categories, while the areas in gray indicate moderate preference for a mix of the three categories. The areas in bright yellow indicate high-preference opportunity for multifamily uses while the areas in light yellow indicate moderate preference for multifamily uses. The areas in dark green indicate high-preference opportunity and the areas in light green indicate moderate preference for commercial uses. The areas in dark blue are high-preference opportunity retail while the areas in light blue are moderately preferred for retail uses. Finally, the red areas have a high preference for a combination of two of the three uses, and the orange areas are moderately preferred for two of the three mixed uses.

Figure from Paul D. Zwick and the Arizona Board of Regents on behalf of the University of Arizona. Airports from the Federal Aviation Administration and the Research and Innovative Technology Administration Bureau of Transportation Statistics National Transportation Atlas Databases (NTAD) 2012; water bodies from the US Geological Survey; state boundary from US Census Bureau TIGER/Line Files; interstates, highways, and major roads from the Florida Department of Transportation; railroads from the Federal Railroad Administration.

CEM allocates the following six primary land uses: (1) residential, (2) commercial, (3) retail, (4) service, (5) industrial, and (6) institutional. These allocations aid planners in land-use decision-making.

Creating three CEMs for residential population allocation

The *greenfield CEM* is for greenfield development. It includes the six primary land-use types. Because its primary land use is single-family residential, it tends to produce less dense development. The greenfield CEM uses the greenfield conflict categories in table 4.5.

The *redevelopment CEM* is for redevelopment. It uses the redevelopment opportunity categories in table 4.6. The redevelopment CEM is based on a fundamental concept. Redevelopment occurs in more densely developed urban areas. It is used to propagate dense urban development. Redevelopment requires the destruction or adaptive reuse of existing structures. These structures are replaced by mixed-use development that combines commercial, multifamily, and retail/service use. The destruction of structures must be thoughtful. It must follow increased reliance on transit-oriented development (TOD), increased urban density where appropriate, and account for the displacement of uses and existing residents or employees. Redevelopment follows either a simple or complex replacement strategy.

The *infill CEM* is for land-use infill. The development of the infill CEM has a conceptual foundation within LUCIS. The infill CEM works on the same six land-use categories but with one fundamental difference. The locations must be identified in the property tax data or future land-use data as existing vacant or platted lands. Essentially, vacant platted lands have been designated for specific uses such as residential, commercial, institutional, or industrial. Additionally, vacant lands are devoid of a structure. That is the essence of the term *vacant*. Finally, the primary difference between the infill CEM and the other two CEMs is that *the infill CEM does not rely on conflict* but rather on the individual land-use suitability for each of the six primary land-use categories. Because the land use has been established for the parcel, the selection does not have to identify the use. It must select locations based on the highest suitability.

Residential mixed-use allocation

Residential allocation within Hillsborough County requires three CEMs because the allocations are based on three separate concepts. The redevelopment CEM concept is discussed first.

As previously calculated, the total additional projected residential population for the county from 2010 to 2045 is 711,300 people. However, the proportion of that new residential population that will be allocated to redevelopment has not been identified. Remember, the difference between trend analysis and scenario allocation can be significant. Trend allocation describes a continuation of an existing process whereas an allocation for an alternative scenario could be a product of planning policy. Policy changes that encourage mixed-use residential development include development tax incentives, identification of mixed-use development zones, and other policy-driven incentives. However, these unknown policy changes do not change the fundamental process of LUCIS^{plus} residential redevelopment allocation. Policy changes can include many variables that LUCIS^{plus} can use to test against the trend.

Trend redevelopment calculation

To identify redevelopment allocations, you can look at time-sequenced land-use change in the property appraisal data or existing land-use data. For example, you can examine the present year's property parcel data. If new residential structures (condominiums or multifamily) are placed on parcels that are either existing residential or were previously another category of land use, redevelopment has occurred. However, if new residential structures are placed on open space categorized as vacant residential property, infill residential development has occurred. Finally, if new residential structures are placed on lands classified as agriculture or open space other than designated for urban use, greenfield development has occurred.

For Hillsborough County, the analysis indicates that 35 percent of residential development during a previous 10-year time period was in redevelopment. Therefore, the LUCIS^{plus} residential redevelopment must account for 35 percent of the 711,300 new projected residents, or 248,955 people.

The number of cities in the county creates an extra complication. You can make some reasonable assumptions to simplify this complication: (1) the new redevelopment proportion for each city remains the same as the existing population distribution for each city (trend); (2) the planning process (policy) determines the percentage of redevelopment that occurs (that is, a city can significantly increase its population portion because of redevelopment incentives or other factors); and (3) one urban city is clearly dominant and gets the lion's share of the redevelopment allocation (again, trend or policy). Ultimately, the analyst must choose the most appropriate fit for the area of interest.

The redevelopment CEM

Table 4.7 shows the pertinent variables for residential redevelopment allocations. The employment allocation variables are also shown and discussed in chapter 5. The location variables may be used as policy indicators (for example, distance to transit stations). However, CEM primarily includes the location variables to accommodate summary information within areas of interest (for example, how many new people or dwelling units were allocated to each individual city, census block, or traffic analysis zone [TAZ]). The conflict/opportunity variable represents the LUCIS mixed-use redevelopment opportunity from table 4.6. Join variables are included to provide additional information that is in separate files, such as property parcel files or zoning files. The joined data can also be used to identify policy (for example, unit density or changes in existing unit density). Finally, the allocation variables are used to allocate residential development. The variable NEWPOP is the potential population available for the CEM raster value attribute table (VAT). Each record calculates the NEWPOP value as (acres * unit density * average household size). If allocated, it would contribute to the new population in the area.

Table 4.7. The redevelopment CEM schema

CEM Structure	Policy Type	Variable	Description
ArcGIS Raster VAT Base		OID	ArcGIS Raster Object ID
		Value	ArcGIS Raster VAT Value
		Count	ArcGIS Raster VAT Cell Count
Location Variables		CENBLKID	Census Block Group ID
	Plan Policy	CITIES	City Identifier
	Policy	CNTERSLCE19	Distance Zone from Major Center City
		TAZID	Traffic Analysis Zone ID
		PARGRID	Parcel ID
Conflict/ Opportunity		REDEVCONFLICT	Redevelopment Mixed-Use Opportunity
		GFCONFLICT	LUCIS Conflict/Preference
Suitability		MFSUIT100I	Multifamily Residential Suitability
		SFSUIT100I	Single-Family Residential Suitability
		RESFSUIT100I	Combined Multifamily and Single- Family Residential Suitability
		INDSUIT100I	Industrial Land-Use Suitability
		COMSUIT100I	Commercial Land-Use Suitability
		SERSUIT100I	Service Land-Use Suitability
		RETSUIT100I	Retail Land-Use Suitability
		INSSUIT100I	Institutional Land-Use Suitability
Join Variables		JUST	Parcel Market Value
		DESCRIPT	Property Use Description
		DOLACRE	Parcel Dollar Value per Acre
		ACT_CENTER	Activity Center ID
		YRBUILT	Structure Year Built
		ACT_DESC	Activity Center Description
		ACREAGE	Property Parcel Acres
	Plan Policy	CITYNAME	City Name
	Plan Policy	AREATYPE	Area Indicator (Urban, Transition, Rural)
	Policy	DENSITY	Future Land-Use Plan Density Units/Acre
	Plan Policy	ZONE_DESC	Zoning Description
	Plan Policy	EZONE	Existing Economic Development Zones

(*continued*)

Table 4.7. The redevelopment CEM schema (*continued*)

CEM Structure	Policy Type	Variable	Description
Allocation Variables	Plan Policy	UDENSITY	Allocation Density Units/Acre
		ALLYRPOP	Allocation Year Residential Population
		NEWPOP	New Population for the Allocation
		ACRES	Available Allocation Acres
		NEWDU	New Residential Dwelling Units Allocated
		ALLYRIND	Allocation Year Industrial Employment
		INDEMPDEN	Industrial Employment Gross Density
		INDEMP	Available Industrial Employment
		ALLYRCOM	Allocation Year Commercial Employment
		COMEMPDEN	Commercial Employment Gross Density
		COMEMP	Available Commercial Employment
		ALLYRSER	Allocation Year Service Employment
		SEREMPDEN	Service Employment Gross Density
		SEREMP	Available Service Employment
		ALLYRRET	Allocation Year Retail Employment
		RETEMPDEN	Retail Employment Gross Density
		RETEMP	Available Retail Employment
		ALLYRINS	Allocation Year Institutional Employment
		INSEMPDEN	Institutional Employment Gross Density
		INSEMP	Available Institutional Employment
Scenario Allocation Variables x = 1, 2, 3, 4 to n		S(X)ALLYR	Allocation Year Residential Population for Scenario (x)
		S(X)INDYR	Allocation Year Industrial Employment for Scenario (x)
		S(X)INDEMPDEN	Industrial Employment Density for Scenario (x)
		S(X)INDEMP	Available Industrial Employment for Scenario (x)
		S(X)COMYR	Allocation Year Commercial Employment for Scenario (x)

(*continued*)

Table 4.7. The redevelopment CEM schema (*continued*)

CEM Structure	Policy Type	Variable	Description
		S(X)COMEMPDEN	Commercial Employment Density for Scenario (*x*)
		S(X)COMEMP	Available Commercial Employment for Scenario (*x*)
		S(X)RETYR	Allocation Year Retail Employment for Scenario (*x*)
		S(X)RETEMPDEN	Retail Employment Density for Scenario (*x*)
		S(X)RETEMP	Available Retail Employment for Scenario (*x*)
		S(X)INSYR	Allocation Year Institutional Employment for Scenario (*x*)
		S(X)INSEMPDEN	Institutional Employment Density for Scenario (*x*)
		S(X)INSEMP	Available Institutional Employment for Scenario (*x*)
		S(X)SERYR	Allocation Year Service Employment for Scenario (*x*)
		S(X)SEREMPDEN	Service Employment Density for Scenario (*x*)
		S(X)SEEMP	Available Service Employment for Scenario (*x*)

The following list of criteria is used for allocations of redevelopment residential population based on land-use selections structured in ArcGIS query language:

- *DESCRIPT*: Property Use Description. Examples include Vacant Commercial, Vacant Industry, Misc. Agriculture, and so forth.
- *ALLYRPOP*: Allocation Year Residential Population. Item used to hold the year when the cell was allocated for residential use. For example, when this item equals 0, the cell has not been allocated. When the item equals 2045, the cell was allocated for residential use by 2045.
- *GFCONFLICT*: LUCIS Conflict/Preference Index for greenfield conflict or opportunity. Item holds the greenfield conflict or opportunity numbers. The index contains 27 unique numbers, ranging from 111 to 333.
- *CNTERSLCE19*: Distance Zone from Major Center City. Item used to indicate a zone of distance from the center of employment activity. The value ranges from 1 to 9, with 9 representing the closest zone of distance.

- *ACRES*: Available Allocation Acres. Item that contains the number of acres represented by the raster zone. A raster zone is a record within an integer raster that contains all the cells in a raster that have a unique value (for example, all the cells that represent single-family residential land use).
- *YRBUILT*: Structure Year Built. Item containing the actual year that a structure on a property parcel was built.
- *CITYNAME*: City Name. Item representing the city the cell is located within: (1) Tampa, (2) Temple Terrace, (3) Plant City, or (0) in the county.
- *REDEVCONFLICT*: Redevelopment Mixed-Use Opportunity. The item is developed, just like the GFCONFLICT land use. However, it represents redevelopment opportunity (in this case, for commercial, multifamily residential, and retail mixed use). It has the same 27 index values.
- *DENSITY*: Future Land-Use Plan Density Units/Acre. This is the residential units per acre.
- *MFSUIT100I*: Multifamily Residential Suitability. As with other suitability raster layers, the numeric values range from 100 to 900. The extrapolated three-digit numbers are used to work with the ArcGIS Combine tool in later sections of the book. The ArcGIS Combine tool works with integers only. Therefore, the 100 to 900 values act like 1.00 to 9.00 values. It retains the suitability sensitivity to three digits. Now you can use the three-digit numbers to query or select suitability.
- *SFSUIT100I*: Single-Family Residential Suitability. As with other suitability raster layers, the numeric values range from 100 to 900.
- *NEWPOP*: New Population Potential. Item contains the population potential in CEM for each record in VAT and the population that is allocated if the record is selected during the allocation query process.
- *ACREAGE*: Property Parcel Acreage. This parcel acreage is from the property files.
- *RESFSUIT100I*: Combined Multifamily and Single-Family Residential Suitability. As with other suitability raster layers, the numeric values range from 100 to 900.

The second allocation variable is ALLYRPOP. It indicates the year the allocation occurs. ALLYRPOP allows time-sequenced population allocations (for example, for every 10 years within the allocation time frame or for only the final year of the allocation). Therefore, 10-year increments from 2005 to 2045 would produce ALLYRPOP values of 0, 2015, 2025, 2035, and 2045, where zero (0) indicates no allocation and is not actually a year. If the allocation was for the entire time increment, the ALLYRPOP values would be 0 and 2045, where 2045 represents the allocated cells for the year 2045 and the 0 cells are cells not yet allocated. The actual allocation is a sequential query, including one, some, or many of the variables in CEM. Suppose, as in query 4.1, you want to allocate only new multifamily residential redevelopment within the City of Tampa. You also want to select locations in which the zoning description is for high or moderate residential density. So you decide to use unit densities greater than or equal to 25 units per acre. You also decide to use only structures that are built in or before 1995. Thus, the selected buildings would be at least 50 years old by 2045. Finally, you decide to look for locations in the top two categories for distances near Tampa's center of employment ("CNTERSLCE19" > 7"). Query 4.1 locates all the cells in the raster layer that meet these requirements. For an explanation of the REDEVCONFLICT values used in the query, see table 4.6.

Query 4.1.
```
"CNTERSLCE19" > 7 AND ("REDEVCONFLICT" = 131 OR "REDEVCONFLICT"
= 132 OR "REDEVCONFLICT" = 231 OR "REDEVCONFLICT" = 232) AND
"YRBUILT" <= 1995 AND "CITYNAME" = 'Tampa' AND "DENSITY" >= 25.
```

Once the ArcGIS Select By Attributes query is complete, determine whether the potential population is greater or less than the needed population (248,955 people). Use ArcGIS to calculate the statistics for the variable NEWPOP in the CEM VAT. Right-click the NEWPOP variable in the table and then click Statistics. If the sum of NEWPOP is less than the needed allocation, the Field Calculator calculates the field ALLYRPOP to the year of the allocations, 2045. The NEWPOP sum equals 11,151 people, the number of new dwelling units equals 5,009, and the selection redevelops 54 acres in Tampa. The unit density for the new development is 93 units per acre. However, you must still allocate an additional 237,804 people.

For the next query, query 4.2, you may want to locate all the cells in the redevelopment CEM with mixed-use opportunity (not only new multifamily residential) for high-preference multifamily residential. You must also continue to fulfill the remainder of the query 4.1 criteria. Finally, select ALLYRPOP = 0 because these cells have not been previously allocated.

Query 4.2.
```
"CNTERSLCE19" > 7 AND ("REDEVCONFLICT" = 131 OR "REDEVCONFLICT"
= 132 OR "REDEVCONFLICT" = 231 OR "REDEVCONFLICT" = 232 OR
"REDEVCONFLICT" = 333 OR "REDEVCONFLICT" = 133 OR "REDEVCONFLICT"
= 233 OR "REDEVCONFLICT" = 331 OR "REDEVCONFLICT" = 332) AND
"YRBUILT" <= 1995 AND "CITYNAME" = 'Tampa' AND "DENSITY" >= 25
AND "ALLYRPOP" = 0.
```

Once query 4.2 is completed, the allocated population is 36,331 more people. There are 15,063 new dwelling units. The extra redevelopment is located on an additional 361 acres with an average unit density of 42 units (table 4.8). Figure 4.11 shows the location of new multifamily residential redevelopment. Query 4.2 also provides opportunities for new mixed-use employment opportunity for commercial, retail, and service employment. Employment allocations are discussed in chapter 5.

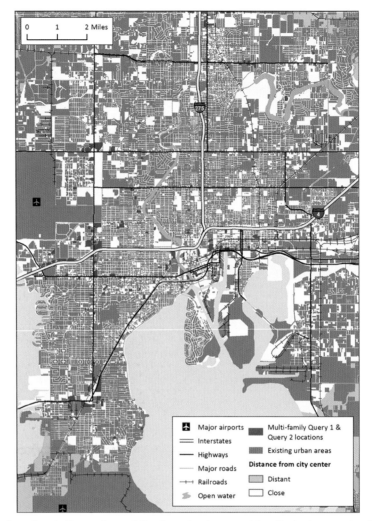

Figure 4.11. Locations in Tampa for multifamily redevelopment identified using queries 4.1 and 4.2. The cells identified for the initial two redevelopment allocations are closest to the city center.

Figure from Paul D. Zwick and the Arizona Board of Regents on behalf of the University of Arizona. Airports from the Federal Aviation Administration and the Research and Innovative Technology Administration Bureau of Transportation Statistics National Transportation Atlas Databases (NTAD) 2012; water bodies from the US Geological Survey; state boundary from US Census Bureau TIGER/Line Files; interstates, highways, and major roads from the Florida Department of Transportation; railroads from the Federal Railroad Administration.

The final redevelopment multifamily residential allocation, query 4.3, identifies additional redevelopment locations for 201,352 people (see table 4.8). It allocates 82,383 new dwelling units on 5,373 more acres of redeveloped land. The average unit density is 15.33 units per acre. Figure 4.12 shows the highest preference locations of multifamily residential and multifamily residential mixed-use resulting from queries 4.1, 4.2, and 4.3. Table 4.8 shows the complete results of the residential redevelopment allocations.

Chapter 4 Analyzing and mapping residential land-use futures

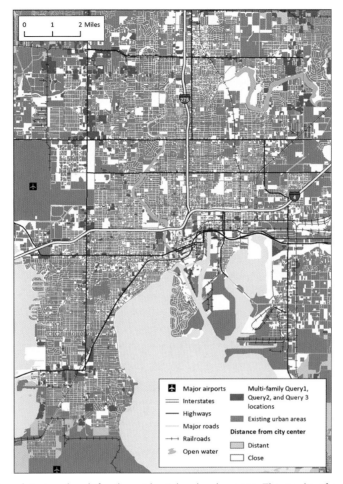

Figure 4.12. Complete trend multifamily residential redevelopment. The results of queries 4.1, 4.2, and 4.3 primarily occur close to the city center.

Figure from Paul D. Zwick and the Arizona Board of Regents on behalf of the University of Arizona. Airports from the Federal Aviation Administration and the Research and Innovative Technology Administration Bureau of Transportation Statistics National Transportation Atlas Databases (NTAD) 2012; water bodies from the US Geological Survey; state boundary from US Census Bureau TIGER/Line Files; interstates, highways, and major roads from the Florida Department of Transportation; railroads from the Federal Railroad Administration.

Query 4.3.
"CNTERSLCE19" >= 5 AND ("REDEVCONFLICT" = 131 OR "REDEVCONFLICT" = 132 OR "REDEVCONFLICT" = 231 OR "REDEVCONFLICT" = 232 OR "REDEVCONFLICT" = 333 OR "REDEVCONFLICT" = 133 OR "REDEVCONFLICT" = 233 OR "REDEVCONFLICT" = 331 OR "REDEVCONFLICT" = 332 OR "REDEVCONFLICT" = 333) AND "YRBUILT" <= 1995 AND ("CITYNAME" = 'Tampa' OR "CITYNAME" = 'Plant City' OR "CITYNAME" = 'Temple Terrace' OR "CITYNAME" = 'County') AND "DENSITY" >= 6 AND "ALLYRPOP" = 0.

Table 4.8. Complete results of trend residential redevelopment allocations

Query	Population	Dwelling Units	Acres	Density
4.1	11,151	5,009	54	92.8
4.2	36,331	15,063	361	41.7
4.3	201,352	82,383	5,373	15.3
Totals	248,834	102,455	5,788	17.7

So how many new people must still be allocated to accommodate the new 2045 population? It is 711,300 people minus the newly allocated redevelopment residential population of 248,834. Or 462,466 people. These people must be allocated in infill and greenfield locations. Infill residential development occupies approximately 40 percent of the total population. Greenfield residential development is 25 percent of the total population. Therefore, the infill population allocation should account for 711,300 * 0.4, which equals 284,520 people. The greenfield allocations account for 177, 825 people.

Queries 4.1, 4.2, and 4.3 complete the allocation of residential redevelopment because the analysis selects all multifamily redevelopment opportunities within the top five areas near the center of Tampa. The analyst also selects all multifamily locations available within the two other metropolitan Tampa cities, Temple Terrace and Plant City (figure 4.12). Plus, the analyst selects other available high-potential sites in the unincorporated county that meet the query's redevelopment opportunity.

Residential infill allocation

Unlike residential redevelopment allocation, the infill residential allocation is completed on lands that have an existing land classification of vacant residential. The allocation uses the LUCIS infill CEM (table 4.9). The difference between the two allocations is that redevelopment requires demolition or adaptive reuse of existing structures. Infill puts new structures on lands available for residential use that do not have existing structures. The vacant residential land-use classification is often designated in the region's property parcel GIS data files or within the future land-use GIS data files. It depends on the region's depth and quality of GIS data. The use of lands classified as vacant residential provides the analyst with opportunity to allocate. In this case, it is residential development on lands already designated in the regional land-use plan and properly zoned for residential development. Developers plat subdivisions with the intent of selling the property. But they realize that it may take many years to build out the development. Vacant lands represent existing capacity dedicated in the region's land-use plan to accommodate all or a portion of the projected residential growth.

One difficulty in assigning projected growth to vacant residential lands is accurately identifying what portion of the growth should be allocated to the vacant lands. A second difficulty is replacing the vacant land capacity that has been allocated. A third difficulty is recognizing that some vacant lands are not appropriate for development. They remain vacant because they are not salable.

Table 4.9. The infill CEM schema

CEM Structure	Policy Type	Variable	Description
ArcGIS Raster VAT Base		OID	ArcGIS Raster Object ID
		Value	ArcGIS Raster VAT Value
		Count	ArcGIS Raster VAT Cell Count
Location Variables		CENBLKID	Census Block Group ID
	Plan Policy	CITIES	City Identifier
	Policy	CNTERSLCE19	Distance Zone from Major Center City
		TAZID	Traffic Analysis Zone ID
		PARGRID	Parcel ID
Suitability		MFSUIT100I	Multifamily Residential Suitability
		SFSUIT100I	Single-Family Residential Suitability
		RESFSUIT100I	Combined Multifamily and Single-Family Residential Suitability
		COMSUIT100I	Commercial Land-Use Suitability
		SERSUIT100I	Service Land-Use Suitability
		RETSUIT100I	Retail Land-Use Suitability
		INSSUIT100I	Institutional Land-Use Suitability
		INDSUIT100I	Industrial Land-Use Suitability
Join Variables		JUST	Parcel Market Value
		ACREAGE	Property Parcel Acres
		DESCRIPT	Property Use Description
		DOLACRE	Parcel Dollar Value per Acre
		ACT_CENTER	Activity Center ID
		YRBUILT	Structure Year Built
		ACT_DESC	Activity Center Description
	Plan Policy	CITYNAME	City Name
	Plan Policy	AREATYPE	Area Indicator (Urban, Transition, Rural)
	Policy	DENSITY	Future Land-Use Plan Density Units/Acre
	Plan Policy	EZONE	Existing Economic Development Zones
	Plan Policy	ZONE_DESC	Zoning Description

(continued)

Table 4.9. The infill CEM schema (*continued*)

CEM Structure	Policy Type	Variable	Description
Allocation Variables	Plan Policy	UDENSITY	Allocation Density Units/Acre
		ALLYRPOP	Allocation Year Residential Population
		NEWPOP	New Population for the Allocation
		ACRES	Available Allocation Acres
		NEWDU	New Residential Dwelling Units Allocated
		ALLYRIND	Allocation Year Industrial Employment
		INDEMPDEN	Industrial Employment Gross Density
		INDEMP	Available Industrial Employment
		ALLYRCOM	Allocation Year Commercial Employment
		COMEMPDEN	Commercial Employment Gross Density
		COMEMP	Available Commercial Employment
		ALLYRSER	Allocation Year Service Employment
		SEREMPDEN	Service Employment Gross Density
		SEREMP	Available Service Employment
		ALLYRRET	Allocation Year Retail Employment
		RETEMPDEN	Retail Employment Gross Density
		RETEMP	Available Retail Employment
		ALLYRINS	Allocation Year Institutional Employment
		INSEMPDEN	Institutional Employment Gross Density
		INSEMP	Available Institutional Employment

(*continued*)

Table 4.9. The infill CEM schema (*continued*)

CEM Structure	Policy Type	Variable	Description
Scenario Allocation Variables $x = 1, 2, 3, 4$ to n		S(X)ALLYR	Allocation Year Residential Population for Scenario (x)
		S(X)INDYR	Allocation Year Industrial Employment for Scenario (x)
		S(X)INDEMPDEN	Industrial Employment Density for Scenario (x)
		S(X)INDEMP	Available Industrial Employment for Scenario (x)
		S(X)COMYR	Allocation Year Commercial Employment for Scenario (x)
		S(X)COMEMPDEN	Commercial Employment Density for Scenario (x)
		S(X)COMEMP	Available Commercial Employment for Scenario (x)
		S(X)RETYR	Allocation Year Retail Employment for Scenario (x)
		S(X)RETEMPDEN	Retail Employment Density for Scenario (x)
		S(X)RETEMP	Available Retail Employment for Scenario (x)
		S(X)INSYR	Allocation Year Institutional Employment for Scenario (x)
		S(X)INSEMPDEN	Institutional Employment Density for Scenario (x)
		S(X)INSEMP	Available Institutional Employment for Scenario (x)
		S(X)SERYR	Allocation Year Service Employment for Scenario (x)
		S(X)SEREMPDEN	Service Employment Density for Scenario (x)
		S(X)SEEMP	Available Service Employment for Scenario (x)

LUCIS[plus] uses three concepts to solve the three vacant-land difficulties. First, the single-family or multifamily suitability raster layers are used to select vacant residential lands for allocation. Thus, the analyst selects the most suitable vacant residential lands. Using residential suitability provides an answer to the first and third allocation difficulties. Because the vacant residential lands are already listed as appropriate for residential development, there is, by definition, no conflict. Conflict is a method of identifying overlapping preferential use. When conflict is removed, you must rely on residential suitability to determine the best vacant residential lands to allocate. Because suitability includes multiple models for proximity, physical characteristics, land-use change, density, and historical growth patterns, vacant lands in one subdivision have different suitability than another subdivision. Using residential suitability reduces the chance of picking vacant lands that are unsuitable. Unsuitability could range from being located far from amenities to being on physically less suitable lands such as wetland areas to being in areas of low residential density or to being in areas that have been historically vacant for long periods of time.

Finally, the second difficulty is the allocation of replacement vacant residential land allocated for the current infill. The allocation of greenfield residential development addresses the need to replace vacant lands allocated for infill residential development. It is discussed in the following section of this chapter, "Greenfield residential allocation."

The allocation of new infill residential using the infill CEM closely resembles the allocations performed for residential redevelopment. The analyst develops queries that select vacant lands with highest suitability that also accommodate policy decisions. One such policy decision could be the location of infill residential uses closest to urban cities. A second policy example might include incentives for affordable-housing opportunities linked to identifying infill multifamily residential development. The single-family suitability model was discussed earlier in this chapter in the "Hillsborough County residential suitability" section. Table 4.10 shows the proportion of infill residential development that must be allocated to multifamily and single-family uses as calculated from the 2010 county uses. Figure 4.13 shows the complete allocation for trend infill residential development.

Table 4.10. Infill residential required with LUCIS^{plus} allocated population

Category	2010 Vacant Infill Acreage	2010 Vacant Infill Percentage	Required Population (portion of 284,520)	Allocated 2045 Population	People per Acre	Allocated 2045 Acreage	Density Units per Acre
Multifamily	43,077	24.7	70,276	70,131[a]	13.55	5,188	4.88
Single family	131,349	75.3	214,244	214,677[b]	11.49	18,677	4.23
Total Residential	174,426	100.0	284,520	284,808	11.92	23,865	4.4 (overall)

[a]Query MF1.
[b]Query SF1.

The following two queries allocate the trend multifamily and single-family infill residential population. Query MF1 selects vacant residential lands, including lands allocated for town houses (Townhomes), within the top five nearest area slices to the center of Tampa. Multifamily residential suitability is in the top 66 percent of suitable locations. Property parcel acreage is greater than four-tenths of an acre. The allocation accounts for 70,131 people within 5,188 acres. The unit density is 4.88 units per acre.

Query SF1 selects the single-family infill residential development. The allocation locates 18,677 residential single-family acres at a dwelling density of 4.23 units per acre. The total

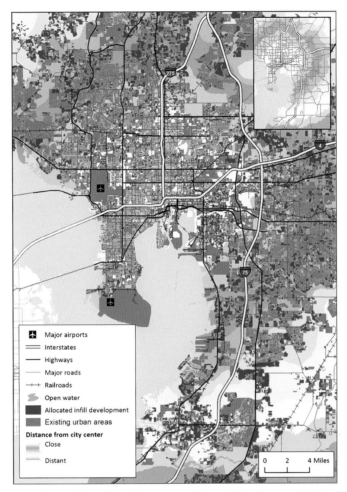

Figure 4.13. Trend infill residential development as allocated for the years 2010 to 2045.

Figure from Paul D. Zwick and the Arizona Board of Regents on behalf of the University of Arizona. Airports from the Federal Aviation Administration and the Research and Innovative Technology Administration Bureau of Transportation Statistics National Transportation Atlas Databases (NTAD) 2012; water bodies from the US Geological Survey; state boundary from US Census Bureau TIGER/Line Files; interstates, highways, and major roads from the Florida Department of Transportation; railroads from the Federal Railroad Administration.

is 79,004 dwelling units housing 214,677 people. The query finds acreage within the entire county for single-family residential suitability in the top 31.3 percent of all opportunities. The selection finds SFSUIT100I greater than 618, which is 31.3 percent of total opportunity for single-family suitability [(900 − 618) / 900]. Again, the "`ALLYRPOP`" = 0 query removes any chance of reselecting lands already allocated for multifamily residential infill. Figure 4.13 shows the allocated areas for infill residential development.

Query MF1.
```
"CNTERSLCE19" > 5 AND "MFSUIT100I" > 300 AND ("DESCRIPT" = 'VACANT
TOWNHOME' OR "DESCRIPT" = 'VACANT RESIDENTIAL') AND "ACREAGE" > 0.4.
```

Query SF1.
```
"SFSUIT100I" > 618 AND ("DESCRIPT" = 'VACANT TOWNHOME' OR "DESCRIPT"
= 'VACANT RESIDENTIAL' OR "DESCRIPT" = 'VACANT ACREAGE') AND
"ACERAGE" >= 0.25 AND "ALLYRPOP" = 0.
```

Greenfield residential allocation

The greenfield residential allocation uses the greenfield CEM. The greenfield CEM uses greenfield conflict, residential suitability, location, and joined variables already described for the other two CEMs. The greenfield CEM is not presented in detail. It has the same basic LUCIS conflict variables for greenfield development areas. Greenfield areas are predominantly rural and agriculture land-use areas. The primary difference between the other residential development CEMs and the greenfield CEM is the use of greenfield conflict shown in table 4.5.

The previous residential allocations allocated 248,834 residents in redevelopment locations and 284,808 residents in infill locations, out of 711,300 residents. Therefore, you must allocate close to 177,625 new residents in greenfield locations.

The following two queries locate the trend greenfield residential allocations for the years 2010 to 2045. Query GF1 selects any cells that have a greenfield urban preference.

Query GF1.
```
("GFCONFLICT" = 113 OR "GFCONFLICT" = 213 OR "GFCONFLICT" = 223 OR
"GFCONFLICT" = 123 OR "GFCONFLICT" = 112) AND "ALLYRPOP" = 0 AND
"ACREAGE" >= 0.26 AND "NEWPOP" > 0 AND "DESCRIPT" <> 'UTILITY'.
```

Query GF2 selects the cells that have high greenfield residential suitability (greater than 6.63 on a scale of 1 to 9). They are also in conflict with areas that have high agriculture suitability. Also, no cells are selected unless the property parcel is greater than or equal to 3.25 acres. The query `"DESCRIPT" <> 'UTILITY'` assures that no cells are selected within utility power line corridors.

Query GF2.
```
"DESCRIPT" <> 'UTILITY' AND "ALLYRPOP" = 0 AND ("GFCONFLICT" = 313
OR "GFCONFLICT" = 323) AND "ACREAGE" >= 3.25 AND "NEWPOP" > 0 AND
"RESFSUIT100I" > 663.
```

The selected locations are inside and outside urban areas. They represent a selection of 12,895 acres. The allocation accommodates 178,157 new residents within 61,222 units. It represents a residential unit density of 4.75 units per acre. All units are on 2010 existing productive agricultural lands. The requirement for high urban preference ensures this condition. It primarily results from the proximity to existing residential land uses (both inside and outside city boundaries). The total greenfield future occupied residential development demographics are presented in table 4.11. They include demographic data for potential new vacant residential occupancy, units, and density discussed in the next section of this chapter, "New allocation of vacant residential lands (future capacity)."

The last requirement for residential location is to identify the new vacant residential capacity. Vacant residential capacity is the result of platted vacant residential parcels. Because this allocation is for trend residential development, the used residential capacity (from infill residential allocation) must be replaced with new residential vacant land.

Figure 4.14 shows the trend greenfield residential allocations for the 177,625 people needed.

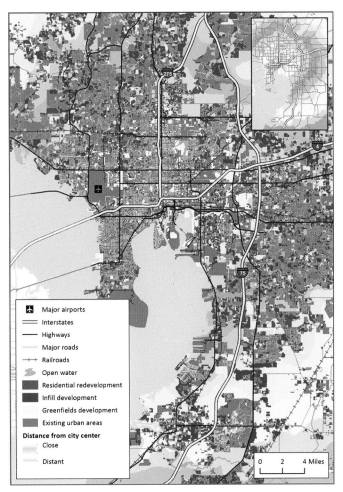

Figure 4.14. Trend total residential development allocation for the years 2010 to 2045. The red areas represent the residential redevelopment allocations. The blue areas represent the infill residential development allocations, and the yellow areas represent the greenfield residential development allocations. Employment allocations are discussed in chapter 5.

Figure from Paul D. Zwick and the Arizona Board of Regents on behalf of the University of Arizona. Airports from the Federal Aviation Administration and the Research and Innovative Technology Administration Bureau of Transportation Statistics National Transportation Atlas Databases (NTAD) 2012; water bodies from the US Geological Survey; state boundary from US Census Bureau TIGER/Line Files; interstates, highways, and major roads from the Florida Department of Transportation; railroads from the Federal Railroad Administration.

New allocation of vacant residential lands (future capacity)

To locate new vacant residential capacity, you must find 23,865 acres of new suitable land. Table 4.10 shows that it will take 23,865 acres to replace the acres allocated for the 2045 trend infill residential development. To replace the total acres allocated for new growth, the analyst may simply use the residential suitability raster (figure 4.15) to allocate 23,865 acres to the highest residential suitable lands.

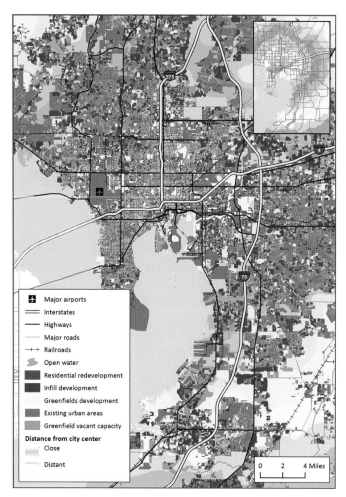

Figure 4.15. Total residential allocations, including replacement capacity in the form of new vacant residential land capacity. The yellow areas are allocated for greenfield development from 2010 to 2045. The green areas are greenfield replacement vacant residential capacity to replace the allocated vacant residential development used for infill residential development. Infill residential is blue, and redevelopment residential is red.

Figure from Paul D. Zwick and the Arizona Board of Regents on behalf of the University of Arizona. Airports from the Federal Aviation Administration and the Research and Innovative Technology Administration Bureau of Transportation Statistics National Transportation Atlas Databases (NTAD) 2012; water bodies from the US Geological Survey; state boundary from US Census Bureau TIGER/Line Files; interstates, highways, and major roads from the Florida Department of Transportation; railroads from the Federal Railroad Administration.

These vacant residential lands are replacement capacity for infill residential allocated lands. The following greenfield residential reserve capacity (GFRRC) query was used to locate the trend infill residential replacement capacity lands for the years 2010 to 2045. Query GFRRC1 selects any cells that have a greenfield urban preference.

Query GFRRC1.
```
"ALLYRPOP" = 0 AND "RESFSUIT100I" > 610 AND "ACRES" > 5.
```

The replacement capacity is a trend allocation because it replaces the vacant platted lands allocated for trend infill development. It is also new capacity based on an increase in the existing population by 2045. Infill allocation uses lands that have already been allocated for future residential capacity. That capacity must be replaced if there is reason to believe the population will continue to grow. Query GFRRC1 selected 23,865 acres. The capacity is for 77,779 new dwelling units housing 226,336 future residents. The unit density is 3.26 units per acre. The equation for the trend infill allocation (acres * unit density * average household size) calculates the dwelling unit numbers and potential population. It is an estimate of the capacity that might be needed for continued growth at the same trend rate.

Table 4.11. Greenfield residential required with LUCIS allocated population

Greenfield Residential	2045 Allocated Greenfield Acreage	Residents (Potential Occupants)	Units (Potential Units)	Unit Density (Potential Density)
New Occupied	12,895	178,157	61,222	4.75
New Vacant	23,865	226,336*	77,779*	3.26*

*Estimated potential occupants, potential units, and potential unit density for the new vacant residential capacity.

Summarizing the data

The reasons for analyzing land-use change are numerous. So are the reasons for spatially visualizing the impacts of land-use and population change. Planners must understand how these land-use changes are distributed within various spatial boundaries to start envisioning the impacts of any land-use change. For example, when the trend residential allocations are summarized within US Census blocks, you can visualize how the changes are distributed. You can also expand the concepts for the census blocks to summarize land-use changes within any boundary or polygonal area. Areas include, but are not limited to, (1) county boundaries, (2) TAZs, (3) other census boundaries, (4) water management districts, (5) historic districts, (6) US House or Senate districts, (7) regional planning districts, and (8) city boundaries.

Census blocks are political boundaries that are familiar to planners and nonplanners alike. They are useful areas that contain many types of demographic information. This information includes cohort population numbers. The census blocks for Hillsborough County contain population numbers for age cohorts 5–17, 18–21, 22–29, 30–39, 40–49, 50–64, and 65 and up. There are 21,851 census blocks in the polygon layer for Hillsborough County. Once the trend population has been allocated, the existing cohort population data can be used to determine what percentage of the new population is within particular population age cohorts. Once again, with the trend allocations, population demographic changes increase or decrease based on the percentages for individual block groups. They can be used to describe the changes based on new population allocations. In alternative geodesign scenarios, the population age cohort demographics must change to better reflect scenario policy. For example, if an alternative scenario were to accommodate professional TOD, you would modify the numbers of elderly and children age cohort percentages from the trend.

Linking CEMs to census blocks

The location data in each of the three CEMs has an attribute for census block ID (CENBLKID). It is a unique link to each census block in the polygon layer. Using that link, you can use the percentage of existing population in each of the seven population age cohorts to aggregate new population allocations within individual census blocks. Table 4.12 shows a small selection of census blocks with the trend infill population age cohort percentage for age 5–17. It shows the total number of dwelling units allocated, the total number of acres used for the allocation, and the newly allocated people in the age cohort of children 5–17. For example, the census block with ID 3 has 210 new residents occupying 103 new dwelling units on 38.63 acres of vacant residential infill lands. There are 37 children in the age population cohort 5–17.

Table 4.12. Sample of selected census blocks with age cohort percentage and newly allocated people

Census Block ID	Infill Residential Acres	New Total Residential Population	New Dwelling Units	Percentage Age 5–17	Number People Age 5–17
3	38.63	210	103	17.75	37
5	2.79	18	9	14.89	3
7	55.08	427	139	25.59	109
11	10.61	213	69	24.17	51

Calculating age cohort population is simple. Determine the existing percentage of each age cohort in individual census blocks. Apply these percentages to allocate newly allocated population to the seven age cohort categories. Interestingly, one of the hardest problems to solve for regional geodesign land-use planning is the small-area population projections.

Mapping age cohort population

Figure 4.16 shows the allocation for individual census blocks for the population age cohort of school-age children age 5–17. Summarizing the spatial allocation of population within specific age cohorts has its advantages. Advantages include school planning; planning for elderly populations, including elderly outpatient care facilities; and employment accessibility for the working population. Each of these advantages can be mapped using a LUCIS^plus geodesign process. Knowing the total number of new trend dwelling units can help with trend identification of the distribution of new school-age children in single-female or single-male

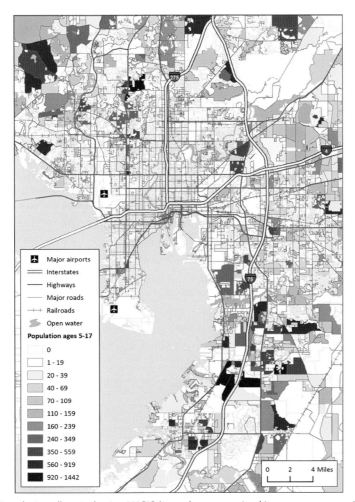

Figure 4.16. Population allocated using LUCIS^{plus} can be summarized in numerous ways, including according to the school-age population cohort (age 5–17) by census block. The spatial distribution and allocation of future populations impacts future school siting or closure policies, and LUCIS^{plus} aids in the decision-making process.

Figure from Paul D. Zwick and the Arizona Board of Regents on behalf of the University of Arizona. Airports from the Federal Aviation Administration and the Research and Innovative Technology Administration Bureau of Transportation Statistics National Transportation Atlas Databases (NTAD) 2012; water bodies from the US Geological Survey; state boundary from US Census Bureau TIGER/Line Files; interstates, highways, and major roads from the Florida Department of Transportation; railroads from the Federal Railroad Administration.

head-of-household families, the distribution of population by age cohort near hospitals or other health-care facilities, and owner-occupied versus rental units.

The total number of new children in the age cohort 5–17 is 27,442 for greenfield development, 59,062 for infill development, and 39,313 for urban redevelopment. The total number of school-age children is 125,817. Although the lower categories shown in the map (figure 4.16) might not require the addition of new schools, the top two or three categories have enough children to indicate a high likelihood that school construction might be needed in those school districts, particularly in those census blocks.

Chapter summary

The following points are discussed in this chapter:

- The five methods discussed to project population are basic linear population projection, basic exponential projection, modified exponential method, proportional population projection, and multivariate regression. Each method uses a different approach to determine population increase or decrease over time. Differences between each method include the rate of population change and the complexity of the calculation.
- The LUCIS^{plus} CEM(s) use customized queries to allocate residential population and its accompanying land uses. The queries reflect policy, individualized scenarios, and physical limitations to accommodate population. The LUCIS^{plus} CEM automates allocation in what was, in the base LUCIS method, a manual approach to population allocation.
- The three LUCIS^{plus} CEMs—the greenfield CEM, redevelopment CEM, and infill CEM—facilitate allocation according to each respective land-use classification. Each CEM combines conflict/opportunity, individual land-use suitability, and other data for making land-use decisions. The CEMs help allocate residential population, and regional and local employment.
- Results of the residential LUCIS^{plus} CEM(s) allocation can be summarized to political, census, and transportation units.

ArcGIS tools referenced in this chapter

Tool Name	Version 10.2 Toolbox/Toolset
Weighted Overlay	Spatial Analyst Tools/Overlay
Zonal Statistics as Table	Spatial Analyst Tools/Zonal

References

Carr, Margaret H., and Paul D. Zwick. 2007. *Smart Land-Use Analysis: The LUCIS Model*. Redlands, CA: Esri Press.

Chapter 5

Analyzing and mapping employment land-use futures

Paul D. Zwick

What this chapter covers

For years, urban planners have understood that employment and jobs are not solely governed by the local economy. Projecting regional employment is not easy, especially when the economy goes through a period like what the world and the United States experienced from 2007 to 2009. Regional employment projections provide a sense of economic opportunities and advantages that lend to satisfying or reevaluating strategic goals. Integrating employment projections into land-use analysis can result in scenarios that balance sound land-use planning with employment potential in the region. This chapter explains how to use employment trends to project future employment demand. Moreover, it describes the challenges of linking land use to employment.

Understanding employment

Understanding and then projecting employment growth or decline in an area can be a daunting task. Many models can be used to project employment and jobs. These models include (1) economic base and shift-share analysis, (2) multivariate regression, (3) curve fitting, and (4) proportional share. Depending on the funding available for economic analysis, more sophisticated economic models include the Regional Economic Model Inc. (REMI). REMI is useful in projecting economic activity and local employment. However, these models can be expensive. They can also require personnel with advanced economic modeling skills to manipulate the programs.

One of the earliest and most often used methods of trying to understand local economic activity is economic base theory. The theory explains that a local economy is less self-sufficient than the national or regional economies because a local economy tends to be more specialized and less diverse (Galambos and Schreiber 2006; Chapin 1965). The essence behind economic base theory is that jobs are tied to the production of export goods and services and the import of other goods and services. For many years, planners have used economic base theory with shift-share analysis to understand the local economy's health and stability. Economic base theory stipulates that if a local economy has few basic employers (subsequently, few jobs producing exports), the economy is less resilient. It is subject to swings based on the health of those few sectors of the economy. Whatever method is chosen to project employment, the economic mix of the region's employment sectors is important.

The US Department of Labor Bureau of Labor Statistics' Quarterly Census of Employment and Wages website (BLS 2012) provides historical employment data. The data is available by North American Industry Classification System (NAICS) employment sectors (figure 5.1). This employment data is useful in developing employment projections. On the BLS web page, click the Location Quotient Calculator. Follow the directions to get national, state, and study region employment data.

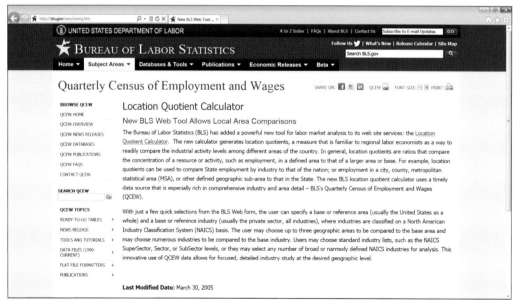

Figure 5.1. US Department of Labor Bureau of Labor Statistics' website providing employment data by NAICS code sectors for calculation of the economic base analysis location quotient and for use with shift-share employment projections.

Figure from the US Department of Labor Bureau of Labor Statistics.

Economic base analysis: Getting the basic import and export employment

The Quarterly Census of Employment and Wages website provides some useful data. Economic base theory compares the diversity and health of a local economy with the national or regional economy within which it exists and competes. It compares various sectors within the local economy with the same sectors within the region and nation. It uses the location quotient (LQ) for this comparison. To get an LQ, equation (5.1) divides the proportion of an individual sector's employment for a local economy by the same sector's proportion of the national or regional economy.

$$LQ = (e_i/e_t)/(E_i/E_t), \quad (5.1)$$

where:
LQ = location quotient for sector i,

e_i = local-sector employment for sector i,

e_t = total local employment,

E_i = national or regional employment for sector i, and

E_t = total national or regional employment.

Table 5.1 shows base employment numbers for 19 sectors in the US economy, Florida's regional economy, and Hillsborough County's local economy gathered from the Quarterly Census of Employment and Wages website. (The null sector 99 is not counted.)

The NAICS codes used here are the two-digit codes. However, data is collected for much more specific categories, down to the five-digit level. Table 5.2 shows the NAICS codes for the wholesale trade sector. The two-digit code is 42 for wholesale trade. It is followed by the three-digit code 423 for merchant wholesalers and durable goods. The four-digit code 4231 is for motor vehicle and motor vehicle parts and supplies merchant wholesalers. The five-digit code 42311 is for automobile and other motor vehicle merchant wholesalers. The following discussion and analysis uses only the two-digit codes for brevity and ease of description. The analysis presented is useful for all levels of NAICS codes. However, the more the two-digit codes are subdivided, the greater the chance of there being an open category or unavailable data.

Table 5.3 shows the percentage of employment by sector. Although the BLS Quarterly Census Location Quotient Calculator website calculates these values, the calculation used in the table is the percentage of employment for an NAICS code individual category calculated by the author. The individual employment sector proportion of employment is based on data in table 5.1. For example, the 2003 NAICS category codes Utilities (22), Construction (23), and Professional and technical services (54) illustrate how employment projections can be made from the labor statistics data. Table 5.1 for US utilities shows 575,877 employees. The total industrial employment for the 19 NAICS categories is 106,857,816 employees for the United States. The US utilities employment of 575,877 divided by 106,857,816 is 0.54 percent (the shaded row in table 5.3).

Table 5.1. 2003 annual average industrial employment for 19 two-digit NAICS codes for the United States, Florida, and Hillsborough County, Florida

NAICS Codes	Category	US Total	Florida Statewide	Hillsborough County
	Base Industry: All Industries	106,857,816	6,149,378	524,341
11	Agriculture, forestry, fishing, and hunting	1,156,242	97,843	12,266
21	Mining, quarrying, and oil and gas extraction	500,103	5,221	80
22	Utilities	575,877	26,805	3,075
23	Construction	6,672,360	444,520	32,649
31–33	Manufacturing	14,459,712	386,800	28,877
42	Wholesale trade	5,589,032	313,088	27,838
44–45	Retail trade	14,930,765	922,170	63,758
48–49	Transportation and warehousing	3,946,170	195,525	15,796
51	Information	3,180,752	171,726	22,357
52	Finance and insurance	5,782,062	330,009	44,437
53	Real estate, and rental and leasing	2,044,868	153,515	10,647
54	Professional and technical services	6,638,679	386,073	35,453
55	Management of companies and enterprises	1,660,137	64,969	4,030
56	Administrative and waste services	7,559,641	807,956	90,265
61	Educational services	2,016,163	85,523	7,097
62	Health care and social assistance	13,721,850	770,226	54,056
71	Arts, entertainment, and recreation	1,816,902	157,007	9,984
72	Accommodation and food services	10,345,336	650,538	41,918
81	Other services, except public administration	4,261,165	234,415	15,293
99	Unclassified			

Source: US Department of Labor Bureau of Labor Statistics' Quarterly Census of Employment and Wages—Location Quotient Calculator.
Note: Although BLS collects data for sector 99 for this search, it does not collect statistics for code 99.

Table 5.2. Example of two- to five-digit NAICS codes for wholesale trade

NAICS	Title
42	Wholesale Trade
423	Merchant Wholesalers, Durable Goods
4231	Motor Vehicle and Motor Vehicle Parts and Supplies Merchant Wholesalers
42311	Automobile and Other Motor Vehicle Merchant Wholesalers

Source: US. Department of Labor Bureau of Labor Statistics' Quarterly Census of Employment and Wages—Location Quotient Calculator.

Table 5.3. Percentage of 2003 industrial employment by two-digit NAICS codes (22, 23, and 54)

NAICS Codes	Category	US Percentage	Florida Percentage	Hillsborough Percentage
	Base Industry: All Industries	100%	100%	100%
22	Utilities	0.54%	0.44%	0.59%
23	Construction	6.24%	7.23%	6.23%
54	Professional and technical services	6.21%	6.28%	6.76%

Source: US Department of Labor Bureau of Labor Statistics' Quarterly Census of Employment and Wages—Location Quotient Calculator.

The calculation for the individual subsector category LQ is again relatively straightforward. In table 5.4, each of the three column LQs is calculated as the ratio of the local percentage compared with the national or regional percentage for 19 individual NAICS categories. For example, for NAICS code 22 (Utilities), the LQ for Florida's utilities compared with the United States is computed as the ratio of Florida percentage to the US percentage. In table 5.3, it is 0.44 percent divided by 0.54 percent, or 0.81. The LQ for Hillsborough County compared with the United States is 0.59 percent divided by 0.54 percent, or 1.09. The LQ for Hillsborough County utilities compared with Florida's utilities is 0.59 percent divided by 0.44 percent, or 1.35.

So what does the LQ say about local economic employment? When the LQ equals 1, the percentage of local employment equals the percentage of regional employment for the same sector. In this situation, local employment and its production of goods and services are enough to accommodate the needs of local residents. When LQ for a sector is greater than 1 (table 5.4, green cells), economic base theory postulates that the local percentage of employment, and therefore production of goods and services, for that sector is greater than the needs of local residents. Excess production is exported to the regional or national economy. Similarly, it follows that the excess employment production of exports, and the sales of those goods and services outside the local economy, will be circulated within the

Table 5.4. The 2003 location quotients of industrial employment by two-digit NAICS codes (22, 23, and 54)

NAICS Codes	Category	Florida/ US	Hillsborough County/US	Hillsborough County/ Florida
22	Utilities	0.81	1.09	1.35
23	Construction	1.16	1.00	0.88
54	Professional and technical services	1.01	1.09	1.08

Source: US Department of Labor Bureau of Labor Statistics' Quarterly Census of Employment and Wages—Location Quotient Calculator.

local economy. Thus, the local economy will become more stable. It means extra dollars for consumption within the local area. These extra dollars from exports produce a multiplier effect for local economic activity.

When the LQ for a local sector is less than 1 (table 5.4, yellow cells), the sector's employment is not producing all the goods and services that local residents require within that sector. Thus, it represents a sector in which the local economy imports goods and services. Because the local sector has a smaller employment percentage than the regional or national economies, it represents a deficit in production for the local economy. The local economy therefore imports from the regional or national economy in that sector.

Finally, what part of the local economy's employment by sector is base employment? Again, for the NAICS utilities sector, the US percentage of NAICS sector 22 is 0.54 percent and the Hillsborough percentage is 0.59 percent. The LQ is 1.09. The excess between 0.59 percent within Hillsborough and 0.54 percent in the United States is the export employment in the Hillsborough study region. If the US percentage is applied to the total employment for the utilities sector in Hillsborough County, it is 0.0054 * 575,877 = 3,110. The actual employment for the sector in Hillsborough County is 3,075 (see table 5.1). The excess base employment is 3,110 − 3,075 = 35. The base employment excess of 35 represents the jobs producing export goods and services for the NAICS utilities sector. The export is goods and services within Hillsborough County exported to the United States. For example, Hillsborough County provides extra electricity, gas, and other utility goods and services to the national utility grid. These exports support utility jobs in Hillsborough County.

Using economic base location quotients to investigate the shifts and shares within local employment

Now that you have an idea of the employment characteristics of Hillsborough County, you can use the data to make employment predictions for the county's local economy. Edgar S. Dunn Jr. and others developed the shift-share technique for modeling employment changes within a local economy in 1960 (Klosterman et al. 1993, 186). The following employment projection process is published in the classic text *Spreadsheet Models for Urban and Regional Analysis* (Klosterman et al. 1993).

How can you predict the 2045 employment for NAICS sectors 22, 23, and 54 based on the change in employment between 2003 and 2010? The top row in table 5.5 shows the change in employment for the 19 NAICS sectors between 2003 and 2010. The table also shows that in 2003, the US utilities sector had 575,877 jobs. By 2010, the sector provided 551,287 jobs. Therefore, the US utilities sector lost 24,590 jobs during that time frame. During the same time frame, Hillsborough County also lost utility jobs (3,075 jobs − 2,574 jobs = 501 jobs). The data shows that the local utilities sector followed the US utilities sector's declining job creation but at a different rate. Hillsborough County's utilities sector jobs were reduced to 83.71 percent of its 2003 employment (2,574 / 3,075 = 0.8371). The decrease in employment for the time frame 2003–10 is 16.29 percent. The US utilities sector jobs decreased to 95.73

Table 5.5. Quarterly census of employment and wage change in historical employment for selected NAICS codes (between 2003 and 2010)

NAICS Codes	Category	US 2003	Hillsborough 2003	US 2010	Hillsborough 2010
	Base Industry: All Industries	106,857,816	524,341	106,048,565	486,978
22	Utilities	575,877	3,075	551,287	2,574
23	Construction	6,672,360	32,649	5,489,499	25,668
54	Professional and technical services	6,638,679	35,453	7,457,913	46,901

percent of its 2003 employment (551,287 / 575,877 = 0.9573). The reduction for the time frame is 4.27 percent. So the reduction in utilities employment was greater in Hillsborough County than in the United States. Hence, the idea of changing, or shifting, shares of the national and local economies within specific sectors. Further, even though the LQ for sector 22 indicates that Hillsborough County has a greater than 1 ratio of employment in utilities, it is losing employment in the sector faster than the US economy. Logically, if the trend continues, at some point the Hillsborough County LQ for the utilities sector will be less than 1. At that point, it will no longer be a basic employment sector for the region.

As the previous discussion has shown, planners need to know how the local economy compares with the regional/national economy. They need to know how each sector's local share of employment has shifted compared with the national rate of change. It is possible to lose jobs in the local economy and still export goods and services to the regional or national economies. Combining economic base analysis with shift-share employment projections for a specified period can show whether the local economy is growing or declining. Planners also need to know whether the projected growth or decline is impacting exports or basic employment. Such a change will affect the local economy's economic health and stability. Comparing Hillsborough County's local economy to the US economy identifies sectors where the growth is strong and where it is weak, or declining. Economic base analysis helps planners identify the strength and weakness in the local economy's NAICS sectors. Shift-share analysis provides a method for projecting growth or decline in the local employment NAICS sectors based on a comparison of the local and national growth rates.

Using shifts and shares by employment sector as a projection technique

Using the shift-share analysis for NAICS sector 22 as an example, the employment projection for utilities employment between 2003 and 2045 for Hillsborough County is displayed in table 5.6. The calculations are presented in the following six-step process. The basic shift-share projection model is shown in equation (5.2).

$$LE_i^{t'} = [1 + R_i^{(t-t')} + S_i^{(t-t')}] * LE_i^t \quad (5.2)$$

where:
t = the base year of employment observation (2003),
t' = the final year of employment observation (2010),
$LE_i^{t'}$ = future local employment within sector i at time t',
LE_i^t = existing local employment within sector i at time t,
$R_i^{(t-t')}$ = regional growth rate for time period $(t-t')$, and
$S_i^{(t-t')}$ = projected employment shift term for sector i for time period $(t-t')$.

Table 5.6. The shift-share factors for Hillsborough County, Florida (showing local employment projections)

NAICS Codes	Category	(1) LCT	(2) NGR	(3) LGR	(4) AGR	(5) S2045	(6) PSE	(7) PCP	(8) LNEC
22	Utilities	0.0053	-0.0427	-0.1629	-0.1960	-0.12023	2,155	5,054	1,031
23	Construction	0.0527	-0.1772	-0.2149	-0.6230	-0.03762	20,152	50,402	9,609
54	Professional and technical services	0.0963	0.1234	0.3235	0.7892	0.20018	62,077	92,096	30,186

Note: LCT = local constant term, NGR = national growth rate, LGR = local growth rate, AGR = adjusted growth rate, S2045 = shift term 2045, PSE = projected shift employment, PCP = population change proportion, LNEC = LUCIS new employment change.

Step 1: Calculating the local constant term

The local constant term (LCT), column (1) in table 5.6, is the proportion of employment in sector i for the final year of the input data values. In this case, it is the 2010 employment for the utilities sector divided by the 2010 employment for the local region, in equation (5.3).

$$LCT_{22}^{2010} = LE_{22}^{2010} / \sum LE_{22}^{2010}$$
$$LCT_{22}^{2010} = 2{,}574 / 486{,}989$$
$$LCT_{22}^{2010} = 0.00528554. \quad (5.3)$$

where:
LE_{22}^{2010} = local employment 2010 for sector 22, and
$\sum LE_{22}^{2010}$ = sum of local employment 2010 for sector 22.

Step 2: Calculating the national (or regional) growth rate

The national growth rate (NGR), column (2) in table 5.6, is the rate of growth for the observed national employment data. For the example, it is the national employment for the utilities sector in the final observed year (2010) minus the national employment for the same sector in the base observed year (2003), divided by the national employment for the sector (2003), in equation (5.4).

$$NGR_{22} = (NE_{22}^{2010} - NE_{22}^{2003})/NE_{22}^{2003}$$
$$NGR_{22} = (551{,}287 - 575{,}877)/551{,}287$$
$$NGR_{22} = -0.04270009, \qquad (5.4)$$

where:
NE_{22}^{2003} = national employment for base observed year (2003), and
NE_{22}^{2010} = national employment for final observed year (2010).

Step 3: Calculating the local growth rate

The local growth rate (LGR), column (3) in table 5.6, is the rate of growth for the observed local employment data. For the example, it is the local employment for the utilities sector in the final observed year (2010) minus the local employment for this sector in the base observed year (2003), divided by the local employment for the sector (2003), in equation (5.5).

$$LGR_{22} = (LE_{22}^{2010} - LE_{22}^{2003})/LE_{22}^{2003}$$
$$LGR^{22} = (2{,}574 - 3{,}075)/3{,}075$$
$$LGR_{22} = -0.162926829, \qquad (5.5)$$

where:
LE_{22}^{2003} = local employment 2003 for NAICS sector 22, and
LE_{22}^{2010} = local employment 2010 for NAICS sector 22.

Step 4: Calculating the adjusted growth rate

The adjusted growth rate (AGR), column (4) in table 5.6, is the NGR adjusted for the difference between the projection time range and the observed range for data collection, in equation (5.6).

$$AGR_{22} = ((1 + NGR_{22})^{(Pm/Pn)}) - 1$$
$$AGR_{22} = ((1 - 0.04270009)^{(35/7)}) - 1$$
$$AGR_{22} = ((0.9573)^{5} - 1$$
$$AGR_{22} = -0.1960, \qquad (5.6)$$

where:
Pm = annual periods from 2010 to 2045, and
Pn = annual periods from 2003 to 2010.

Step 5: Calculating the shift term 2045

The shift term 2045 (S2045), column (5) in table 5.6, is the employment growth rate shift term (in this case, for the year 2045), in equation (5.7).

$$S_{22}^{(t-t')} = LGR_{22}^{2010} - NGR_{22}^{2010}$$
$$S_{22}^{(t-t')} = -0.162926829 - (-0.04270009)$$
$$S_{22}^{(t-t')} = -0.12023. \tag{5.7}$$

Step 6: Calculating the projected shift employment 2045

Projected shift employment (PSE), column (6) in table 5.6, is the projected shift term for local employment by sector based on the national growth rate and sector shift term, in equation (5.8).

$$PSE2045_{22} = (1 + NGR_{22}^{2010} + S_{22}) * LE_{22}^{2010}$$
$$PSE2045_{22} = (1 - 0.0427 - 0.12023) * 2{,}574$$
$$PSE2045_{22} = 2{,}155. \tag{5.8}$$

The previous six steps complete the shift-share employment projection for utilities. Interestingly, the equations in this process present the possibility that the projected shift-share employment could be a negative number. But that outcome is impossible. Column (6) is adjusted to zero if the calculation produces negative employment. The change in employment for an NAICS sector can be negative, but the projected total-sector employment cannot be negative.

Columns (7) and (8) in table 5.6 are the population change proportion (PCP) and LUCIS new employment change (LNEC), respectively. They are an adjustment to the shift-share employment projections used in Land-Use Conflict Identification Strategy planning land-use scenarios (LUCISplus). As discussed previously, the shift-share projections are less accurate when the time frame is long. Shift-share projections usually rely on a short observation period. (In NAICS sector 22, the observation period was only seven years.) That period is then extrapolated for the long projection period. The LUCIS process uses the study region population projection to adjust the employment sector projections because population projections usually rely on a longer observation period.

To get the population proportion adjustment, determine the proportion of total employment to total population for the years 2003 and 2010. In 2003, the population was 883,522, and the total employment was 520,746. The total 2003 employment proportion was approximately 59 percent of the 2003 population. The proportion of employment to population in 2010 was approximately 39.5 percent. The resulting mean employment ratio for Hillsborough County thus equals 49 percent.

Column (7) in table 5.6 shows the 2045 NAICS sector 22 proportion of employment to population, or 5,054 projected jobs in the utilities sector. Column (8) indicates that the mean change in employment NAICS sector 22 is shift-share new 2045 employment plus the new 2045 employment to population proportion divided by 2. It is an increase of 1,031 new utilities sector jobs by 2045.

Step 7: Calculating the population change proportion 2045

Population change proportion (PCP), column (7) in table 5.6, for 2045 is shown in equation (5.9).

$$PCP_{22}^{2045} = POP^{2045} * LCT_{22}^{2010} * AEPR^{(2003-2010)}$$
$$PCP_{22}^{2045} = 1{,}940{,}526 * 0.00528544 * 0.49$$
$$PCP_{22}^{2045} = 5{,}054, \tag{5.9}$$

where:
LCT_{22}^{2010} = local constant term from table 5.6, column (1), and

$AEPR^{(2003-2010)}$ = average employment to population ratio for the observed data.

Step 8: Calculating the LUCIS new employment change 2045

LUCIS new employment change (LNEC) for 2045, column (8) in table 5.6, is shown in equation (5.10).

$$LNEC_{22}^{2045} = \left[PCP_{22}^{2045} + PSE_{22}^{2045}\right)/2\right] - LE_{22}^{2010}$$
$$LNEC_{22}^{2045} = \left[(5{,}054 + 2{,}155)/2\right] - 2{,}574$$
$$LNEC_{22}^{2045} = 1031, \tag{5.10}$$

where:
PCP_{22}^{2045} = population change proportion for NAICS sector 22 in 2045,

PSE_{22}^{2045} = projected shift employment for NAICS sector 22 in 2045, and

LE_{22}^{2010} = local sector employment for NAICS sector 22 in 2010.

LUCISplus employment land-use allocation for Hillsborough County, Florida

In the previous section, "Using shifts and shares by employment sector as a projection technique," economic base theory, location quotients, and shift-share analysis were used to calculate the total and NAICS sector employment projections for 2045. The economic base employment projections were then adjusted by the employment to population ratio. Next, you must assign land-use classifications for the types of employment projections. Table 5.7 shows 237,824 new jobs projected for 2045 in the New Emp. column. The total employment in 2045 is the base industry employment in 2003 of 524,341 jobs shown in table 5.1 plus the new employment of 237,824 jobs from table 5.7. So the total employment is 762,165 jobs for 2045.

Table 5.7. The shift-share factors for Hillsborough County, Florida (showing local employment projections)

NAICS Codes	Category	Land-Use Category	Shift Share	Emp. Prop.	New Emp.
11	Agriculture, forestry, fishing, and hunting	Agriculture	9,743	21,814	4,670
21	Mining, quarrying, and oil and gas extraction	Industry	994	554	492
22	Utilities	Industry	2,155	5,054	1,031
23	Construction	Industry	20,152	50,402	9,609
31–33	Manufacturing	Industry	18,049	44,830	8,609
42	Wholesale trade	Service	24,890	51,689	11,967
44–45	Retail trade	Retail	63,798	125,236	30,739
48–49	Transportation and warehousing	Industry	11,274	26,205	5,394
51	Information	Service	13,101	33,606	6,239
52	Finance and insurance	Commercial	38,882	81,622	18,685
53	Real estate, and rental and leasing	Service	9,852	20,111	4,740
54	Professional and technical services	Commercial	62,077	92,096	30,186
55	Management of companies and enterprises	Service	16,068	15,801	7,888
56	Administrative and waste services	Service	24,104	91,593	11,204
61	Educational services	Institutional	13,696	19,359	6,669
62	Health care and social assistance	Institutional	85,071	133,159	41,302
71	Arts, entertainment, and recreation	Service	15,550	24,467	7,548
72	Accommodation and food services	Service	48,897	88,899	23,625
81	Other services, except public administration	Service	15,004	29,745	7,227
99	Unclassified	Service	0	22	0
	Totals		493,357	956,264	237,824

NAICS categories and land-use classifications

The NAICS categories are not a perfect match with land-use classifications, even at the five-digit level. The two-digit NAICS categories are even more difficult to link to land-use classifications. However, table 5.7 is the authors' "crosswalk" for NAICS codes to the property parcel land-use classifications for Hillsborough County, Florida.

Besides showing the projected new employment for the year 2045, table 5.7 also shows the number of new jobs by major employment land-use category. For example, there are 30,739 total new retail jobs, 48,871 total new commercial jobs, and 80,438 total new service-related jobs. The projected new employment is an average of the shift-share and population proportion projections.

Employment density for Hillsborough County, Florida

Once the study region employment projections are determined, the next step is to calculate the employment density for the six major employment land-use categories. The categories are agriculture, commercial, industrial, institutional, retail, and service. What is "employment density"? Within LUCIS, employment density refers to the acres of existing lands used to support the known employment for an individual employment category during the base year (2010). For example, industrial employment density is the number of industrial employees in the 2010 base year divided by the total number of industrial acres in the region for the same year. Table 5.8 includes employment densities for the six major categories. These individual employment densities are used in the following sections of the chapter, when applicable, to spatially allocate employment for the employment land-use categories.

Using employment densities and the information gathered from historical property analysis, you can calculate the necessary industrial land-use acreage for 2045. Table 5.7

Table 5.8. Base year (2010) employment density for the six major employment categories

Employment Land Use Category of Change	Acres	Percent	Jobs	Occupied Density	Gross Density with Vacant Proportion
Agriculture	180,665.24	54.51%	11,109	0.061	0.061
Commercial	10,535.00	3.18%	88,468	8.40	7.14
Industrial	17,680.74	5.33%	64,474	3.65	3.23
Institutional	27,307.94	8.24%	77,672	2.84	2.84
Retail	7,087.27	2.14%	63,778	9.00	6.26
Service	13,830.21	4.17%	181,263	13.10	9.73
Vacant Commercial	5,520.50	1.67%			
Vacant Industrial	2,270.46	0.68%			
Totals	264,897.36	79.92%	486,764		

Note: The employment densities for industrial and commercial include vacant land for these categories in 2010, thus decreasing the employment densities for these categories.

indicates that the five industry land-use sectors will produce 25,135 new industrial jobs. Using the industrial employment densities of 3.65 and 3.23 employees per acre (table 5.8), the number of acres needed for new industrial employment is as follows:

25,135 employees / 3.65 employees per occupied acre = 6,886 acres, or
25,135 employees / 3.23 employees per gross acre = 7,782, respectively.

There are implications regarding these two employment density calculations. The industrial occupied employment density shows how many acres are required for the new employment without additional future capacity. The use of gross employment density allocates an additional 896 acres (7,782 acres – 6,886 acres) of future industrial capacity beyond that needed for 2045. Because future land-use plans typically allocate lands to include some excess employment capacity, the industrial allocation for 2045 within the Hillsborough County study area will be the 7,782 acres that include the 896 extra acres.

You can make LUCIS employment land allocations in three ways: (1) infill allocation using existing 2010 vacant industrial parcels to accommodate some of the future employment, (2) redevelopment of some existing 2010 developed lands, and (3) greenfield allocation in primarily 2010 existing agricultural lands.

Table 5.9 indicates that between 2003 and 2010, 88,616.19 acres of land changed use from some form of agriculture, existing urban/suburban use, or vacant platted use. The development of existing 2003 agricultural acres accounted for 55.41 percent of the total new land-use change during the seven-year period. During the same period, 23.81 percent of 2003 existing urban lands changed to a different category of urban use. This change ultimately represents the area's redevelopment. Finally, 20.78 percent of the area's 2003 vacant platted parcels were developed. They represent new occupied infill residential, commercial, or industrial use for the period.

Combining the results from tables 5.7 and 5.8, the 25,135 new industrial jobs require 7,782 additional acres. That is, 25,135 people / 3.23 people per acre = 7,782 acres. Assuming that the distribution of development shown in table 5.9 applies to industrial development (a more complicated assumption to be used later), 1,617 acres of vacant industrial parcels will need to

Table 5.9. The change in acres for agriculture, redevelopment, and vacant platted lands (between 2003 and 2010)

Category	Changed Acres	Unchanged Acres	Column Changed %	Column Unchanged %
Greenfield lands (Agriculture)	49,099.27	118,631.60	55.41%	27.76%
Redevelopment (Existing Uses)	21,099.56	299,652.24	23.81%	70.11%
Infill (Vacant Parcels)	18,417.36	9,139.94	20.78%	2.14%
Totals	88,616.19	427,423.78	100.00%	100.00%

Table 5.10. New employment allocations for Hillsborough County by 2045 (in acres)

Category	Industrial	Commercial	Retail	Institutional	Service	Row Totals
New Employment 2045	25,135.00	48,871.00	30,739.00	47,971.00	87,107.00	**239,823.00**
Greenfield (Agriculture)	4,311.86	3,792.63	2,720.84	9,359.41	4,960.53	**25,145.27**
Redevelopment (Existing Uses)	1,852.83	1,629.72	1,169.16	4,021.79	2,131.57	**10,805.07**
Infill (Vacant Parcels)	1,617.04	1,422.32	1,020.38	3,509.99	1,860.31	**9,430.04**
Column Totals	**7,781.73**	**6,844.67**	**4,910.38**	**16,891.19**	**8,952.41**	**45,308.38**

be allocated for industrial use by 2045 (table 5.10). That is, 20.78 percent of 7,782 acres = 1,617 acres. Redevelopment of existing urban uses will result in 1,853 acres of existing urban lands changing to new industrial uses. That is, 23.81 percent of 7,782 acres = 1,853 acres. And 4,312 acres of existing agricultural lands will also be converted for new industrial development. That is, 55.41 percent of 7,782 acres = 4,312 acres.

Following the same process for retail, institutional, and service employment, the new land-use allocations are presented in table 5.10. Commercial employment projections indicate that the area can expect 48,871 new commercial jobs. Using the gross commercial density of 7.14 (table 5.8) indicates a need for 6,844.68 acres of new commercial lands by 2045. The proportional distribution of these new commercial lands is determined from table 5.9. It indicates that 3,792.63 acres of existing agricultural lands will need to be developed. Plus, 1,629.72 acres of existing urban development will change to new commercial use. Finally, 1,422.32 acres of vacant commercial lands will be developed for new commercial use.

Industrial land-use allocation

Industrial employment allocation uses the same three LUCIS^{plus} criteria evaluation matrices (CEMs) used in chapter 4 to allocate new 2045 residential development. The CEMs are for redevelopment, infill, and greenfield development. Figure 5.2 shows industrial suitability using a red to green color ramp. The darkest green areas indicate the highest industrial suitability. The red areas indicate the lowest suitable locations. The dark-gray parcels are existing industrial parcels. Clearly, the wet areas and areas close to rivers, streams, and wetlands have the lowest suitability for industrial uses. The green areas are usually in or near the urbanized areas. The green areas are aggregated around major transportation corridors.

A trend industrial land-use allocation begins with the concept of the trend allocation adequately mimicking the historical record. Future land-use planning maps already envision the allocation of future industrial use. The maps show these acres as industrial.

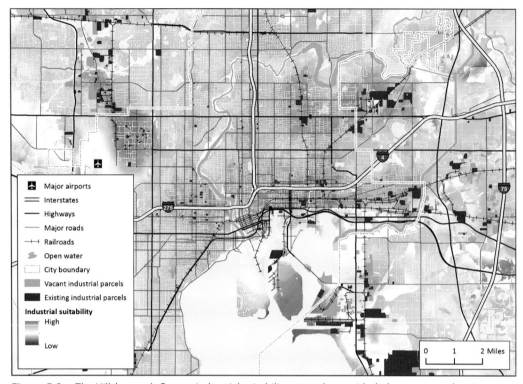

Figure 5.2. The Hillsborough County industrial suitability raster layer with dark-gray areas showing existing 2010 industrial parcels and lavender parcels showing vacant 2010 industrial parcels.

Figure from Paul D. Zwick and the Arizona Board of Regents on behalf of the University of Arizona. Airports from the Federal Aviation Administration and the Research and Innovative Technology Administration Bureau of Transportation Statistics National Transportation Atlas Databases (NTAD) 2012; water bodies from the US Geological Survey; state boundary from US Census Bureau TIGER/Line Files; interstates, highways, and major roads from the Florida Department of Transportation; railroads from the Federal Railroad Administration; parcels from the Florida Department of Revenue.

The property parcel base identifies them as vacant industrial property. Following that logic, the LUCIS^{plus} allocation of new industrial acreage begins with the allocation of 1,617.04 acres (table 5.10) of existing vacant industrial property. Table 5.8 shows 2,270.46 acres of vacant industrial property are available in 2010. Therefore, the process must identify 1,617 acres in the infill CEM that have high industrial suitability. These acres must also be classified as existing vacant industrial property. The allocation process uses vacant industrial property parcels because parcel definitions also adhere to existing zoning requirements.

The following list of criteria is used for allocations of employment based on land-use selections structured in ArcGIS query language:

- *DESCRIPT*: The Property Use Description. Examples include Vacant Commercial, Vacant Industry, Misc. Agriculture, and so forth.
- *ALLYRPOP*: Allocation Year Residential Population. Item used to hold the allocation year when the cell was allocated for residential use. For example, when this item

equals 0, the cell has not been allocated. When the item equals 2045, the cell was allocated for residential use by 2045.
- *INDSUIT100I*: The LUCIS industrial land-use suitability value. The values for this item range from 100 to 900. The values are a three-digit integer representation of suitability decimal values for use in a combined raster layer (see the ArcGIS Combine tool in the Local toolset within the Spatial Analyst toolbox). For example, when the value 2.3456 is converted to a three-digit integer value, the result is 235.
- *ALLYRIND*: Allocation Year Industrial Use. Item used to hold the allocation year when the cell was allocated for industrial use. For example, when this item equals 0, the cell has not been allocated. When the item equals 2045, the cell was allocated for industrial use by 2045. This variable uses the same concept as ALLYRPOP to identify which year the allocation of residential population occurs. Only here, it indicates which year the allocation of industrial employment occurs.
- *GFCONFLICT*: LUCIS Conflict/Preference. The greenfield conflict index of LUCIS conflict/preference categories contains 27 unique values that represent the conflict level between the major land-use types: agriculture, conservation, and urban.
- *CNTERSLCE19*: Distance Zone from Major Center City. Item used to indicate a zone of distance from the center of employment activity. The value ranges from 1 to 9, with 9 representing the closest zone of distance.
- *ACRES*: Available Allocation Acres. Item that contains the number of acres represented by the raster zone. A raster zone is a record within an integer raster that contains all the cells in a raster that have the same unique value. For example, it could be all the cells that represent single-family residential land use.
- *YRBUILT*: Structure Year Built. Item containing the actual year a structure on a property parcel was built.
- *EZONE*: Economic Development Zone. Item represents a cell within an economic development zone.
- *CITYNAME*: City Name. Item representing the city the cell is located within: (1) Tampa, (2) Temple Terrace, (3) Plant City, or (0) in the county.
- *ACREAGE*: Property Parcel Acres. Item containing the property parcel acreage.
- *COMSUIT100I*: The LUCIS commercial land-use suitability value. The values for this item range from 100 to 900. Allocating commercial employment follows the same concept as using MFSUIT100I and SFSUIT100I for residential population allocation in chapter 4.
- *SERSUIT100I*: The LUCIS service land-use suitability value. The values for this item range from 100 to 900. Allocating service employment follows the same concept as using MFSUIT100I and SFSUIT100I for residential population allocation in chapter 4.
- *REDEVCONFLICT*: Redevelopment Mixed-Use Opportunity. The item is structured just like the GFCONFLICT index of LUCIS Preference/Conflict categories. However, it represents redevelopment opportunity (in this case, for commercial, multifamily residential, and retail mixed use). It has the same 27 index values.
- *ALLYRCOMR*: Allocation Year Commercial Use. Item used to hold the allocation year when the cell was allocated for commercial use. For example, when this item

equals 0, the cell has not been allocated. When the item equals 2045, the cell was allocated for commercial use by 2045. Allocating commercial employment follows the same concept as using ALLYRPOP for residential population allocation in chapter 4.

SQL query 5.1 identifies the most suitable existing vacant industrial property.

Query 5.1.
"DESCRIPT" = 'VACANT INDUSTRY' AND "ALLYRPOP" = 0 AND "INDSUIT100I" >= 326.

Query 5.1 searches for cells in the infill CEM that are classified as vacant residential parcels (DESCRIPT = "VACANT INDUSTRY"). These cells have not been allocated for future residential use (ALLYRPOP = 0) and have moderate to high industrial suitability (INDSUIT100I >= 326). Industrial suitability ranges from a low value of 100 to the highest value of 900. Values of 330 and greater are in the moderate to high range of suitability. Query 5.1 results in the identification of 1,617.04 acres of existing vacant industrial land that will support future industrial employment. The extra benefit is that all the acres selected are moderate to high LUCIS suitability. Future land-use maps do not necessarily incorporate suitability when mapping future land use, but the LUCIS allocation does. Next, allocate the rows within the infill CEM that fulfill query 5.1. Set the item ALLYRIND in the value attribute table (VAT) to 2045 to identify the selected cells for future industrial employment. The ALLYRIND variable assures that other land-use allocations do not double-count any previously allocated acreage, especially if the historical property data does not indicate mixed-use allocations that include industrial parcels.

Figure 5.3 shows the selected 2010 vacant industrial parcels. The selected areas are in lavender. Clearly, these vacant parcels are close to the existing industrial properties. This close proximity is typical for future land-use plans. A close inspection of figures 5.1 and 5.2 shows that many of the vacant 2010 industrial parcels are developed for industrial use by 2045.

Once the correct portion of the 2010 vacant industrial property parcels (in red) has been allocated for infill industrial development, the next allocation is for greenfield industrial development. The process of allocating future greenfield industrial acres is like that for industrial infill but occurs within existing agriculture property. However, one major part of the allocation process does not occur for infill allocation. The allocation of future industrial infill does not use LUCIS conflict because Hillsborough County already set aside the lands for future industrial use. However, LUCIS conflict is used for all greenfield employment allocations. Therefore, conflict is used for allocation of future industrial acres in greenfield areas. Again, a trend concept should be employed for the allocation. In Hillsborough County's case, it means trying to locate industrial acreage that is near or close to the center of urban activity in the area. Weighting the existing industrial activity by acreage identifies the center of urban activity in Hillsborough County. A quick analysis of the historical property parcel base indicates that, for practical purposes, future industrial parcels should be larger than one acre in size.

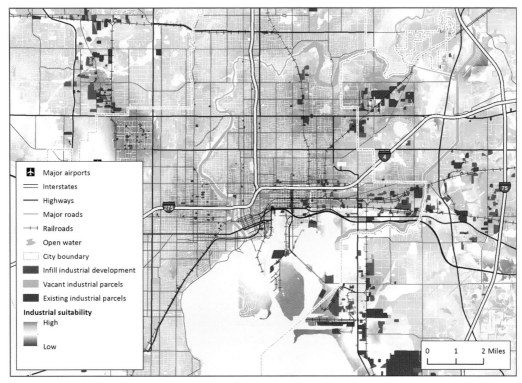

Figure 5.3. Future 2045 infill industrial properties. The red areas are the infill industrial parcels selected for new development. The dark-gray areas are existing industrial parcels. The lavender parcels are available vacant industrial parcels.

Figure from Paul D. Zwick and the Arizona Board of Regents on behalf of the University of Arizona. Airports from the Federal Aviation Administration and the Research and Innovative Technology Administration Bureau of Transportation Statistics National Transportation Atlas Databases (NTAD) 2012; water bodies from the US Geological Survey; state boundary from US Census Bureau TIGER/Line Files; interstates, highways, and major roads from the Florida Department of Transportation; railroads from the Federal Railroad Administration; parcels from the Florida Department of Revenue.

Query 5.2 uses LUCIS conflict values to allocate greenfield industrial development.

Query 5.2.
```
("DESCRIPT" = 'MISC AG' OR "DESCRIPT" = 'PASTURE') AND "ALLYRPOP"
= 0 AND ("GFCONFLICT" = 113 OR "GFCONFLICT" = 123 OR "GFCONFLICT"
= 213 OR "GFCONFLICT" = 223 OR "GFCONFLICT" = 323 OR "GFCONFLICT"
= 313) AND "CNTERSLCE19" >= 5 AND "ACRES" >= 1.4.
```

Query 5.2 starts by selecting existing 2010 agricultural lands for new industrial land use. It selects two categories of agricultural lands: miscellaneous agricultural lands and pasture lands [("DESCRIPT" = 'MISC AG' OR "DESCRIPT" = 'PASTURE')]. As with the infill industrial allocations, the greenfield allocation process removes any chance of selecting previously allocated 2045 residential lands ("ALLYRPOP" = 0). Next, the greenfield conflict values for urban preference are selected [("GFCONFLICT" = 113 OR "GFCONFLICT" = 123 OR "GFCONFLICT" = 213 OR "GFCONFLICT" = 223 OR "GFCONFLICT" = 323 OR "GFCONFLICT" = 313)]. The final two criteria in the query include proximity to the center of urban activity ("CNTERSLCE19" >= 5) and setting the minimum parcel size for industrial greenfield parcels ("ACRES" >= 1.4). Query 5.2 results in 4,306 acres of greenfield pasture with high urban preference in reasonably close proximity to the urban center of Hillsborough County. The minimum parcel size is 1.4 acres. Setting the item ALLYRIND to 2045 in the greenfield CEM allocates the 4,306 acres.

Figure 5.4 shows the greenfield industrial parcel allocation. The lavender areas show vacant industrial parcels. These parcels include allocated infill industrial parcels. Inspection of the query shows that the minimum acreage for industrial parcel development in greenfield property is 1.4 acres. The only conflict within the allocation process occurs between high agriculture preference and high urban preference. Only lands that were categorized as high

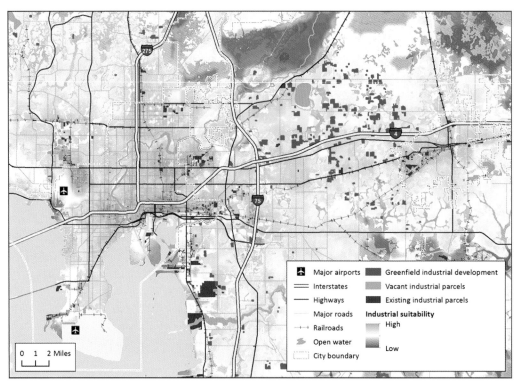

Figure 5.4. Greenfield 2045 industrial land-use allocation. The blue properties have been allocated for industrial use by the year 2045. The dark-gray areas represent the existing 2010 industrial parcels. The lavender industrial parcels remain vacant after this allocation process.

Figure from Paul D. Zwick and the Arizona Board of Regents on behalf of the University of Arizona. Airports from the Federal Aviation Administration and the Research and Innovative Technology Administration Bureau of Transportation Statistics National Transportation Atlas Databases (NTAD) 2012; water bodies from the US Geological Survey; state boundary from US Census Bureau TIGER/Line Files; interstates, highways, and major roads from the Florida Department of Transportation; railroads from the Federal Railroad Administration; parcels from the Florida Department of Revenue.

urban preference were allocated to greenfield industrial uses. Moreover, the new greenfield industrial properties have above average suitability for proximity to the center of urban activity within the study region. Finally, a query of the allocated cells indicates that no miscellaneous agricultural lands were selected. The entire 4,306 acres of future greenfield industrial development will be on pasture land. This selection seems logical, considering that pasture lands are the easiest to develop. To start, there are fewer clearing costs for development. By using pasture land, the allocation does not remove any cropland from productivity. The new industrial properties selected from the greenfield development raster layer are primarily in the northern part of the county between Tampa, to the west, and Plant City, to the northeast. These locations have excellent major roadway access. They are close to Interstate 4, the major highway connection between Tampa and Orlando. Access to these major roadways provides for more efficient movement of exported industrial goods and services out of the county. It also provides greater accessibility for importing raw materials, decreasing production costs. However, the industrial suitability raster layer shows that all the new industrial parcels selected from the greenfield allocation have moderate suitability.

The final industrial allocation is industrial redevelopment acreage. Again, it uses the same redevelopment CEM as chapter 4. It includes some variable substitutions to accommodate employment allocations rather than residential population allocations. The process is similar to both infill and greenfield allocations. However, there is one nuance for redevelopment allocations. The analyst must first determine the areas within which redevelopment can occur. Redevelopment areas may be (1) defined by the county planning department as special planning areas, (2) identified as areas designated for future economic redevelopment, or (3) selected based on land values and age of existing structures (low land values with older structures). LUCIS^{plus} combines all three options to accommodate redevelopment opportunity. It also incorporates existing county economic or policy directives.

The trend concept for redevelopment allocation starts with the identification of redevelopment opportunity. By definition, it is within the major urban cities in the county. Again, it is on parcels that are larger than an acre in size. The redevelopment mask for this allocation selects parcels with structures that are 50 years or older by 2045. The parcels also are not eligible for historic preservation status and have below average existing market value. Figure 5.5 displays the selection of vacant industrial properties (lavender areas). The black-and-white hashed outline polygons are the city boundary for the three cities in Hillsborough County: Tampa, Temple Terrace, and Plant City. The yellow areas are the parcels selected for industrial redevelopment. Query 5.3 fulfills the allocation.

Query 5.3.
"YRBUILT" <= 1986 AND ("EZONE" = 1 OR "CITIES" > 0) AND "ALLYRPOP" = 0 AND "PARACRES" >= 1.7.

Query 5.3 selects cells in the redevelopment CEM raster with structures that were built in or before 1986 ("YRBUILT" <= 1986), not including historic structures. The query assures the minimum age of structures redeveloped in 2045 is 59 years or older. The selection also includes cells within identified existing economic development zones ("EZONE" = 1) or in

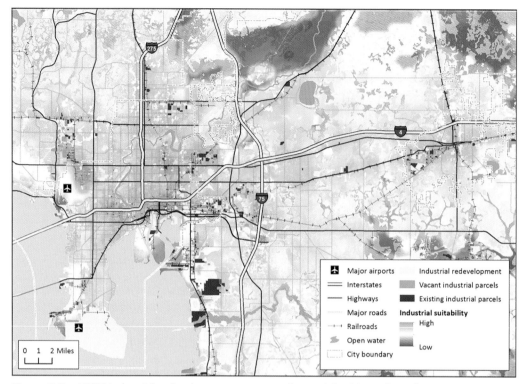

Figure 5.5. LUCIS industrial redevelopment property allocated for 2045. The yellow properties were selected for future industrial uses that require redevelopment of other existing urbanized uses.

Figure from Paul D. Zwick and the Arizona Board of Regents on behalf of the University of Arizona. Airports from the Federal Aviation Administration and the Research and Innovative Technology Administration Bureau of Transportation Statistics National Transportation Atlas Databases (NTAD) 2012; water bodies from the US Geological Survey; state boundary from US Census Bureau TIGER/Line Files; interstates, highways, and major roads from the Florida Department of Transportation; railroads from the Federal Railroad Administration; parcels from the Florida Department of Revenue.

any of Hillsborough's three major urban cities ("CITIES" > 0), where Tampa = 1, Temple Terrace = 2, and Plant City = 3. Finally, cells have not previously been identified in chapter 4 as residential opportunity for 2045 ("ALLYRPOP" = 0). The query assures no double allocation of uses. The minimum parcel acreage is 1.7 acres ("PARACRES" >= 1.7).

Figure 5.6 shows the entire allocation of 2045 industrial parcel allocations. The newly allocated parcels are shaded in yellow (future redevelopment industrial parcels), red (future infill industrial parcels), and blue (future greenfield industrial properties). The yellow cells are located within the three city boundaries; the red parcels are dispersed inside and outside the cities of Tampa, Temple Terrace, and Plant City; and the blue parcels are all in greenfield areas (on 2010 existing pasture parcels).

To summarize the allocation of new locations for industrial land uses by 2045, three separate SQL queries selected appropriate cell locations from within the infill, greenfield, and redevelopment CEMs to complete the allocation. The allocation process is relatively straightforward. Future infill industrial uses are all located on existing vacant industrial

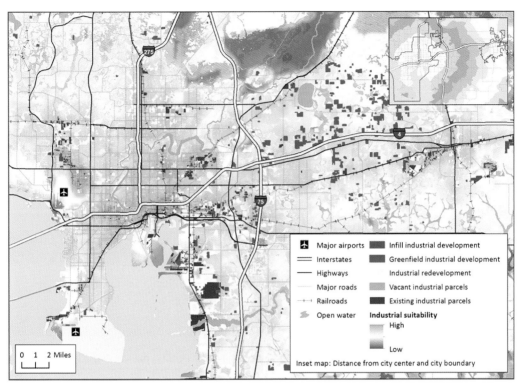

Figure 5.6. Industrial land-use 2045 allocations for infill, greenfield, and redevelopment. The dark-gray areas are existing 2010 industrial parcels. The red, blue, and yellow areas are newly allocated industrial property, and the lavender areas are remaining vacant industrial properties.

Figure from Paul D. Zwick and the Arizona Board of Regents on behalf of the University of Arizona. Airports from the Federal Aviation Administration and the Research and Innovative Technology Administration Bureau of Transportation Statistics National Transportation Atlas Databases (NTAD) 2012; water bodies from the US Geological Survey; state boundary from US Census Bureau TIGER/Line Files; interstates, highways, and major roads from the Florida Department of Transportation; railroads from the Federal Railroad Administration; parcels from the Florida Department of Revenue.

properties. These properties were already identified within the county's future land-use plans. Greenfield future industrial uses are on existing pasture lands, with a minimum parcel size of 1.4 acres. All parcels have moderate to high industrial suitability values. Future redevelopment industrial acres are on parcels with a minimum size of 1.7 acres. They are located within one of the three urban areas in the county. They all require the demolition or adaptive reuse of nonhistoric structures built before 1987. Table 5.11 shows the allocation details and a few selected impacts of these allocations.

Commercial employment in Hillsborough County is projected to grow by 48,871 new jobs for the time frame 2010 to 2045. Table 5.7 indicates an increase of 30,186 new commercial jobs in the professional and technical services sector and 18,685 jobs in the finance and insurance sector. The total increase is 48,871 new commercial jobs.

As with industrial employment, the three LUCIS[plus] CEMs are used to spatially allocate land uses that support these commercial job opportunities. However, using suitability and

Table 5.11. Future industrial allocation details with select impacts of the allocations

Category	Allocated Acres	Tampa Acres	Temple Terrace Acres	Plant City Acres	County Acres	Industrial Jobs
Greenfield (Agriculture)	4,306	0	0	0	4,306	13,915
Redevelopment (Existing Uses)	1,863	1,336	119	408	0	6,011
Infill (Vacant Parcels)	1,616	357	0	341	918	5,220
Column Totals	7,785	1,693	119	749	5,224	25,146
Category	County Major Habitat Change	Tampa Major Use Change	Temple Terrace Major Use Change	Plant City Major Use Change	County Major Use Change	
Greenfield (Agriculture)	Pasture	No Change	No Change	No Change	Pasture	
Redevelopment (Existing Uses)	Urban High Density	Municipal 278 Acres 896 Jobs	Warehousing 107 Acres 344 Jobs	Multifamily 22 Acres 71 Jobs	No Change	
Infill (Vacant Parcels)		Vacant Industry	Vacant Industry	Vacant Industry	Vacant Industry	

Note: The table shows the percentages from 2003–10. These percentages are needed for allocations for 2045. The time frame 2003–10 is used to develop the information needed for the trend allocations for 2045. The trend extrapolates what has occurred from 2003 to 2010.

conflict to determine the most appropriate new locations for future acreage of commercial businesses is more complicated than industrial job allocations. The entanglement that arises from the lack of categories for vacant retail, service, and institutional land uses complicates infill commercial land-use allocation. As a result, you must complete the allocation of commercial, retail, service, and institutional future land uses simultaneously.

Commercial, retail, service, and institutional land-use suitability

Table 5.9 shows the percentages of the existing 2003 greenfield, redevelopment, and vacant lands that changed to developed land uses by 2010. Employing these change percentages, table 5.10 shows the greenfield, redevelopment, and infill lands required to accommodate the 2045 future commercial, retail, institutional, and service development. Figures 5.7 through 5.12 display the commercial infill, greenfield, and redevelopment allocations over the LUCIS combined commercial, retail, service, and institutional land-use suitability raster layer for the Hillsborough County study region. The red to green color ramp marks suitability. In each figure, the allocated commercial, retail, service, and institutional acres are displayed in either yellow (infill or redevelopment), blue (greenfield), bright purple (mixed use), or black for additional allocations in or near city boundaries.

As in residential and industrial land-use allocations, the allocation of future commercial, retail, service, and institutional future acres begins with infill use on parcels already identified for future commercial use. Table 5.12 indicates that by 2045, there will be 7,813 new acres of land supporting these four job-generating sectors. The 7,813 acres represent 20.78 percent of the 37,599 acres of land necessary to support the projected employment. The concept for future infill commercial, retail, service, and institutional employment is to locate new land uses on general vacant acreage. This acreage can be vacant commercial, vacant commercial/condominium mixed use, and vacant professional parks near the center of urban activities. The acreage should all have moderate to high commercial and service LUCIS suitability. Because the vacant lands have already been approved through the planning process, no minimum acreage is assigned. Nor are there any double land-use allocations for future and existing industrial or residential uses.

Next, query 5.4 allocates future commercial, retail, service, and institutional acreage in Hillsborough County for 2045. These acres are selected from property already defined by the county government as available for each respective development type.

Query 5.4.
```
"CNTERSLCE19" >= 5 AND ("COMSUIT100I" >= 395 OR "SERSUIT100I" >= 395) AND ("DESCRIPT" = 'VACANT ACREAGE' OR "DESCRIPT" = 'VACANT COMM' OR "DESCRIPT" = 'VACANT COMM CONDO' OR "DESCRIPT" = 'VACANT PRO PARK') AND "ALLYRIND" = 0 AND "ALLYRPOP" = 0.
```

Query 5.4 locates the infill vacant commercial cells in the infill CEM by selecting cells that are close to the urban center of the county (`"CNTERSLCE19" >= 5`). The cells also have moderate to high LUCIS suitability for commercial and service land use (`"COMSUIT100I" >= 395 OR "SERSUIT100I" >= 395`). The selection then looks for vacant commercial lands defined as vacant acreage, vacant commercial, vacant commercial and condominium, or vacant professional park [(`"DESCRIPT" = 'VACANT ACREAGE' OR "DESCRIPT" = 'VACANT COMM' OR "DESCRIPT" = 'VACANT COMM

Table 5.12. New employment allocations for Hillsborough County by 2045 (in acres)

Category	Commercial	Retail	Institutional	Service	Percentage	Row Totals
Greenfield (Agriculture)	3,792.64	2,720.84	9,359.41	4,960.53	55.41%	20,833.42
Redevelopment (Existing Uses)	1,629.72	1,169.16	4,021.79	2,131.57	23.81%	8,952.24
Infill (Vacant Parcels)	1,422.32	1,020.38	3,509.99	1,860.31	20.78%	7,813.00
Column Totals	6,844.68	4,910.38	16,891.19	8,952.41	100.00%	37,598.66

CONDO' OR "DESCRIPT" = 'VACANT PRO PARK')]. Finally, the selection removes the chance that cells already allocated for infill residential or industrial land uses were reselected for one of these uses ("ALLYRIND" = 0 AND "ALLYRPOP" = 0). Query 5.4 allocates 7,889.43 acres of existing vacant commercial lands for infill commercial, retail, service, and institutional uses. These acres support new 2045 jobs for these sectors (figure 5.7 and table 5.13).

Table 5.13 shows the allocation of existing vacant commercial acres by location and distance for future commercial, retail, service, and institutional use. Most of the allocated acreage is located outside the urban areas and within the county (6,545.64 acres). It represents approximately five to six times the allocation of new commercial, retail, service, or institutional jobs on vacant platted acreage within the three major urban areas (1,343.79 acres). Among the three major cities, Tampa has approximately 10 times the acreage (1,110.59 acres) compared with the allocated acreage in Temple Terrace (86.81 acres) and Plant City (146.39 acres).

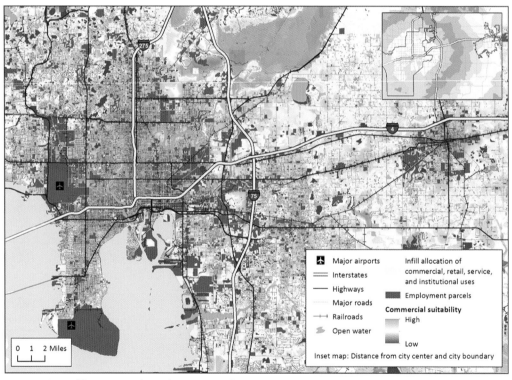

Figure 5.7. Infill commercial, retail, service, and institutional acres allocated by query 5.4 are drawn in yellow and presented as an overlay on the commercial suitability raster layer. The dark-gray areas are existing commercial, retail, service, industrial, and institutional properties.

Figure from Paul D. Zwick and the Arizona Board of Regents on behalf of the University of Arizona. Airports from the Federal Aviation Administration and the Research and Innovative Technology Administration Bureau of Transportation Statistics National Transportation Atlas Databases (NTAD) 2012; water bodies from the US Geological Survey; state boundary from US Census Bureau TIGER/Line Files; interstates, highways, and major roads from the Florida Department of Transportation; railroads from the Federal Railroad Administration; parcels from the Florida Department of Revenue.

Table 5.13. Infill commercial, retail, service, and institutional development acreages by distance from the central urban core in Hillsborough County

Location	Closest (9)	Very Close (8)	Close (7)	Moderately Close (6)	Relatively Near (5)	Row Totals
County	994.41	1,813.26	1,253.47	1,510.86	973.64	6,545.64
Tampa	632.48	56.68	215.85	132.22	73.36	1,110.59
Temple Terrace	4.20	55.57	27.04	0.00	0.00	86.81
Plant City	0.00	0.00	0.00	0.00	146.39	146.39
Column Totals	1,631.09	1,930.51	1,496.36	1,643.08	1,193.39	7,889.43

The distance characteristics range from "Closest" to "Relatively Near." These descriptions refer to the distance of the individual cell locations in the infill CEM to the major center of urban activity. The closest distance ranking is 9, and the farthest distance ranking is 1. Table 5.13 indicates that all the cells, and thus the acres selected, range from 5 (relatively near) to 9 (closest). The allocated existing vacant commercial lands are evenly distributed according to distance from the major center of urban activity. The greatest number of allocated acres (1,930.51) is rated as "Very Close," and the lowest number of acres (1,193.39) is in the category "Relatively Near." The range of allocations includes a small difference of approximately 800 to 900 acres between "Relatively Near" and "Closest."

The next allocation is greenfield acreage for the four employment classifications. As with industrial employment, query 5.5 uses the LUCIS greenfield preference/conflict categories.

Query 5.5.
```
"ALLYRIND" = 0 AND "ALLYRPOP" = 0 AND "CNTERSLCE19" >= 5 AND
("COMSUIT100I" >= 418 OR "SERSUIT100I" >= 418) AND "ACREAGE" > 2
AND ("GFCONFLICT" = 113 OR "GFCONFLICT" = 123 OR "GFCONFLICT" = 213
OR "GFCONFLICT" = 223 OR "GFCONFLICT" = 313 OR "GFCONFLICT" = 323
OR "GFCONFLICT" = 333 OR "GFCONFLICT" = 112 OR "GFCONFLICT" = 133
OR "GFCONFLICT" = 233 OR "GFCONFLICT" = 212 OR "GFCONFLICT" = 222
OR "GFCONFLICT" = 232 OR "GFCONFLICT" = 312 OR "GFCONFLICT" = 322).
```

Following the same process used by previous queries, query 5.5 assures that no previously allocated acreage was used. The same terms used in query 5.4 assure that no previously allocated industrial employment or residential acres were allocated ("ALLYRIND" = 0 AND "ALLYRPOP" = 0). The distance zone from major center city field in the allocation query ("CNTERSLCE19" >= 5") identifies locations of commercial, retail, service, and institutional employment land uses near the existing urban center of the county. The commercial and service suitability fields in the allocation query [("COMSUIT100I" >= 418 OR "SERSUIT100I" >= 418)] select cells from CEM in which commercial and service suitability are again moderate to high suitability lands that are still available. The greenfield conflict index statements in the query [("GFCONFLICT" = 113 OR "GFCONFLICT" = 123 OR "GFCONFLICT" = 213 OR "GFCONFLICT" = 223 OR "GFCONFLICT" = 313 OR "GFCONFLICT" = 323 OR "GFCONFLICT" = 333 OR "GFCONFLICT" = 112 OR "GFCONFLICT" = 133 OR "GFCONFLICT" = 233 OR "GFCONFLICT" = 212 OR "GFCONFLICT" = 222 OR "GFCONFLICT" = 232 OR "GFCONFLICT" = 312 OR "GFCONFLICT" = 322)] identify conflict values appropriate for allocating commercial, retail, service, and institutional employment.

The greenfield conflict index statements in query 5.5 show that cells were selected that clearly are in conflict among the three major land-use categories: agriculture, conservation, and urban. Table 5.14 describes the conflict that occurred to complete query 5.5. The selection of cells with LUCIS greenfield conflict values 112, 113, 123, 213, and 223 all have high or moderate urban preference. By definition, they are not in conflict with other uses. They are therefore preferred for the future allocation of greenfield acres for new employment development. The LUCIS greenfield conflict categories included 313, 323, 133, 233, and 333, which all share a high urban preference. But they are in conflict with either

Table 5.14. Acres of lost agricultural production or ecologically significant lands

LUCIS Greenfield Conflict	Description	Acres
313	High Agriculture and Urban Preferences; Low Ecological Significance	481.42
323	High Agriculture and Urban Preferences; Moderate Ecological Significance	3,769.68
333	High Agriculture, Ecological Significance, and Urban Preferences	6,696.60
133	Low Agriculture; High Ecological Significance and Urban Preferences	4.71
233	Moderate Agriculture; High Ecological Significance and Urban Preferences	405.88
232	Moderate Agriculture and Urban Preferences; High Ecological Significance	1,266.63
222	Moderate Agriculture, Ecological Significance, and Urban Preferences	180.28
312	High Agriculture, Low Ecological Significance, and Moderate Urban Preferences	136.96
322	High Agriculture; Moderate Ecological Significance and Urban Preferences	7,828.05

high agriculture (313 and 323) or high conservation (133 and 233), or both (333). The logic of this part of the allocation is that even though conflict exists, the selected cells all have high urban preference. Therefore, they are clearly valued for new urban development. Moreover, the allocation spatially identifies the cost (acres of lost agriculture production and ecological services) of allocating these future greenfield employment sites. Finally, the LUCIS conflict numbers 232, 222, 312, and 322 indicate that some cells selected for future greenfield employment development are more preferable for agriculture (312 and 322) or conservation (232), or equally preferred for all three major uses (222). Of all the categories in conflict, the categories that require the use of high-preference agriculture property or lands that have been identified as having high conservation significance for new employment allocation cause the most concern. These issues are addressed in chapters 6 and 7, which deal with mixed-use development, increased density, and conservation allocation.

Table 5.14 identifies allocations that infringe on agriculture or ecologically significant areas. It also shows that the number of acres with high agriculture preference and moderate urban preference is 7,965.01. Following the trend development patterns in Hillsborough County, it will produce a loss of these highly preferred agriculture acres. Also, 1,266.63 acres of high-preference ecologically significant lands are lost for lower preference employment development. Finally, 11,538.57 acres of land are in some form of conflict with urban preference, producing lost agriculturally productive or ecologically significant lands that are *equally* preferred for employment development. Figure 5.8 displays the location selected for future greenfield allocation of commercial, retail, service, and institutional employment/jobs.

The next allocation for commercial, retail, service, and institutional employment is for redevelopment. The following queries, beginning with query 5.6, allocate future commercial, retail, service, and institutional acreage in Hillsborough County. By 2045, this acreage will be mixed use in the urban areas, sharing a location with multifamily residential. Analysis of the parcel base in Hillsborough County indicates opportunity for redevelopment in mixed-use development. Therefore, the concept for query 5.6 is only slightly different from the other queries in this chapter. The difference focuses on two items in the query, ALLYRPOP and REDEVCONFLICT. As previously explained for the industrial employment allocations, "ALLYRPOP" = 0 is used in queries 5.1, 5.2, and 5.3 to assure that future industrial jobs are not spatially allocated in the same cells as residential population. If, as in the case of query 5.6, you want to allocate residential population (multifamily residential units) and jobs—commercial, retail, service, and institutional—in shared locations, the allocation query must be able to identify the cells that share colocation opportunity with future multifamily residential. Query 5.6 uses "ALLYRPOP" = 2045 instead of "ALLYRPOP" = 0 to identify cells that have been allocated for future residential use. The variable REDEVCONFLICT helps in the mixed-use allocation by selecting cells with high multifamily, commercial, and retail use (see table 4.6). Query 5.6 looks for mixed-use opportunity with future multifamily residential acres and future commercial, retail, service, and institutional employment. The query selects 1,300 acres for this type of mixed-use opportunity (figure 5.9).

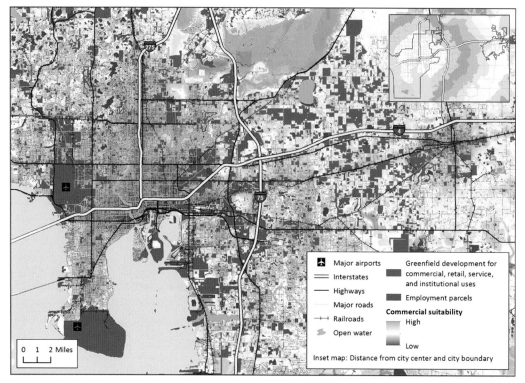

Figure 5.8. Greenfield development properties for commercial, retail, service, and institutional employment land use in blue were selected by query 5.5. Existing employment properties are the areas in dark gray.

Figure from Paul D. Zwick and the Arizona Board of Regents on behalf of the University of Arizona. Airports from the Federal Aviation Administration and the Research and Innovative Technology Administration Bureau of Transportation Statistics National Transportation Atlas Databases (NTAD) 2012; water bodies from the US Geological Survey; state boundary from US Census Bureau TIGER/Line Files; interstates, highways, and major roads from the Florida Department of Transportation; railroads from the Federal Railroad Administration; parcels from the Florida Department of Revenue.

Query 5.6.
"CITIES" >= 1 AND ("REDEVCONFLICT" = 333 OR "REDEVCONFLICT" = 332 OR "REDEVCONFLICT" = 331 OR "REDEVCONFLICT" = 323 OR "REDEVCONFLICT" = 322 OR "REDEVCONFLICT" = 321 OR "REDEVCONFLICT" = 233 OR "REDEVCONFLICT" = 232 OR "REDEVCONFLICT" = 223 OR "REDEVCONFLICT" = 222) AND "ALLYRPOP" = 2045 AND ("DESCRIPT" = 'MIXED USE MH PARK' OR "DESCRIPT" = 'MIXED USE MULTI FAM' OR "DESCRIPT" = 'MIXED USE OFFICE' OR "DESCRIPT" = 'MIXED USE RES').

As in the other redevelopment queries in this chapter, "CITIES" >= 1 contains the spatial search to areas inside the three cities of Tampa, Temple Terrace, and Plant City. Finally, the section [("DESCRIPT" = 'MIXED USE MH PARK' OR "DESCRIPT" = 'MIXED USE MULTI FAM' OR "DESCRIPT" = 'MIXED USE OFFICE' OR "DESCRIPT" = 'MIXED USE RES')] uses property parcels with use descriptions that include mixed-use opportunities.

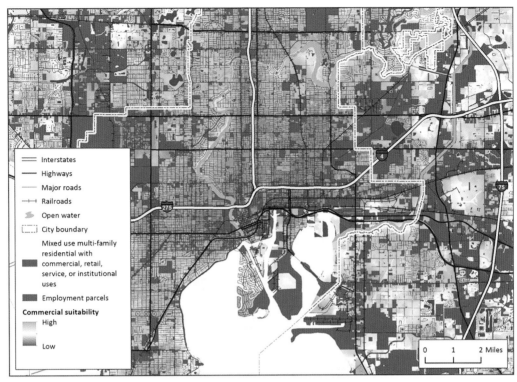

Figure 5.9. Mixed-use multifamily residential with commercial, retail, service, or institutional uses. The locations in bright purple represent the 1,300 acres of future mixed-use allocations selected using query 5.6.

Figure from Paul D. Zwick and the Arizona Board of Regents on behalf of the University of Arizona. Water bodies from the US Geological Survey; state boundary from US Census Bureau TIGER/Line Files; interstates, highways, and major roads from the Florida Department of Transportation; railroads from the Federal Railroad Administration; parcels from the Florida Department of Revenue.

The next redevelopment allocation for commercial, retail, service, and institutional employment is query 5.7. Query 5.7 selects redevelopment lands within the three cities that have not previously been selected for future residential, commercial, retail, service, institutional, or industrial employment. The lands must be within the redevelopment mask that identifies potential using the REDEVCONFLICT index. As with query 5.6, the concept is to understand what the redevelopment mask requires to most closely fulfill the trend indicated by the historical information. Query 5.7 selects 3,759 acres within the city limits of Tampa, Temple Terrace, and Plant City. The lands have commercial, retail, service, and institutional potential to support employment/jobs (figure 5.10).

So far, the process of locating new commercial, retail, service, and institutional employment within the redevelopment areas has resulted in allocating only 5,059 acres of the required 8,952.24 acres, or 56.51 percent. To allocate the appropriate amount of space for redevelopment employment, look to redevelopment areas external to the three major cities. Query 5.8 must allocate an extra 3,893 acres of land outside the cities to support new employment/jobs for these categories. This situation is not unusual, but query 5.8 must find redevelopment acres that are at least near the urbanized areas with high potential for these types of employment (figure 5.11).

Query 5.7.
```
"ALLYRPOP" = 0 AND "ALLYRCOMR" = 0 AND "ALLYRIND" = 0 AND "CITIES"
>= 1 AND ("REDEVCONFLICT" = 333 OR "REDEVCONFLICT" = 332 OR
"REDEVCONFLICT" = 331 OR "REDEVCONFLICT" = 323 OR "REDEVCONFLICT"
= 322 OR "REDEVCONFLICT" = 321 OR "REDEVCONFLICT" = 233 OR
"REDEVCONFLICT" = 232 OR "REDEVCONFLICT" = 223 OR "REDEVCONFLICT"
= 222 OR "REDEVCONFLICT" = 212 OR "REDEVCONFLICT" = 123 OR
"REDEVCONFLICT" = 311 OR "REDEVCONFLICT" = 213).
```

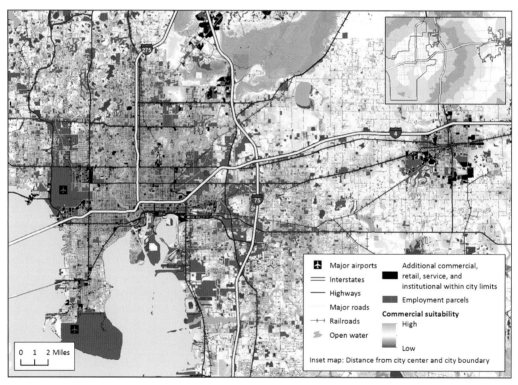

Figure 5.10. Extra commercial, retail, service, and institutional acres within the urbanized areas of Hillsborough County that meet the redevelopment criteria selected using query 5.7 are shown in black.

Figure from Paul D. Zwick and the Arizona Board of Regents on behalf of the University of Arizona. Airports from the Federal Aviation Administration and the Research and Innovative Technology Administration Bureau of Transportation Statistics National Transportation Atlas Databases (NTAD) 2012; water bodies from the US Geological Survey; state boundary from US Census Bureau TIGER/Line Files; interstates, highways, and major roads from the Florida Department of Transportation; railroads from the Federal Railroad Administration; parcels from the Florida Department of Revenue.

Query 5.8.
`"CNTERSLCE19" >= 8 AND "YRBUILT" <= 1992 AND ("REDEVCONFLICT" = 313 OR "REDEVCONFLICT" = 323 OR "REDEVCONFLICT" = 212 OR "REDEVCONFLICT" = 113 OR "REDEVCONFLICT" = 311 OR "REDEVCONFLICT" = 211 OR "REDEVCONFLICT" = 112 OR "REDEVCONFLICT" = 333 OR "REDEVCONFLICT" = 222) AND "ALLYRPOP" = 0 AND "ALLYRCOMR" = 0 AND "ALLYRIND" = 0.`

Query 5.8 is more complex. It selects properties outside of, but very close to, the three major cities (`"CNTERSLCE19" >= 8`). The structures have been built in or before 1992 (`"YRBUILT" <= 1992`). Redevelopment opportunity is for commercial, retail, service, or institutional uses [(`"REDEVCONFLICT" = 313` OR `"REDEVCONFLICT" = 323` OR `"REDEVCONFLICT" = 212` OR `"REDEVCONFLICT" = 113` OR `"REDEVCONFLICT" = 311`

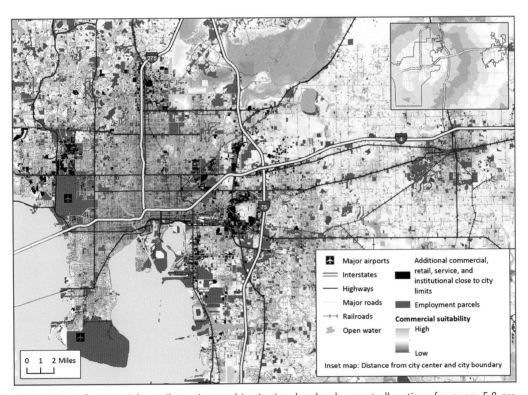

Figure 5.11. Commercial, retail, service, and institutional redevelopment allocations for query 5.8 are shown in black. These areas are allocated outside the urban cities and close to the center of employment activities.

Figure from Paul D. Zwick and the Arizona Board of Regents on behalf of the University of Arizona. Airports from the Federal Aviation Administration and the Research and Innovative Technology Administration Bureau of Transportation Statistics National Transportation Atlas Databases (NTAD) 2012; water bodies from the US Geological Survey; state boundary from US Census Bureau TIGER/Line files; interstates, highways, and major roads from the Florida Department of Transportation; railroads from the Federal Railroad Administration; parcels from the Florida Department of Revenue.

OR "REDEVCONFLICT" = 211 OR "REDEVCONFLICT" = 112 OR "REDEVCONFLICT" = 333 OR "REDEVCONFLICT" = 222)]. Again, none of the selected cell locations have previously been used for another future use ("ALLYRPOP" = 0 AND "ALLYRCOMR" = 0 AND "ALLYRIND" = 0).

The query 5.8 allocation involves an important structural limitation regarding employment land uses. The limitation regards the redevelopment of property close to the center of urban employment activity. It is evident in figure 5.11 that the query essentially allocates new redevelopment, as directed, close to urban employment activity, near the city edges of Tampa and Temple Terrace. The allocation provides an interesting result. It shows that a pattern of trend growth, continuing with new commercial, retail, service, and institutional uses as constricted, leaves out opportunities near Plant City. However, it aggregates new redevelopment employment activity close to major existing infrastructure and other services. It constricts new redevelopment using a single proximity variable (CNTERSLCE19) in the redevelopment CEM. The use of this variable in query 5.8 is an example of the inclusion of trend policy within redevelopment employment/jobs allocation. Finally, the query shows its sensitivity to redevelopment allocation created by the LUCIS suitability raster layers. First, the allocations to the northeast are directly linked to Tampa's international airport. Second, the allocations to the east are located along the major transportation corridors for Interstate 75 for north–south access and Interstate 4 for east–west access. Finally, the allocations to the north, although external to the city limits, follow a major transportation connector, locally identified as Dale Mabry Highway.

Figure 5.12 presents the total allocation for infill (yellow areas), greenfield (blue areas), mixed-use redevelopment (bright purple areas), and additional allocations for commercial, retail, service, and institutional (black areas). The figure shows that the trend allocation requirement for 45,175 acres, even including the use of redevelopment opportunity, produces urban sprawl. Significant land-use changes primarily occur in a north–south corridor between Tampa and Temple Terrace to the west of Plant City. Query statements 5.1 through 5.8 all use moderate to high suitability for the allocations. The use of proximity to the center of existing employment activity varies for each query. Some use moderate to high proximity for their allocations, while one (query 5.7) uses only high values for proximity. The results produce a trend that, overall, is not suitable, even when suitability is used as a foundation. This situation brings us to an important conclusion: suitability alone, without a vision for the future or the inclusion of planning policy, can help show how a trend progresses. However, in most locations of the world, it is the trend allocation that depicts a trend that planners, the general public, administrators, and policy makers want to change for the better.

Chapters 6 and 7 use other mixed-use development opportunities, increased-density options, and conservation alternatives combined within the LUCIS^{plus} CEMs to develop alternatives for making better decisions regarding future land-use allocations. The remainder of this book explores the power of LUCIS^{plus} as a regional geodesign methodology for urban planning, land development, and regional policy analysis, and as a mechanism for integrating the facets of comprehensive geodesign.

Chapter 5 Analyzing and mapping employment land-use futures

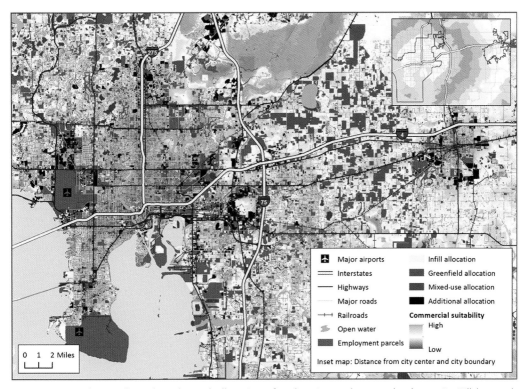

Figure 5.12. The total combined trend allocations for the six employment land uses in Hillsborough County, Florida, through 2045.

Figure from Paul D. Zwick and the Arizona Board of Regents on behalf of the University of Arizona. Airports from the Federal Aviation Administration and the Research and Innovative Technology Administration Bureau of Transportation Statistics (RITA/BTS) National Transportation Atlas Databases (NTAD) 2012; water bodies from the US Geological Survey; state boundary from US Census Bureau TIGER/Line files; interstates, highways, and major roads from the Florida Department of Transportation; railroads from the Federal Railroad Administration; parcels from the Florida Department of Revenue.

Chapter summary

The following points are discussed in this chapter:

- Economic base theory analysis and location quotient calculations can determine the demand for and diversity of industrial sectors in a region and serve as a factor in predicting future employment by sector. Combined with shift-share employment projections and the LUCIS^{plus} CEMs, LUCIS^{plus} can help planners, elected officials, and other professionals in land-use planning visualize the growth or decline of a local economy.
- The greenfield, redevelopment, and infill LUCIS^{plus} CEMs play a crucial role in the spatial allocation of projected employment. Complications arise in most employment category allocations because many land-use designations do not include categories for "vacant employment."
- The LUCIS^{plus} CEMs consider proximity to existing activity centers to allocate employment. CEMs ensure that industrial jobs are not spatially allocated in the same locations as previous residential allocations. They maximize colocation opportunities with nonconflicting land uses.

ArcGIS tools referenced in this chapter

Tool Name	Version 10.2 Toolbox/Toolset
Extract by Attributes	Spatial Analyst Tools/Extraction
Select By Attributes	Attribute Table/Table Menu tool

References

BLS (US Department of Labor Bureau of Labor Statistics). 2012. *Quarterly Census of Employment and Wages.* http://www.bls.gov/cew/cewlq.htm.

Chapin Jr., F. S. 1965. *Urban Land Use Planning,* 2nd ed. Urbana, IL: University of Illinois Press.

Galambos, E. C., and A. F. Schreiber, project directors. 2006. *Making Sense Out of Dollars: Economic Analysis for Local Governments.* Washington, DC: National League of Cities.

Klosterman, R. E., R. K. Basil, and E. G. Bossard, eds. 1993. *Spreadsheet Models for Urban and Regional Analysis.* New Brunswick, NJ: Rutgers University, Center for Urban Policy.

Part II

Land-use analysis and alternative futures

Chapter 6

Identifying and mapping an alternative urban mixed-use opportunity

Paul D. Zwick

What this chapter covers

For many years, planners, engineers, landscape architects, architects, and public administrators have proposed mixed-use or other increased-density alternatives as a solution to urban sprawl. Some problems associated with urban sprawl or low-density development include (1) the increased cost of providing urban services such as roads, utilities, and emergency services; (2) the loss of productive agricultural lands and environmentally sensitive habitats; and (3) increased travel costs and time to work. Proponents have proposed mixed-use land development alternatives to increase urban density and, at the same time, reduce the expansion of urban services. Mixed-use development will also reduce the cost of installing and maintaining urban services.

Chapters 4 and 5 developed trend allocations for residential and employment development in the study area. These trend allocations set the stage for a comparison with this chapter's allocation scenarios. This chapter explores three allocation scenarios for increased residential and employment density: increased-density allocation scenarios 3 and 4 and urban mixed-use allocation scenario 5. The intent is to explore possibilities for reducing the trend sprawl produced in chapters 4 and 5. Allocation scenario 3 tests a policy to increase the density of new infill and greenfield residential development, leading to a small change in existing density. Allocation scenario 4 explores the possibility of increasing residential density of new development. It also restricts new residential infill and greenfield development based on its proximity to the center of urban employment. Finally, allocation scenario 5 investigates a policy to increase residential and employment density for new redevelopment and infill. It uses mixed-use opportunities while again restricting new development based on proximity to the center of urban employment, similar to allocation scenario 4. All three allocation scenarios investigate reducing trend residential sprawl through increased residential density. Only allocation scenario 5 explores the opportunity for mixed-use development that colocates commercial, retail, service, and institutional employment with multifamily residential development.

Identifying alternative land-use allocations to reduce trend sprawl

Table 6.1 presents the basic information for the trend residential development demographics developed from the allocations chapter 4 presents. Without the development of a trend scenario, it is impossible to determine whether future scenarios are an improvement on the problems observed with the trend development. Essentially, the trend scenario is the allocation of population and employment land use through time using historical development rates that have occurred within the region. Interestingly, a future scenario can produce a dual effect. That is, the scenario may provide a solution to a few existing problems, but at the same time it may cause new problems. A classic example of this duality is a scenario that reduces transportation congestion but simultaneously creates new urban sprawl. The value of Land-Use Conflict Identification Strategy planning land-use scenarios (LUCIS^{plus}) in modeling for geodesign is that LUCIS^{plus} can preview beneficial or problematic impacts of a new land-use policy. It can provide insights that may otherwise be hidden or counterintuitive. A benefit of using LUCIS^{plus} modeling is its ability to test what is believed to be a good idea or an attractive policy opportunity. Many good ideas or well-intended policies have unintended consequences. LUCIS^{plus} modeling can reveal the beneficial and problematic impacts before an idea or a policy is implemented. At a minimum, LUCIS^{plus} modeling can raise alternative questions about a policy or scenario that are worth exploring. If LUCIS^{plus} shows that a land-use policy is capable of generating problematic impacts, you can use it to test either a modification of that policy or an alternative policy. The ability to rapidly generate multiple land-use alternatives using LUCIS^{plus} modeling provides planners and administrative officials a strategy to investigate policies before they are implemented.

Exploring increased density as an alternative land-use allocation concept

The purpose of the following three allocation scenarios is structural. It provides a method for easing into the complexity of scenario development. It is also an opportunity to use the

Table 6.1. Summary descriptive statistics for trend 2045 residential development allocations

Allocation Type	Allocated Acres	Allocated Residents or Jobs	New Total Dwelling Units	New Dwelling Units/Acre or Employees/Acre
Trend Residential Redevelopment	5,788	248,834	102,454	17.70
Trend Infill Residential	23,865	284,521	104,965	4.40
Trend Greenfield Residential	12,788	177,371	60,969	4.77
Trend Totals	**42,441**	**710,726**	**268,388**	**6.32**
Trend Redevelopment Industry	7,782	25,135		3.23
Trend Redevelopment Commercial	6,845	48,871		7.14
Trend Redevelopment Retail	4,910	30,739		6.26
Trend Redevelopment Institutional	16,891	47,971		2.84
Trend Service Employment	8,952	87,107		9.73
Trend Employment Totals	**45,380**	**239,823**		**5.24**

three LUCIS[plus] criteria evaluation matrices (CEMs) as a land-use allocation tool. The three LUCIS[plus] CEMs provide a structure for redevelopment, infill, and greenfield allocations. You can use the three CEMs to explore land-use change through residential and employment allocations.

Allocation scenario 3: Testing a small increase in residential density for infill and greenfield development

Allocation scenario 3 tests a simple increase in residential density. The scenario increases residential density by one person per acre of new residential development. This across-the-board policy only slightly changes the area's residential unit density (residential units per acre). The use of a simple policy scenario often helps identify and present a fundamental concept more clearly. The question is, Can small changes in new residential development densities change the pattern of development, reduce sprawl, and thus reduce the impacts of sprawl? Because residential land-use occupies, by far, the largest area of land within the study area, scenario 3 tests the effects of a policy that impacts only one sector of the county's development community, residential development.

This section of the chapter presents the scenario's concept and results. Then it provides an explanation of the query used for the residential allocation. Next, it presents the tabular data summarizing the allocations.

Spatially, what does an increase of one person per acre mean for new infill and greenfield residential development? This policy is basic to analyze. For infill residential development, it would accommodate an additional 23,865 residents on the same infill residential acres allocated in chapter 4. Table 4.10 shows that the infill trend multifamily residential density is 13.55 people per acre (70,276 people / 5,188 acres). The unit density is 4.88 units per acre. The table also shows that the single-family residential density is 11.49 people per acre (214,244 people / 18,677 acres). The unit density is 4.23 units per acre. The average trend residential density is 11.92 people per acre, and the unit density is 4.40 units per acre. Therefore, a simple policy decision to increase residential density by one person per acre would increase scenario 3's infill residential density to 12.92 people per acre (308,385 people / 23,865 acres) compared with the residential trend allocation of 11.92 people per acre in chapter 4. Scenario 3 would split the allocation to accommodate 5,188 acres of multifamily residents and 18,677 acres of single-family residents. At the same time, it would not increase the need for any new infill residential development acreage. If the scenario was implemented, it would result in an increase of 23,865 residents (308,385 people instead of 284,520 people) allocated within the same infill residential acres. It would increase the county's infill residential unit density only slightly, from 4.40 residential units per acre to 4.77 residential units per acre (see table 6.2). The multifamily residential density increased from 4.88 units per acre to 5.37 units per acre. The single-family residential density increased from 4.23 units per acre to 4.60 units per acre.

Since the new infill density increase (one person per acre) did not change the base footprint of the 2045 infill residential development, how does it affect the trend's residential sprawl? For greenfield residential development, the impact of allocation scenario 3 would first be felt as a reduction of acreage within the greenfield residential allocation. This reduction results because the scenario removes the need to allocate 23,865 future residents within greenfield residential development. They have been allocated within the trend infill area instead. Therefore, as a result of the altered infill density, this scenario's allocation for new greenfield residential development will change to accommodate the new allocation. It reduces the need for residential acreage in greenfield development. Again turning to the residential allocations in chapter 4, the average density for greenfield residential is 13.82 people per acre. The existing average household size in greenfield residential development is 2.91 people per household. The average unit density is 4.75 units per acre for single-family residential development. The resulting reduction in greenfield residential acreage is 1,729 acres [23,895 residents / (4.75 units per acre * 2.91 residents per acre)]. This reduction is significant: reducing 1,729 acres of greenfield development saves 2.70 square miles of agricultural production.

Second, you can increase the reduction of 1,729 acres in new greenfield residential development even further. Table 6.2 shows that the new greenfield population for scenario 3 is 154,081 people. A simple increase in residential density of one person per acre in the scenario greenfield residential development results in an increase from 13.82 people per acre to 14.82 people per acre. As a result of the scenario adjustments to infill residential development, there are 154,081 more people to be allocated for the scenario's greenfield residential allocation. Dividing 154,081 people by 14.82 people per acre indicates that 10,397 acres are needed for the greenfield residential development. The reduction in scenario 3 greenfield acres from the trend greenfield acres is 2,498 acres (12,895 acres - 10,397 acres). The 2,498 acres represents 1,729 acres because of increased infill density and 769 acres because of increased greenfield density. Converting 2,498 acres to square miles results in a total savings of 3.903 square miles of agricultural productive land not needed for scenario 3 greenfield residential development.

So what does a reduction of 3.903 square miles of residential development really mean in regard to the trend residential sprawl shown in chapter 4? Clearly, the decreased new residential acreage produced by allocation scenario 3 would not significantly impact the county's residential sprawl. However, scenario 3 shows the need to significantly increase residential density to reduce the impacts of chapter 4 trend residential sprawl and protect agricultural productivity or ecologically significant conservation areas. Ultimately, scenario 3 shows that a significant reduction in urban/suburban sprawl requires high levels of increased residential density or development. Either that, or there must be some other spatial policy combined with increased density to reduce the impact of trend residential sprawl.

Query 6.1 identifies the increased residential density allocation described for allocation scenario 3:

Query 6.1.
```
"ALLYRPOP" = 2045 AND "RESFSUIT100I" >= 676.
```

The following list of criteria is used for allocations of increased residential density based on land-use selections structured in ArcGIS query language:

- *ALLYRPOP*: Allocation Year Residential Population. This value equals 0 if the selected cells have not been previously allocated.
- *RESFSUIT100I*: Combined Multifamily and Single-Family Residential Suitability. Reclassified for integers in a range of 100 to 900. This reclassification is completed to provide a sufficient range of suitability for CEM. For example, a floating point suitability number 2.3456 would be reclassified to an equivalent three-digit integer value of 235.
- *CNTERSLCE19*: Distance Zone from Major Center City. Item used to indicate a zone of distance from the center of employment activity. The value ranges from 1 to 9, with 9 representing the closest zone of distance.
- *S3ALLYR*: Allocation Year Residential Population for Scenario 3. This value equals 0 if the selected cells have not been previously allocated.
- *S3INDYR*: Allocation Year Industrial Employment for Scenario 3. This value equals 0 if the selected cells have not been previously allocated.
- *S53COMYR*: Allocation Year Commercial Employment for Scenario 3. This value equals 0 if the selected cells have not been previously allocated.
- *S3RETYR*: Allocation Year Retail Employment for Scenario 3. This value equals 0 if the selected cells have not been previously allocated.

- *S3INSYR*: Allocation Year Institutional Employment for Scenario 3. This value equals 0 if the selected cells have not been previously allocated.
- *S3SERYR*: Allocation Year Service Employment for Scenario 3. This value equals 0 if the selected cells have not been previously allocated.
- *ACRES*: Available Allocation Acres. This item contains the number of acres represented by the raster zone. A raster zone is a record within an integer raster that contains all the cells in a raster that have the same value (for example, all the cells that represent single-family residential land use).
- *REDEVCONFLICT*: Redevelopment Mixed-Use Opportunity. This item is structured just like the GFCONFLICT index of LUCIS preference/conflict categories. However, it represents redevelopment opportunity (in this case, for commercial, multifamily residential, and retail mixed use). It has the same 27 index values.
- *CITYNAME*: City Name. This item represents the city in which the cell is located: (1) Tampa, (2) Temple Terrace, (3) Plant City, or (0) in the county.
- *DESCRIPT*: Property Use Description. Examples include Vacant Commercial, Vacant Industry, Misc. Agriculture, and so forth.
- *INSSUIT100I*: Institutional Land-Use Suitability. This is the same as residential suitability described in chapter 4 but for locations supporting institutional employment.
- *COMSUIT100I*: Commercial Land-Use Suitability. This is the same as residential suitability but for locations supporting commercial employment.
- *RETSUIT100I*: Retail Land-Use Suitability. This is the same as residential suitability but for locations supporting retail employment.
- *INDSUIT100I*: Industrial Land-Use Suitability. This is the same as residential suitability but for locations supporting industrial employment.
- *SERSUIT100I*: Service Land-Use Suitability. This is the same as residential suitability but for locations supporting service employment.

Conceptually, query 6.1 selects cells from the greenfield CEM that were originally selected for trend residential development. These cells have higher values for the single-family residential suitability than the cells allocated for the trend. By now, you should be familiar with the variable ALLYRPOP. It is used as a placeholder in chapters 4 and 5 to identify residential cells or acres of land that have been allocated for trend 2045 greenfield residential development. The variable RESFSUIT100I is single-family residential suitability, which ranges from a low value of 100 to a high value of 900. Query 6.1 processes the selection of cells iteratively. It ultimately selects cells with suitability values greater than or equal to 676 that were previously allocated for residential trend development. Yet the query results in removing only a small range of residential suitability values. In fact, it removes a total of 20 values for residential cell suitability, beginning with the value of 656 and ending with 675. The iteration begins with the lowest single-family suitability of 656 and iterates to 657, then 658, then 659, and so forth. It continues iterating until it reaches the desired acres to support the new residential development. As query 6.1 shows, the final value removed is 675. The iterative process used for query 6.1 results in a reduction of 2,649.40 acres for future greenfield residential development.

The italic numeric values in table 6.2 represent the scenario's allocation statistics. Compare them with table 6.1 to see how an increase of one person per acre for residential development alters the allocation demographics. Table 6.2 shows that an increase of one person per acre results in 308,385 new residents in this same "new infill" acreage. The extra increase for the number of residents occurs on the same 23,865 acres of infill. It also results in less acreage needed for future greenfield residential development. Not surprisingly, the resulting decrease in greenfield acres from 12,895 acres for trend development to 10,397 acres for scenario 3 also decreases the total number of housing units allocated from 61,222 to 52,920 units. Finally, even though allocation scenario 3 reduces the total number of residents allocated to greenfield acres, it increases the residential unit density for infill and greenfield residential development. Another benefit of scenario 3 is increased average residential densities measured in units per acre. The overall density for scenario 3 residential development is 6.70 units per acre while the trend overall residential development is 6.30 units per acre. To reiterate, the result of scenario 3 shows that a policy decision to use a small increase in residential density (one person per acre) can support the same number of residents and slightly reduce sprawl. However, small density changes are not adequate to significantly reduce the county's sprawl.

Table 6.2. Summary descriptive statistics for allocation scenario 3 (one person per acre increase in residential development allocations)

Allocation Type	Allocated Acres	Allocated Residents	Total Dwelling Units	People / Acre	Dwelling Units/Acre
Redevelopment Multifamily	5,788	248,834	102,455	42.99	17.70
Existing Infill (Table 4.10)	23,865	284,520	104,387	11.92	4.40
Multifamily (Table 4.10)	5,188	70,276	25,318	13.55	4.88
Single Family (Table 4.10)	18,677	214,244	79,069	11.49	4.23
Greenfield	12,895	177,946	61,222	13.82	4.75
Totals	**42,548**	**711,300**	**268,064**	**16.73**	**6.30**
Redevelopment Multifamily	*5,788*	*248,834*	*102,455*	*42.99*	*17.70*
New Infill	*23,865*	*308,385*	*113,774*	*12.92*	*4.77*
Multifamily	*5,188*	*75,464*	*27,860*	*14.55*	*5.37*
Single Family	*18,677*	*232,921*	*85,914*	*12.49*	*4.60*
Greenfield	*10,397*	*154,081*	*52,920*	*14.82*	*5.09*
New Totals	***40,050***	***711,300***	***269,149***	***17.76***	***6.70***

Note: Green is trend and its totals. Yellow is scenario and its totals. The multifamily and single-family numbers are both summed for existing infill and new infill.

Allocation scenario 4: Testing increased density and adding a new proximity restriction

Allocation scenario 4 tests a policy to increase density and simultaneously restrict new occupied and vacant residential development to areas near the center of urban employment activity. Although increased density reduces land consumption, which was the concept tested with allocation scenario 3 in the previous section of this chapter, the restriction on the location of new development should further improve the reduction of residential sprawl.

A restriction on location of development is not a new concept. However, testing that concept for future residential development might reveal some unintended consequences of instituting such a policy. The decision to increase density while restricting the location of that denser development nearer urban employment should reduce sprawl. However, redistricting the location of development is not without controversy. Using a defined urban services boundary (USB) or urban growth boundary (UGB) is one method to restrict the location of new development. Since 1979, Portland, Oregon, has used USB to preserve rural character and agricultural lands, and thus enhance agricultural productivity. Portland is one of 240 USBs in Oregon (Oates 2006).

A USB is designed to limit the necessity of providing urban services at ever increasing distances from the urban area without some control over sprawl development into the countryside. The concept of restricting urban services (such as roads, water, sewer, fire, and schools) to an area near, or very near, the center of urban activity is based on the assumption that it reduces the cost of providing such services to residents and businesses. Because the urban form generally has higher densities near its core, a linear mile of urban infrastructure can serve more residents at a reduced cost of construction and maintenance. However, any USB policy has its winners and losers. Land values increase more rapidly in the USB area, and so, the price of housing increases. It can place some at a disadvantage for homeownership. Some individual landowners endure development restrictions as the land development of their private property is restricted because they are outside the USB. However, the problem of restrictions on property outside the USB is not permanent, and agricultural productivity and rural character are preserved. As the urban area grows, new property is added to the USB to support new growth. Also, mixed-use development provides multifamily opportunities for less-expensive development with options for homeownership within the lower price ranges.

Scenario 4 development allocations begin by using the median center of urban activity/employment. The CNTERSLCE19 attribute represents the median center of urban employment in all three LUCIS^{plus} CEMs. The employment properties used to locate the median center are the same employment properties discussed in chapter 5 and include (1) industrial, (2) commercial, (3) retail, (4) service, and (5) institutional parcels. Agriculture jobs are not part of the urban employment mix.

Figure 6.1 shows the trend redevelopment and infill areas from chapter 4 in blue and red, respectively. The trend greenfield areas are shown in yellow as an overlay on the network equal-area zones for distance from the center of employment.

Allocation scenario 4 is being tested to remove the infill and greenfield residential development in the zones farther from the urban center and reallocate these trend residential allocations closer to the major urban core. The goal is for the relocation to show a decrease in sprawl within the county by allocating growth closer to or inside urban areas.

Table 6.3 shows the 2045 trend residential demographic acreage summaries, number of residents, number of residential dwelling units, residential density in dwelling units per acre, and average household size for the nine distance zones from the median center of employment. These zones are shown in figure 6.1. The first section in the table presents descriptive demographic statistics for trend greenfield residential allocations. The second

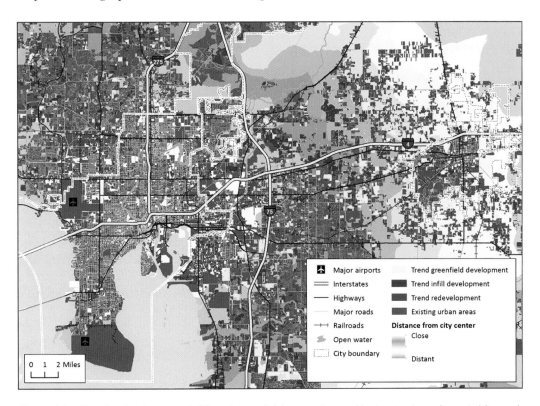

Figure 6.1. Trend redevelopment, infill, and greenfield areas allocated in chapter 4 are shown in blue, red, and yellow, respectively. The development areas are shown as an overlay on the network equal-area zones based on distance from the center of employment.

Figure from Paul D. Zwick and the Arizona Board of Regents on behalf of the University of Arizona. Airports from the Federal Aviation Administration and the Research and Innovative Technology Administration Bureau of Transportation Statistics National Transportation Atlas Databases (NTAD) 2012; water bodies from the US Geological Survey; city and state boundary from US Census Bureau TIGER/Line Files; interstates, highways, and major roads from the Florida Department of Transportation; railroads from the Federal Railroad Administration; parcels from the Florida Department of Revenue.

section presents the same descriptive statistics for the trend infill residential development, and the third section presents the statistics for the trend redevelopment allocations.

The average household size seems to be larger for the greenfield development. Perhaps it is because the families are larger in the suburbs whereas the household size is smaller in the redevelopment area, reflecting a trend for young singles to live in urban areas. Table 6.3 illustrates that the average household size is not significantly different between infill, redevelopment, and greenfield development. A significant difference clearly is present for the number of dwelling units per acre, with urban zones (zones 9 and 8) having a greater residential unit density.

Table 6.3. Summary descriptive statistics for allocation scenario 4

Trend Greenfield Zones	Allocated Acres	Number of Residents	New Total Dwelling Units	New Dwelling Units/Acre	Average Household Size
9	203.7	3,321	1,157	5.69	2.87
8	1,500.11	23,737	8,532	5.69	2.78
7	1,642.27	14,784	5,095	3.1	2.9
6	2,909.33	45,501	15,282	5.25	2.98
5	2,223.77	34,071	11,482	5.16	2.97
4	2,452.26	33,566	11,730	4.78	2.86
3	1,345.33	18,608	6,422	4.77	2.9
2	465.26	3,371	1,131	2.43	2.98
1	45.76	412	137	3.0	3.0
Totals	12,787.79	177,371	60,968	4.77	2.92
Trend Infill Zones	Allocated Acres	Number of Residents	New Total Dwelling Units	New Dwelling Units/Acre	Average Household Size
9	1,020.39	19,043	7,074	6.93	2.69
8	2,937.30	39,789	15,466	5.27	2.57
7	4,637.05	56,015	20,273	4.37	2.76
6	3,616.86	43,195	15,656	4.33	2.76
5	4,187.32	53,108	20,113	4.8	2.64
4	2,806.29	32,165	11,758	4.19	2.74
3	2,042.47	23,635	8,712	4.27	2.71
2	1,228.88	12,375	4,309	3.51	2.87
1	572.79	5,195	1,604	2.63	3.24
Totals	23,049.35	284,520	104,965	4.55	2.78

(*continued*)

Table 6.3. Summary descriptive statistics for allocation scenario 4 (*continued*)

Trend Redevelopment Zones	Allocated Acres	Number of Residents	New Total Dwelling Units	New Dwelling Units/Acre	Average Household Size
9	1,915.09	102,796	44,077	23.02	2.33
8	2,041.86	104,962	43,503	21.31	2.41
7	736.93	16,847	6,516	8.84	2.59
6	741.28	16,013	5,552	7.49	2.88
5	352.9	8,216	2,807	7.95	2.93
4	0	0	0	0	0
3	0	0	0	0	0
2	0	0	0	0	0
1	0	0	0	0	0
Totals	5,788.06	248,834	102,455	17.7	2.63

Note: The trend residential development (chapter 4) allocations for greenfield, infill, and redevelopment are summarized for the equal-area zones from the median center of employment.

The planning department exploring these scenarios believes the information presented in table 6.3 for allocation scenario 4 warrants testing a concept of increasing density for residential development and restricting location. In particular, it wants to test the concept of restricting development outside zones 8 through 6, where zone 9 is closest to the urban center of employment. It wants to increase the density in zones 8 through 6 as a means of reallocating 100 percent of the 2045 occupied trend infill and greenfield residential development from zones 5 through 1. In this way, it will accommodate the units previously allocated for zones 5 through 1. Second, the planning department staff wants to relocate the 2045 trend vacant residential allocations from zones 5 through 1 into zones 8 through 6. By increasing the unit densities in zones 8 through 6 and reallocating the trend vacant residential capacity, the planning staff believes the scenario model will show a significant reduction in suburban sprawl.

A quick review of the 2045 trend acreage allocations from table 6.3 in zones 5 through 1 shows 6,531 acres of occupied residential development in greenfield areas and 10,837 acres of occupied infill residential development. The trends allocated for zones 8 through 6 are 6,052 and 11,191 acres, respectively. Table 6.4 shows the integration of the trend residential allocations for zones 5 through 1 with the trend allocations for zones 8 through 6, which results in a unit density increase of 207 percent for greenfield development. Unit density increases 190 percent for infill residential development. Essentially, the infill and greenfield residential unit density will double from the trend residential density for allocation scenario 4. The proposed unit density increases are based on the average household size. They will remain very close to or the same as the trend average household size. The increased units per acre densities would increase to an average 9.88 units per acre for new residential greenfield development in zones 8 through 6. They would increase to 8.85 units per acre for new infill residential development in the same zones. The increased allocations for greenfield and infill future residential development from new allocation scenario 4 are shown in the shaded columns in table 6.4. Pay particular attention to the increased number of units per zone and, thus, the increased units per acre densities.

The final relocation of lands for allocation scenario 4 is for the future vacant (unoccupied) residential capacity. Table 6.4 shows a reduction of 10,837 acres of occupied infill development in zones 5 through 1. The residents that would have been placed on these 10,837 acres were accommodated in zones 8 through 6 because of the increased density. Of further interest is

Table 6.4. Residential allocations for scenario 4 increased density with location restrictions

Category	Trend Zones 5–1 Data	Trend Zones 8–6 Data	Zone 6 New Allocations	Zone 7 New Allocations	Zone 8 New Allocations	New Totals
Future Greenfield Residential						
Acres	6,532	6,052	2,910	1,642	1,500	**6,052**
Residents	90,028	84,022	94,254	30,625	49,171	**174,050**
Units	30,902	28,909	31,618	10,541	17,652	**59,811**
Units/Acre	4.73	4.78	10.87	6.42	11.77	**9.88**
Mean HH Size*	2.92	2.92	2.98	2.91	2.79	**2.91**
Future Infill Residential						
Acres	10,837	11,191	3,262	4,389	4,250	**11,901**
Residents	126,478	138,999	82,499	106,984	100,304	**289,787**
Units	46,496	51,395	29,783	36,622	38,878	**105,283**
Units/Acre	4.29	4.59	9.13	8.34	9.14	**8.85**
Mean HH Size*	2.72	2.72	2.77	2.92	2.58	**2.75**
Future Vacant Residential in Greenfield for Zones 6, 7, and 8						
Acres	23,865		2,969	1,443	783	**5,195**

*The Average Household Size (Mean HH Size) values are not spatially weighted averages and should not be used to calculate the Units/Acre or Residents columns. The values are provided for general information and for use only as approximations.

the change in density for infill property in zones 8 through 6. The new density for these zones is 8.85 units per acre, again approximately double the trend density. Taking into account the new doubling in unit density, the requirement for 10,837 acres of new vacant lands can be cut in half. As a result, scenario 4 must allocate only 5,195 acres. The allocation of the 5,195 acres is shown in the table section "Future Vacant Residential in Greenfield for Zones 6, 7, and 8" in each respective zone column and in the New Totals column.

Figure 6.2 displays the spatial allocations for scenario 4. The scenario increases residential density by approximately doubling the number of units per acre. It restricts the urban expansion beyond the three zones closest to the major urban employment center, not allowing new residential development in zones 5 through 1. The result is as expected: (1) an increase in density measured in units per acre, (2) an increase in the density of people per developed acre, (3) a decrease in the distance to the major urban center of employment, and (4) a decrease in the number of acres developed. Allocation scenario 4

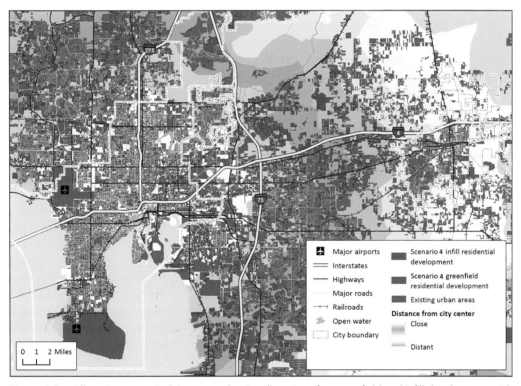

Figure 6.2. Allocation scenario 3 increases density allocations for greenfield and infill development with a priority of allocating new development near urban employment centers. As compared with the trend scenario, scenario 4 decreases the number of acres developed and directs development within the first three urban employment zones. A summary of scenario 4 allocations is described in table 6.4.

Figure from Paul D. Zwick and the Arizona Board of Regents on behalf of the University of Arizona. Airports from the Federal Aviation Administration and the Research and Innovative Technology Administration Bureau of Transportation Statistics National Transportation Atlas Databases (NTAD) 2012; water bodies from the US Geological Survey; state boundary from US Census Bureau TIGER/Line Files; interstates, highways, and major roads from the Florida Department of Transportation; railroads from the Federal Railroad Administration; parcels from the Florida Department of Revenue.

reduces residential sprawl, in that distance measured in network miles decreased distance to work by an average of approximately five miles per unit. It also may have decreased travel time to work. And it reduced the cost for increased new infrastructure—the roads, pipelines, and other infrastructure needed to support new development in zones 5 through 1.

Allocation scenario 4 shows that increasing density and controlling location can reduce sprawl, which again is not surprising. The LUCIS^{plus} modeling process identifies viable spatial locations that help reduce sprawl and, further, illustrates the results on a map.

Allocation scenario 5: Testing increased density using mixed-use development

For the third scenario in this chapter, allocation scenario 5, the planning department wants to test the concept of increasing residential density combined with a greater proportion of mixed-use development. The planning department believes that scenario 5 should significantly decrease suburban sprawl compared with allocation scenarios 3 and 4. Mixed-use development, as the name implies, colocates employment with residential development. Again, the intent is to reduce suburban sprawl and its associated costs. Allocation scenario 5 will also restrict or relocate the future residential development and employment from zones 5 through 1 to alternative areas in zones 9 through 6. They will also be within the city boundary of Plant City, Florida. Plant City is outside zones 9 through 6 but is an urbanizing city within the county. It lies within the Interstate 4 corridor between Orlando and Tampa.

Steps 1, 2, and 3: Allocating redevelopment residential

The first step in completing allocation scenario 5 is to use the trend multifamily residential redevelopment allocated in zones 9 through 6. The trend allocations for these zones have created adequate residential unit density. Query 6.2a selects 240,618 residents within these zones. Table 6.5 presents the demographic summary by zone for the allocations the query selects. These 240,618 residents are allocated within 99,648 units on 5,435 acres. The average residential unit density is 18.64 units per acre.

    Query 6.2a.
    `"CNTERSLCE19" >= 6 AND "ALLYRPOP" = 2045`.

Table 6.5. Results for query 6.2a scenario 5 residential mixed-use redevelopment allocations

Query 6.2a	Zone 9	Zone 8	Zone 7	Zone 6	Totals
Population	102,796	104,962	16,847	16,013	**240,618**
Dwelling Units	44,077	43,503	6,516	5,552	**99,648**
Acres	1,915	2,042	737	741	**5,435**
Unit Density	23.02	21.30	8.84	7.49	**18.64 (overall)**

The second step is to locate other redevelopment opportunities not selected by query 6.2a within the preferred zones 9 through 6. Query 6.2b selects the extra mixed-use opportunity. The selection finds areas with high or moderately high multifamily preference combined with high or moderately high commercial or retail preference. The concept of query 6.2b is to select areas with appropriate mixed-use opportunity for multifamily residential development. In the REDEVCONFLICT variable, the middle digit is for multifamily preference. The first digit is commercial preference, and the third digit is retail preference. Using the REDEVCONFLICT variable is appropriate because the selection is within the high commercial employment opportunity zones. Table 6.6 presents the demographics associated with the allocations from query 6.2b.

Query 6.2b.
```
"CNTERSLCE19" >= 6 AND ("REDEVCONFLICT" = 333 OR "REDEVCONFLICT"
= 332 OR "REDEVCONFLICT" = 233 OR "REDEVCONFLICT" = 232 OR
"REDEVCONFLICT" = 322 OR "REDEVCONFLICT" = 233) AND "S5ALLYR" = 0
AND "ACRES" >= 2.
```

Table 6.6 shows that an additional 94,708 residents could be allocated within zones 9 through 6 in an extra 39,230 residential multifamily units. This allocation would require an extra 2,811.23 acres near the county's center of employment activity. It would create an average residential unit density of 13.95 units per acre.

Table 6.6. Results for query 6.2b allocation scenario 5 residential redevelopment mixed-use opportunity

Query 6.2b	Zone 9	Zone 8	Zone 7	Zone 6	Totals
Population	52,472	31,077	9,402	1,757	94,708
Dwelling Units	22,080	12,644	3,819	687	39,230
Acres	947.31	1,205.44	529.69	128.79	2,811.23
Unit Density	23.31	10.49	7.21	5.33	13.95 (overall)

The third step in the process is to locate other mixed-use redevelopment opportunity that has not already been allocated and allocate it within the boundary of Plant City. Query 6.3 selects the extra mixed-use redevelopment opportunity confined within the Plant City boundary. Plant City lies within a major transportation corridor along Interstate 4 between Tampa and Orlando. It is an urbanizing area with mixed-use development opportunity. Again, the concept is to find areas with high or moderately high multifamily residential preference combined with high or moderately high commercial and retail preference.

Query 6.3.
```
("REDEVCONFLICT" = 333 OR "REDEVCONFLICT" = 332 OR "REDEVCONFLICT"
= 233 OR "REDEVCONFLICT" = 232 OR "REDEVCONFLICT" = 322 OR
"REDEVCONFLICT" = 233) AND "S5ALLYR" = 0 AND "ACRES" >= 2 AND
"CITYNAME" = 'Plant City'.
```

Table 6.7 displays the combined residential development allocation information for queries 6.2a, 6.2b, and 6.3 in Plant City, Florida. Allocation query 6.3 accommodates only 2,843 new multifamily residents in 895 units on 160 acres. The average residential density is 5.6 units per acre. A total of 338,169 residents would be accommodated within 8,406.04 acres in 139,773 mixed-use residential units. The average density would be 16.63 units per acre. The average total residential density does not preclude lower-unit density in particular areas. But a policy to contain the expansion of suburban sprawl, which preserves agricultural production and environmentally sensitive habitats, requires average residential density to increase in all areas. The increased-density policy requires the reduction of single-family suburban low-density residential development in favor of higher density urban multifamily residential mixed-use development.

Table 6.7. Results for queries 6.2a, 6.2b, and 6.3 allocation scenario 5 residential redevelopment mixed-use opportunities

Category	Zone 9	Zone 8	Zone 7	Zone 6	Plant City	Totals
Population	155,268	136,039	26,249	17,770	2,843	**338,169**
Dwelling Units	66,157	56,147	10,335	6,239	895	**139,773**
Acres	2,862.31	3,247.44	1,266.69	869.79	159.81	**8,406.04**
Unit Density	23.11	17.29	8.16	7.17	5.60	**16.63 (overall)**

Step 4: Allocating infill residential

Step 4 allocations are for infill residential development. Query 6.4 selects infill locations that were part of the trend infill residential development but restricts the location to zones 9 through 6. Table 6.8 presents the selection results. The selection locates 16,901.40 acres that support 247,769 residents in 80,516 dwelling units. The residential density is 4.76 units per acre.

Query 6.4.
"CNTERSLCE19" >= 6 AND "RESSUIT100I" > 600 AND ("DESCRIPT" = 'VACANT ACREAGE' OR "DESCRIPT" = 'VACANT RESIDENTIAL').

Table 6.8. Results for query 6.4 allocation scenario 5 residential infill opportunities

Category	Zone 9	Zone 8	Zone 7	Zone 6	Totals
Population	37,043	61,389	72,859	76,478	**247,769**
Dwelling Units	11,377	18,646	24,236	26,257	**80,516**
Acres	1,722.91	3,809.97	5,649.11	5,719.41	**16,901.40**
Unit Density	6.60	4.89	4.29	4.59	**4.76 (overall)**

Steps 5 and 6: Allocating greenfield residential

Step 5 uses query 6.5 to select locations with high residential development suitability on agricultural lands. It restricts the location to zones 9 through 6. The selections locate 3,460.06 acres that support 114,013 multifamily residents in 38,755 dwelling units. The residential density is 11.2 units per acre.

Step 6 uses query 6.6 to select locations with mixed-use opportunities that have service and institutional uses. Table 6.9 shows the selection results for queries 6.5 and 6.6. The selection allocated 12,186 residents on 444.87 acres within 4,685 dwelling units. The residential density is 10.53 units per acre. The New Totals column in table 6.9 is the summations of selections generated by queries 6.5 and 6.6. The individual zonal summations are shown in the zones 8 through 6 columns. The overall density is the total units divided by the total acres.

Query 6.5.
"CNTERSLCE19" >= 6 AND "RESSUIT100I" >= 677.

Query 6.6.
"CNTERSLCE19" >= 6 AND "RESSUIT100I" >= 368 AND "S5INSTYR" = 2045.

Table 6.9. Results for queries 6.5 and 6.6 allocation scenario 5 residential infill opportunities

Category	Zone 9	Zone 8	Zone 7	Zone 6	Totals
Population	0	37,145	23,761	65,294	**126,200**
Dwelling Units	0	13,176	8,262	22,003	**43,441**
Acres	0	1,090.50	1,166.98	1,647.46	**3,904.94**
Unit Density	0	12.08	7.08	13.36	**11.12 (overall)**

Table 6.10 presents allocation scenario 5 residential allocations by zone and category. Compared with the trend residential development from chapter 4, allocation scenario 5 requires 29,212.38 acres. The trend residential allocation requires 41,625.20 acres. (Neither acreage value includes the replacement infill acres.) Therefore, scenario 5 saves 12,412.82 acres of residential development, or 19.4 square miles of urban/suburban residential sprawl. Figure 6.3 displays the scenario 5 residential allocations.

Next, the LUCIS^{plus} process allocates employment for scenario 5. The goal is to save greenfield development acreage based on mixed-use development.

Table 6.10. Results for allocation scenario 5 residential redevelopment mixed-use opportunities

Category	Zone 9	Zone 8	Zone 7	Zone 6	Plant City	Totals
Population	192,311	234,573	122,869	159,542	2,843	**712,138**
Dwelling Units	77,534	87,969	42,833	54,499	895	**263,730**
Acres	4,585.22	8,147.91	8,082.78	8,236.66	159.81	**29,212.38**
Unit Density	16.91	10.80	5.30	6.62	5.60	**9.03 (overall)**

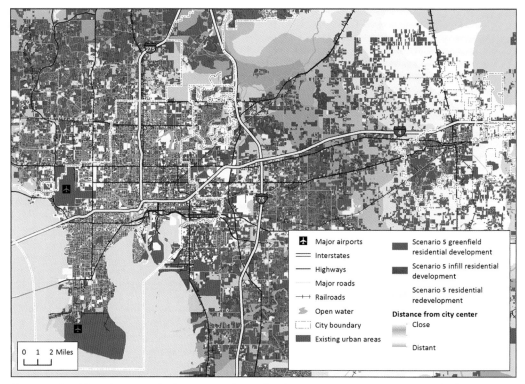

Figure 6.3. Allocation scenario 5 uses mixed-use development opportunities as a method to increase density. Redevelopment, infill, and greenfield allocations consider high suitability for residential land use as a means to identify mixed-use opportunities. Compared with the chapter 4 trend residential scenario in which allocations were based on single uses, the combined use of increased densities and mixed-use development saves almost 20 square miles of urban/suburban residential sprawl.

Figure from Paul D. Zwick and the Arizona Board of Regents on behalf of the University of Arizona. Airports from the Federal Aviation Administration and the Research and Innovative Technology Administration Bureau of Transportation Statistics National Transportation Atlas Databases (NTAD) 2012; water bodies from the US Geological Survey; state boundary from US Census Bureau TIGER/Line Files; interstates, highways, and major roads from the Florida Department of Transportation; railroads from the Federal Railroad Administration; parcels from the Florida Department of Revenue.

Step 7: Allocating employment

In LUCIS[plus], redevelopment refers to new development as a result of new construction that replaces existing structures or an adaptive reuse of an existing structure. The concept of mixed-use redevelopment employment in allocation scenario 5 follows the same strategy as the residential mixed-use allocations. However, the selection is directed at the employment preference digits in the REDEVCONFLICT variable. Before making the allocations, adjust the scenario to employment density. (Employment density in chapter 5 used the mean density for employment—that is, employees per acre within the major employment categories.)

Employment density adjustment

To adjust the employment density and remain consistent with the use of mean employment density requires a fairly simple transformation. This transformation is based on proximity to the center of employment within the county. The concept is to increase the employment density near the allocated land-use location, the center of employment, without increasing the mean employment density. The transformation allocates greater employment near the existing employment activity. The reason to keep the mean the same, or at least close to the same, is to compare apples to apples. If the mean employment density remains the same or very close to the same, the county's employment trend or historical employment dynamic remains unchanged. Therefore, the only change in employment results from a selection of location—within zones 9 through 6. These zones are rewarded increased density because of their proximity. The remaining proximity zones 5 through 1 are then penalized with density reductions because of low proximity. The county planning department would also view the penalty as an increased use of land for employment in those zones (that is, more acres required per employee).

Table 6.11 presents the mean normalized adjusted employment factor (MNAEF). It is calculated by normalizing the employment base factor (EBF) by its mean value. Employment base factors (column A) are created based on the concept that employment will drop to one-third of its value for zone 1 and increase by a factor of 3 for zone 9. The values increase in a nonlinear curve from one-third of the trend base employment to three times the trend base employment. A problem arises with this technique because the mean value of employment will increase on average for the county. Therefore, normalize the factor to keep the trend mean equal to the zone adjusted mean. The result is the MNAEF (column B).

Table 6.11. Base employment proximity factor normalized to keep the mean base employment densities consistent with trend employment densities

Proximity Zone	Employment Base Factor (A)	Mean Normalized Adjusted Employment Factor (B)
1	0.333	0.251848739
2	0.400	0.302521008
3	0.500	0.378151261
4	0.667	0.504453782
5	1.000	0.756302521
6	1.500	1.134453782
7	2.000	1.512605042
8	2.500	1.890756303
9	3.000	2.268907563
Mean Normalizing Factor	1.322	1.000000000

Table 6.12 presents the employment densities for each zone by category. To get the employment for each category, multiply MNAEF by the trend base employment for each zone. For example, the trend industrial base employment is 3.23 employees per acre, and MNAEF for zone 4 is 0.504453782. So the zone employment per acre is 1.629 employees per acre for zone 4.

Table 6.12. Adjusted employment densities by proximity zone for the five major urban employment categories

Proximity Zone	Adjustment Factor from Column (B) Table 6.11	Industrial Base = 3.23	Commercial Base = 7.14	Service Base = 9.73	Retail Base = 6.26	Institutional Base = 2.84
1	0.251848739	0.813	1.798	2.450	1.576	0.7152
2	0.302521008	0.977	2.160	2.943	1.893	0.8591
3	0.378151261	1.221	2.700	3.679	2.367	1.0739
4	0.504453782	1.629	3.601	4.908	3.157	1.4326
5	0.756302521	2.442	5.400	7.358	4.734	2.1478
6	1.134453782	3.664	8.100	11.038	7.101	3.2218
7	1.512605042	4.885	10.800	14.717	9.468	4.2957
8	1.890756303	6.107	13.500	18.397	11.836	5.3697
9	2.268907563	7.328	16.200	22.076	14.203	6.4436
Mean		3.230	7.140	9.730	6.260	2.8400

Step 8: Allocating industrial employment

As in chapter 5, the allocation of industrial employment begins with infill allocation. Query 6.7 selects vacant industrial property within zones 9 through 6 and within the city boundary of Plant City. The query selects 1,799.24 acres and assigns 10,071 industrial employees within that acreage. The industrial employment density is 5.6 industrial jobs per acre.

Query 6.7.
"DESCRIPT" = 'VACANT INDUS' AND ("CNTERSLCE19" >= 6 OR "CITYNAME" = 'Plant City').

Next, the redevelopment CEM is used to allocate redevelopment industrial jobs. In chapter 5, query 5.3 selected the best available employment allocations for industrial redevelopment based on the age category, parcel size, and location. Query 6.5 will use the existing redevelopment allocation from chapter 5 ("ALLYRPOP" = 2045) but modify it ("ALLYRIND" = 2045) so that the selected areas are within zones 9 through 6 or in Plant City. Therefore, the selection set for query 6.8 comprises 4,410 industrial jobs on 1,852.83 acres. The average employment density is 2.4 jobs per acre.

Query 6.8.
`"ALLYRIND" = 2045 AND ("CNTERSLCE19" >= 6 OR "CITYNAME" = 'Plant City')`.

The final query for industrial employment must allocate greenfield industrial job acres. Query 6.9 again follows the concept of selecting acreage for industrial jobs within the area defined by zones 9 through 6. Industrial suitability is moderately high to high (`"INDSUIT100I" > 300`). The selection places 10,583 jobs within 2,309.62 acres. It produces an average employment density of 4.58 jobs per acre (figure 6.4).

Query 6.9.
`"ALLYRIND" = 2045 AND "CNTERSLCE19" >= 6 AND "INDSUIT100I" > 300`.

The combined selections for queries 6.7, 6.8, and 6.9 result in the placement of 25,334 total industrial jobs on 5,961.69 acres (table 6.13, scenario 5). Figure 6.4 shows the industrial allocations for scenario 5.

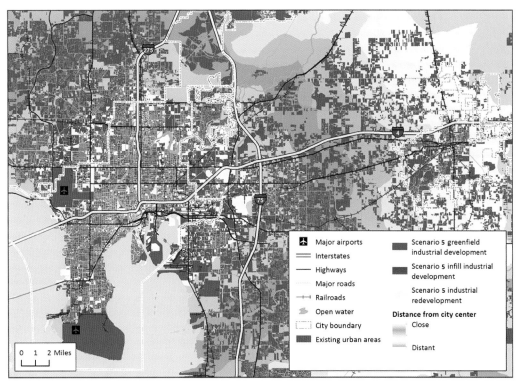

Figure 6.4. Allocated acres for industrial employment are selected based on available infill and redevelopment areas far from the city center.

Figure from Paul D. Zwick and the Arizona Board of Regents on behalf of the University of Arizona. Airports from the Federal Aviation Administration and the Research and Innovative Technology Administration Bureau of Transportation Statistics National Transportation Atlas Databases (NTAD) 2012; water bodies from the US Geological Survey; city and state boundary from US Census Bureau TIGER/Line Files; interstates, highways, and major roads from the Florida Department of Transportation; railroads from the Federal Railroad Administration; parcels from the Florida Department of Revenue.

Commercial, retail, multifamily residential mixed-use employment

After allocating industrial jobs and their associated lands, allocate the redevelopment employment mixed-use allocations. The first allocation, query 6.10, selects locations previously selected for mixed-use residential development. The locations must also possess high commercial and retail redevelopment preference.

> **Query 6.10.**
> ("S3ALLYR" = 20451 OR "S3ALLYR" = 20452) AND ("REDEVCONFLICT" = 232 OR "REDEVCONFLICT" = 233 OR "REDEVCONFLICT" = 322 OR "REDEVCONFLICT" = 323 OR "REDEVCONFLICT" = 331 OR "REDEVCONFLICT" = 332 OR "REDEVCONFLICT" = 133 OR "REDEVCONFLICT" = 333) AND "CNTERSLCE19" >= 6.

Query 6.10 identifies previously allocated mixed-use residential (S3ALLYR). The selection values 20451 and 20452 represent the first two mixed-use residential allocations (queries 6.1 and 6.2a and b). The item REDEVCONFLICT is used to select mixed-use commercial and retail opportunity using the values 232, 233, 332, 323, 332, or 333. The REDEVCONFLICT values 133 and 331 select locations that are suitable for multifamily residential with moderate or high retail or commercial preference, respectively. The item CNTERSLCE19 again is used to select locations within the zones nearest the employment center. Query 6.10 locates 29,105 commercial jobs and 23,785 retail jobs on 6,435.34 acres of redevelopment lands. The employment density is 8.22 employees per acre. These 6,435.34 acres also have residential units on the property. The residential units may be multistory, town house, or clustered housing units, depending on the architectural design of the development. Also, the residential unit density and employment density may vary as long as the mean densities do not fall below the allocated densities for the individual zones.

The next allocation must locate the remaining commercial and retail jobs. Query 6.11 selects 19,750 commercial jobs and 6,955 retail jobs on 1,853.56 acres. The combined employment density is 14.41 employees per acre (figure 6.5).

> **Query 6.11.**
> "ALLYRCOMR" = 2045 AND "ALLYRRET" = 2045 AND "CNTERSLCE19" = 9 AND "COMSUIT100I" >= 400 AND "RETSUIT100I" >= 400.

Step 9: Allocating service and institutional employment

Now that the industrial, commercial, and retail employment has been allocated, next up is the service and institutional employment. Query 6.12 selects the redevelopment service employment locations in zones 9 through 6 with high service suitability. These locations were allocated for the trend scenario as well. The selection excludes locations previously allocated for industrial, commercial, retail, or residential in allocation scenario 5.

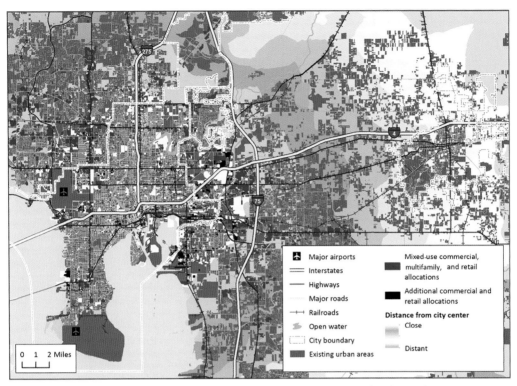

Figure 6.5. Mixed-use commercial, multifamily, and retail allocations. The bright-purple areas represent the acreage of mixed-use residential employment allocations, while the black areas are the additional non-mixed-use commercial (multifamily or single-family) residential, and retail allocations.

Figure from Paul D. Zwick and the Arizona Board of Regents on behalf of the University of Arizona. Airports from the Federal Aviation Administration and the Research and Innovative Technology Administration Bureau of Transportation Statistics National Transportation Atlas Databases (NTAD) 2012; water bodies from the US Geological Survey; state boundary from US Census Bureau TIGER/Line Files; interstates, highways, and major roads from the Florida Department of Transportation; railroads from the Federal Railroad Administration; parcels from the Florida Department of Revenue.

Query 6.12.
```
"CNTERSLCE19" >= 6 AND "ALLYRSERVR" = 2045 AND "S5INDYR" = 0 AND
"S5COMYR" = 0 AND "S5RETYR" = 0 AND "S5ALLYR" = 0.
```

Query 6.12 allocates 58,590 service jobs within 5,891.46 acres. The employment density is 9.8 jobs per acre.

Next, the second part of the process allocates institutional acres within zones 9 through 6 that were originally allocated for institutional jobs. Query 6.13 also excludes any acres previously allocated in scenario 5 for industrial, commercial, retail, or service employment and any acres previously allocated for residential development.

Query 6.13.
```
"S5INDYR" = 0 AND "S5COMYR" = 0 AND "S5RETYR" = 0 AND "S5SERYR" = 0
AND "CNTERSLCE19" >= 6 AND "S5INSYR" = 2045.
```

The allocation selects 3,343.41 acres and 20,298 institutional jobs. The average employment density is 6.07 institutional jobs per acre. However, because the query does not restrict the selection for S5ALLYR, it also colocates 8,338 of these institutional jobs with residential development on 2,448.76 acres.

The third part of the service and institutional employment allocation process selects the rest of the service jobs required for the county by choosing places where the trend scenario allocated infill service employment jobs but only within the area of zones 9 through 6. Query 6.14 selects 28,217 service jobs within 1,769.18 acres. The average service employment density is 15.9 jobs per acre.

Query 6.14.
"S5INDYR" = 0 AND "S5COMYR" = 0 AND "S5RETYR" = 0 AND "CNTERSLCE19" >= 6 AND "S5ALLYR" = 0 AND "S5SERYR" = 2045 AND "SERSUIT100I" > 454.

The service suitability value was greater than 454, which selects highly suitable employment opportunity. The entire service employment selection (86,807 jobs) is therefore accomplished within zones 9 through 6 and within redevelopment opportunity or in infill locations.

The final part of step 9 of the employment allocation is to select the remaining institutional jobs.

Query 6.15.
"CNTERSLCE19" >= 6 AND "S5INDYR" = 0 AND "S5INDYR" = 0 AND "S5COMYR" = 0 AND "S5RETYR" = 0 AND "S5SERYR" = 0 AND "S5INSYR" = 0 AND "INSSUIT100I" >= 461.

Query 6.15 finds 19,335 institutional job locations on 4,496.67 acres. The institutional employment density is 4.3 jobs per acre. The entire institutional employment selection (47,971 jobs) is therefore made within zones 9 through 6. All are within infill opportunity or in greenfield locations with better proximity to the center of county employment activities (figure 6.6).

Table 6.13 compares trend employment acreage versus scenario 5 employment acreage allocations. As the table indicates, the use of mixed-use development opportunity saved 13,090.32 acres, or 20.45 square miles, of productive agricultural lands or ecologically significant habitat (45,380.39 trend acres − 32,290.07 scenario 5 acres).

The increased-density opportunities developed for allocation scenario 5, combined with the requirement that development remain within zones 9 through 6, work to save productive lands and decrease urban sprawl. The 12,412.82 acres, or 19.4 square miles, of reduced residential sprawl development and the 13,090.32 acres, or 20.45 square miles, of reduced employment sprawl allocations as a result of the same population growth and job production scenario represent a significant decrease in urban/suburban development (figure 6.7).

Ultimately, scenario 5 saved approximately 40 square miles of productive agricultural lands or significant native habitat in the study area. However, there is a significant caveat

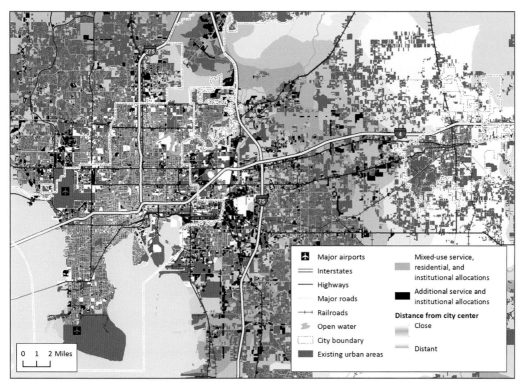

Figure 6.6. Mixed-use service, residential, and institutional allocations. The orange areas represent the acreage of mixed-use service, residential, and institutional employment allocation while the black areas are the additional non-mixed-use service, (multifamily or single-family) residential, and institutional allocations.

Figure from Paul D. Zwick and the Arizona Board of Regents on behalf of the University of Arizona. Airports from the Federal Aviation Administration and the Research and Innovative Technology Administration Bureau of Transportation Statistics National Transportation Atlas Databases (NTAD) 2012; water bodies from the US Geological Survey; state boundary from US Census Bureau TIGER/Line Files; interstates, highways, and major roads from the Florida Department of Transportation; railroads from the Federal Railroad Administration; parcels from the Florida Department of Revenue.

present in scenario 5. It is the traffic congestion that might occur as a result of this scenario. Testing or modeling transportation alternatives is outside the scope of this book. However, a joint team of researchers and professional transportation and land-use consultants tested the concept of scenario 5 mixed use for transit-oriented development (TOD). They used elevated bus toll lanes in Hillsborough County (figure 6.8). The Tampa Hillsborough Expressway Authority project (2013) used LUCIS^{plus} modeling to analyze the land-use patterns for a proposed transit network of bus toll lanes. The project would be constructed using private funding and supported through bus fees and automobile tolls. The land-use and transportation modeling indicated that (1) roadway congestion in the area would be dramatically reduced, (2) transit access to the City of Tampa would be greatly improved, (3) land-use densification would reduce urban sprawl, and (4) future maintenance cost would be covered by user fees and tolls, thus removing public costs for maintenance of this infrastructure.

Table 6.13. Basic trend employment acreage allocations for 2045 produced in chapter 5

Category	Industrial	Commercial	Retail	Institutional	Service	Row Totals
Trend Employment Acreage Allocations						
Greenfield Acres	4,311.86	3,792.64	2,720.84	9,359.41	4,960.53	**25,145.28**
Redevelopment Acres	1,852.83	1,629.72	1,169.16	4,021.79	2,131.57	**10,805.07**
Infill Acres	1,617.04	1,422.32	1,020.38	3,509.99	1,860.31	**9,430.04**
Trend Acreage Totals	**7,781.73**	**6,844.68**	**4,910.38**	**16,891.19**	**8,952.41**	**45,380.39**
Scenario 5 Employment Jobs Allocations						
New Scenario 5 Mixed-Use Employment 2045 (Jobs)	25,334	48,855	30,740	47,971	86,807	**239,707**
Scenario 5 Employment Acreage Allocations						
Greenfield Acres	2,309.62	0	0	4,496.67	0	**6,806.29**
Redevelopment (Commercial & Retail) Acres	0	6,435.34*		0	0	**6,435.34**
Redevelopment Acres	1,852.83	0	0	0	5,981.46	**7,843.29**
Infill Acres	1,799.24	1,853.56	0	3,343.41 2,448.76*	0	**9,444.97**
Infill (Service & Institutional) Acres	0	0	0	0	1,769.18	**1,769.18**
Trend Reduction Acreage Created by Scenario 5 Employment Allocations						
Scenario 5 Mixed-Use Employment Total Acres	5,961.69	8,288.90		10,288.84	7,750.64	**32,290.07**
Scenario 5 Mixed-Use Acreage Reduction	1,820.04	3,466.16		6,602.35	1,201.77	**13,090.32**

*Mixed-use residential and employment by category. Cells showing cross-employment categories have a mixture of employment for the categories identified.

Chapter 6 Identifying and mapping an alternative urban mixed-use opportunity

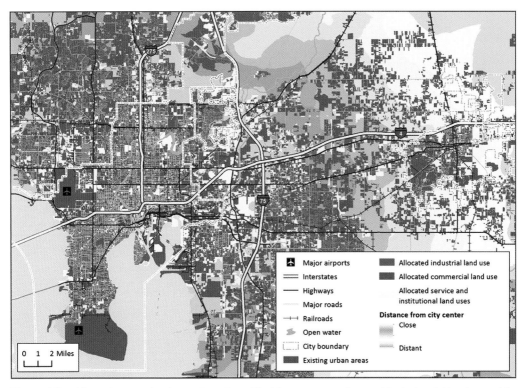

Figure 6.7. All employment land-use allocations. The blue areas are allocated for industrial land use, while the red areas are allocated for commercial and retail land use. The yellow areas are allocated for institutional and service land use.

Figure from Paul D. Zwick and the Arizona Board of Regents on behalf of the University of Arizona. Airports from the Federal Aviation Administration and the Research and Innovative Technology Administration Bureau of Transportation Statistics National Transportation Atlas Databases (NTAD) 2012; water bodies from the US Geological Survey; state boundary from US Census Bureau TIGER/Line Files; interstates, highways, and major roads from the Florida Department of Transportation; railroads from the Federal Railroad Administration; parcels from the Florida Department of Revenue.

Figure 6.8. The Tampa Hillsborough Expressway Authority website presents reports on the use of bus toll lanes to fulfill TOD goals.

Tampa Hillsborough Expressway Authority.

Chapter summary

The following points are discussed in this chapter:

- The CEMs in LUCIS^{plus} can test land-use scenarios for redevelopment, infill development, employment, and regional urban form. The results illustrate increased or saved land and adjusted densities based on the conditions used during the allocation. Conditions can include specific jurisdictions, distance from urban centers, and so forth.
- Manipulating densities and employing contemporary planning methods such as TOD, mixed use, and increased densities near urban centers can reduce the physical impacts from lower-density development patterns and ultimately sprawl development. These strategies can also help maintain employment allocation goals.

ArcGIS tools referenced in this chapter

Tool Name	Version 10.2 Toolbox/Toolset
Median Center	Spatial Statistics Tools/Measuring Geographic Distributions

References

Oates, David. 2006. *The Oregon Encyclopedia*. Portland, OR: Portland State University. http://oregonencyclopedia.org/articles/urban_growth_boundary/#.VO-Xj_nF-JU.

Online resources

Tampa Hillsborough Expressway Authority. 2013. Bus Toll Lanes Documentation. http://www.tampa-xway.com/Projects/BusTollLanes.aspx.

Chapter 7

Analyzing and mapping conservation and agriculture preservation and protection

Paul D. Zwick

What this chapter covers

Chapters 4 and 5 developed trend allocations for residential and employment development in the study area of Hillsborough County, Florida. These trend allocations set the stage for a comparison with the chapter 6 increased-density and urban mixed-use scenarios that showed urban/suburban densification can decrease sprawl development. In fact, allocation scenario 5 in chapter 6 decreases urban/suburban sprawl. It also removed the necessity for developing an additional 40 square miles of agriculturally productive lands. This chapter examines opportunities to integrate conservation with environmental planning to preserve and protect quality habitats, agricultural productivity, and scenic views.

Criteria to be evaluated

The following list of criteria is used for allocating agriculture and conservation lands based on land-use selections structured in ArcGIS query language:

- *ALLYRPOP*: Allocation Year for Residential Population. This value equals 0 if the selected cells have not been previously allocated.
- *RESFSUIT100I*: Combined Multifamily and Single-Family Residential Suitability. This value is reclassified for integers ranging from 100 to 900. This reclassification is

completed to provide a sufficient range of suitability for the criteria evaluation matrix (CEM). For example, a floating point suitability number 2.3456 would be reclassified to an equivalent three-digit integer value of 235.

- *CNTERSLCE19*: Distance Zone from Major Center City. This Item is used to indicate a zone of distance from the center of employment activity. The value ranges from 1 to 9, in which 9 represents the closest zone of distance.
- *S5ALLYR*: Allocation Year Residential Population for Scenario 5. This value equals 0 if the selected cells have not been previously allocated.
- *S5INDYR*: Allocation Year Industrial Employment for Scenario 5. This value equals 0 if the selected cells have not been previously allocated.
- *S5COMYR*: Allocation Year Commercial Employment for Scenario 5. This value equals 0 if the selected cells have not been previously allocated.
- *S5RETYR*: Allocation Year Retail Employment for Scenario 5. This value equals 0 if the selected cells have not been previously allocated.
- *S5INSYR*: Allocation Year Institutional Employment for Scenario 5. This value equals 0 if the selected cells have not been previously allocated.
- *S5SERYR*: Allocation Year Service Employment for Scenario 5. This value equals 0 if the selected cells have not been previously allocated.
- *ACRES*: Available Allocation Acres. This item contains the number of acres represented by the raster zone. A raster zone is a record within an integer raster that contains all the cells in a raster that have the same value (for example, all the cells that represent single-family residential land use).
- *GFCONFLICT*: LUCIS Conflict/Preference. The greenfield conflict index of LUCIS conflict/preference categories contains 27 unique values that represent the conflict level between the major land-use types: agriculture, conservation, and urban.
- *REDEVCONFLICT*: Redevelopment Mixed-Use Opportunity. This item is structured just like the GFCONFLICT index of LUCIS preference/conflict categories. However, it represents redevelopment opportunity (in this case, for commercial, multifamily residential, and retail mixed use). It has the same 27 index values.
- *CITYNAME*: City Name. This item represents the city in which the cell is located: (1) Tampa, (2) Temple Terrace, (3) Plant City, or (0) in the county.
- *DESCRIPT*: Property Use Description. Examples include Vacant Commercial, Vacant Industry, Misc. Agriculture, and so forth.
- *INSSUIT100I*: Institutional Land-Use Suitability. This is the same as residential suitability but for locations supporting institutional employment.
- *COMSUIT100I*: Commercial Land-Use Suitability. This is the same as residential suitability but for locations supporting commercial employment.
- *RETSUIT100I*: Retail Land-Use Suitability. This is the same as residential suitability but for locations supporting retail employment.
- *INDSUIT100I*: Industrial Land-Use Suitability. This is the same as residential suitability but for locations supporting industrial employment.
- *SERSUIT100I*: Service Land-Use Suitability. This is the same as residential suitability but for locations supporting service employment.

Acquisition and land-use options for conservation or agriculture preservation

J. Randolph (2004, 88–90) provides an excellent discussion on land conservation in "Rights and Interests in Property That Can Be Acquired" and "Techniques for Acquiring Land Title or Property Rights by Land Trust or Agency." His presentation includes a discussion of the advantages and disadvantages of each option, which is summarized as follows.

Randolph (2004) begins with a brief discussion of *fee simple acquisition*, which is the purchase of the land area of interest and makes the state or land trust the owner of the property. Randolph describes fee simple acquisition as providing all the rights of, and guaranteeing protection of, ownership. The advantages include full control of the property, full access to the property, and permanent protection of the property. Disadvantages include a costly purchase price. Once purchased for conservation, the property is removed from the tax base. He advises that fee simple acquisition should be used for large tracts of land at a lower cost per acre.

Conservation easements or *development rights* restrict the property owner from specified development uses. Conservation easements and development rights become part of the property deed and remain intact as the land is bought and sold. A partial interest in the property is transferred to the nonprofit or government agency, either by gift or purchase. Advantages include (1) the easement is less expensive than fee simple purchase, (2) the easement can be tailored to the protection desire of the owner and the property, (3) the landowner retains ownership, and (4) the property remains as part of the tax base but at a reduced rate because of restrictions. Disadvantages include (1) reduced, restricted, or no public access; (2) restricted use must be enforced; and (3) resale value may be lowered because of development restrictions. This option is well suited for agriculture preservation.

Purchase with lease-back has a more limited use than the previous options. It results in the land being purchased by a conservation entity and then leased back to the seller, subject to restrictions of use. Advantages include income generated from leasing for the conservation entity and removal of the responsibilities of management and liability, which are assigned to the lessee. Disadvantages include the possible restriction of public access, and land-use availability must be appropriate for lessee use such as agriculture.

Lease, as described by Randolph (2004), is the short- or long-term rental of land for conservation. The advantages include a low cost for the use of the land, and the landowner receives income and retains control of the property. A lease provides some type of conservation or preservation option. The disadvantage results primarily from the conservation entity's lack of equity in the property and limited control of the property's use.

Undivided interest is the divided ownership of the property among multiple owners. Either each owner has a fractional ownership in the property, or each owner has equal ownership. With equal ownership, one owner is prevented from acting on a land-use development option without consent of the other owners. The disadvantage arises from the complications of property management issues, including payment of property taxes.

Fee simple acquisition is the most costly form of acquisition. Eminent domain and foreclosure are less expensive forms of acquisition but are restricted to use by government agencies. These actions also have political repercussions. Thus, donations make a good source of conservation lands. Randolph (2004, 89) states: "Bargain sales and donations save considerable expenses by the trust or agency and can provide large tax benefits. Another option for acquisition is a bequest or donation with reserved life estate, which are similar because the trust or agency doesn't gain use of the property until the death of the landowner."

Green infrastructure and land-use planning

Green infrastructure can mean different things to different people. However, as the following examples indicate, green infrastructure is regarded as a connected network of linkages and conservation hubs for the protection of local, state, and regional natural resources, including water, species, and habitat, and the provision of recreational opportunity within the networked lands.

- The US Environmental Protection Agency (EPA) definition of green infrastructure is primarily based on water quality and storm water management for sustainable communities. Green infrastructure is the general name given to methods that use environmentally friendly techniques to manage storm water and support sustainable communities. Unlike single-purpose gray storm water infrastructure, which uses pipes to dispose of rainwater and relies on dilution as the solution to pollution, green infrastructure uses vegetation and soil to manage rainwater where it falls. By weaving the natural process of water movement and percolation into the built environment, green infrastructure provides storm water management, flood mitigation, air quality management, and more (EPA 2014).
- The Florida Greenways and Trails System, managed by the Office of Greenways and Trails in the Florida Department of Environmental Protection, was established to develop a statewide system of greenways and trails for recreational and conservation purposes. The Florida Greenways and Trails System envisions that the statewide network of greenways and hubs will help conserve wildlife and protect Florida's native biological diversity. The system also offers multiuse trails over the length and breadth of the state. These trails promote appreciation of the state's natural and working landscapes, provide routes for alternative transportation, and protect cultural and historical sites (Florida Department of Environmental Protection 2013).
- T. Hoctor and J. Wood (2002), in a report titled *Identification of Critical Linkages within the Florida Ecological Greenways Network*, provide a historical summary of the project's beginning and the University of Florida's role in the development of the plan. Since 1995, the University of Florida has been working with the Florida Department of Environmental Protection to help develop the Florida Greenways and Trails System Plan. The University of Florida was asked to develop a decision support model to

help identify the best opportunities to protect ecological connectivity statewide. A research team from the university used geographic information system (GIS) software to analyze all the best available data on land use and significant ecological areas, including habitats for native species, natural communities, wetlands, roadless areas, floodplains, and aquatic ecosystems. The process then integrated all this information to identify a statewide Ecological Greenways Network. The network contains all the largest areas of ecological and natural resource significance and the landscape linkages necessary to link these areas together in one functional statewide network. The process was collaborative and overseen by three separate state-appointed greenway councils. During the development of the model, researchers obtained technical input from the Florida Greenways Commission; Florida Greenways Coordinating Council; state, regional, and federal agencies; scientists; university personnel; conservation groups; planners; and the general public in over 20 sessions. When the modeling was complete, public meetings statewide thoroughly reviewed the results as part of the development of the Greenways Implementation Plan. The plan was completed in 1999. The results indicated that approximately 50 percent of the state is potentially suitable for inclusion within a statewide ecological greenway system (Hoctor et al. 2000). To focus protection efforts, the University of Florida research team has worked on developing and applying a process to assess the relative significance of features within the Florida Ecological Greenways Network.

- The Maryland Greenways Commission (2000) developed the 2000 edition of the *Maryland Atlas of Greenways, Water Trails, and Green Infrastructure*. A fruitful partnership with several units within the Department of Natural Resources' Chesapeake and Coastal Watershed Services has greatly expanded this edition from previous versions. Most significant is the addition of *green infrastructure maps*. Unlike the greenway maps, which focus on protected lands, green infrastructure maps are based on a statewide scientific and systematic assessment of ecologically significant lands, regardless of ownership or protection status. These maps can include greenways, agricultural lands, state-owned conservation lands, and open space within a community. The green infrastructure shows a larger context from which a protected greenway system can be carved.
- Greenways are natural corridors set aside to connect larger areas of open space. Greenways provide for the conservation of natural resources, protection of habitat, and movement of plants and animals. They offer opportunities for linear recreation, alternative transportation, and nature study. Maryland has over 1,500 miles of protected greenway corridors, including over 600 miles of trails. Most of the land in the system is publicly owned. All Maryland greenways provide some level of ecological benefit, and most serve multiple purposes. Especially in urban areas, Maryland's stream valley parks tend to take on multiple functions. They act as stream buffers, flood control areas, wildlife corridors, and recreation areas. However, most corridors can be classified as primarily ecological or recreational (Maryland Greenways Commission 2000).

- Maryland's land conservation GreenPrint Program identifies natural unprotected lands in the state. It links these lands through a system of corridors and saves them through targeted acquisitions and easements (Maryland DNR 2014). The state has used general obligation bonds to fund its GreenPrint Program protecting the state's green infrastructure. It allocates 25 percent of the funds to the Maryland Agricultural Land Preservation Foundation (MALPF) to protect green infrastructure in established agricultural districts. The original goal of the program was to increase the amount of protected green infrastructure network land by 10,000 acres per year.
- Maryland uses its Program Open Space funding to buy some lands in the green infrastructure network model. Land purchases through this program account for the bulk of the more than $1.2 billion Maryland has spent on public land protection since 1969. The program is funded through an annual appropriation by state legislators. The money is derived mostly from the state real estate transfer tax and occasional supplements from state general obligation bonds. Most funding is split roughly in half between state acquisitions and funding to each of Maryland's counties and the City of Baltimore (Benedict and Drohan 2004).

Why green infrastructure?

One of the best descriptions in support of green infrastructure is provided in "Maryland's Green Infrastructure Assessment and GreenPrint Program" in the Green Infrastructure—Linking Lands for Nature and People Case Study Series (Benedict and Drohan 2004). The series describes Maryland's green infrastructure network as providing the following benefits:

- It reverses past trends of "haphazard conservation" by identifying for state agencies, land planners, citizens, and developers the most valuable and vulnerable lands for protection. It also leverages public and private investments in land conservation.
- It provides a focal point to coordinate existing conservation programs with one another and increase their overall effectiveness.
- It conserves and connects large contiguous land areas with multiple natural resource features.
- It ensures the preservation of natural resources in each region that help clean the air and water.
- It provides urgently needed extra funding so that agencies can act immediately to protect vulnerable lands.
- It addresses commitments in the Chesapeake Bay Agreement to protect 20 percent of the watershed and reduce the rate of sprawl development by 30 percent.
- It enhances property values.
- It produces a tangible improvement in the quality of life.

- It supports the diverse economy of the state, especially natural resource-based industries such as fisheries, forestry, and tourism, and the jobs they create.
- It identifies and protects lands that
  - serve as natural filter systems for trapping pollutants before they reach the Chesapeake Bay;
  - provide cover and passage for wildlife;
  - supply a "greenway" for the enhancement of biological diversity through the provision of important wildlife habitat and corridors that link existing habitat areas;
  - serve as an outdoor classroom for teaching about Maryland's natural environment; and
  - provide public access to, and recreational opportunities in, the natural world, including the Chesapeake Bay.

LUCIS ecological significance goals

LUCIS uses the identification of conservation opportunity as part of its ecological significance models. The models to evaluate ecological significance include the following major goals:

- Support biodiversity (species and habitat)
- Improve water quality (surface and subsurface)
- Protect ecological processes (fire, floodplain, and wetlands)
- Enhance existing connectivity between conservation areas

Each of the major goals relies heavily on work completed for the Critical Lands and Waters Identification Project (CLIP) (Oetting et al. 2012). CLIP is a GIS database and analysis of statewide natural resource conservation priorities in Florida. CLIP was originally designed as a tool for identifying the state's most important conservation priorities. It supports the efforts of the Century Commission for a Sustainable Florida and the Florida Fish and Wildlife Conservation Commission's Cooperative Conservation Blueprint (CCB). The results and methodology of the CLIP Version 2.0 update in 2011 are summarized as follows. The full technical report, data tutorial, data request form, and online map viewer are available on the Florida Natural Areas Inventory website. CLIP data is also available for free download from the Florida Geographic Data Library. (See "Online resources" at the end of this chapter.)

Figure 7.1 shows the final ecological significance suitability raster layer generated from the combination of the four ecological significance goals. The ecological significance suitability values range from the standard LUCIS 1 to 9 (red to green), in which 9 is the highest suitability. The raster layer combines the highest values for each of the four goal

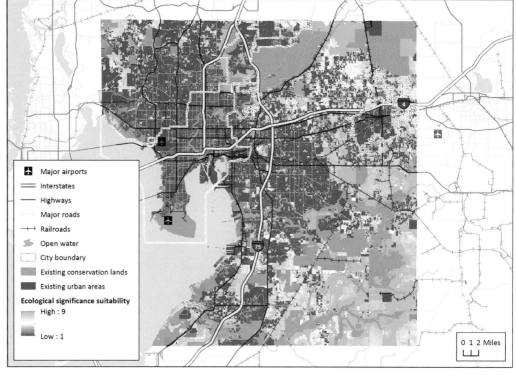

Figure 7.1. Ecological significance suitability ranging from red (lowest suitability) to green (highest suitability). The dark-gray areas are existing urban lands while the orange areas are existing conservation lands.

Figure from Paul D. Zwick and the Arizona Board of Regents on behalf of the University of Arizona. Airports from the Federal Aviation Administration and the Research and Innovative Technology Administration Bureau of Transportation Statistics National Transportation Atlas Databases (NTAD) 2012; water bodies from the US Geological Survey; state boundary from US Census Bureau TIGER/Line Files; interstates, highways, and major roads from the Florida Department of Transportation; railroads from the Federal Railroad Administration; parcels from the Florida Department of Revenue.

layers. This method was chosen because the ecological significance of an area is based on the highest significance in the process.

Equal-interval reclassification converts the ecological significance suitability index into collapsed preference. Figure 7.2 shows the ecological significance preference raster layer using the standard LUCIS values of 1 to 3 (red to green), in which 3 is the highest preference. LUCIS conflict is the combination of agriculture, conservation or ecological significance, and urban preference. In this chapter, the LUCIS conflicts of interest are the values for agriculture and ecological significance. However, urban preference plays an important role in determining whether there is an urban development pressure for areas considered to be important for conservation or agriculture preservation opportunity.

Figure 7.2 shows that, as expected, the ecologically significant areas are all within the greenfield areas. They are located in the eastern, outlying areas of the county, with a distinct separation created by the Interstate 4 corridor between the cities of Tampa and Plant City.

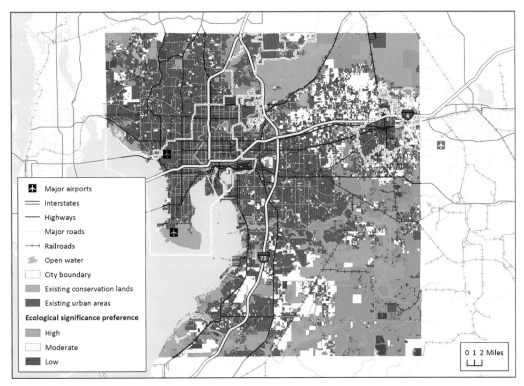

Figure 7.2. Ecological significance collapsed preference ranging from 1 (red—lowest preference) to 3 (green—highest preference). The dark-gray areas are existing urban lands while the orange areas are existing conservation areas.

Figure from Paul D. Zwick and the Arizona Board of Regents on behalf of the University of Arizona. Airports from the Federal Aviation Administration and the Research and Innovative Technology Administration Bureau of Transportation Statistics National Transportation Atlas Databases (NTAD) 2012; water bodies from the US Geological Survey; state boundary from US Census Bureau TIGER/Line Files; interstates, highways, and major roads from the Florida Department of Transportation; railroads from the Federal Railroad Administration; parcels from the Florida Department of Revenue.

Using LUCIS to identify conservation opportunities in the landscape

As the previous section, "LUCIS ecological significance goals," indicated, the protection of lands for conservation is a complex undertaking. It involves many methods of acquisition; various management techniques; and many complex interactions with local economies, including impacts to the tax base. This chapter does not address all these implications but focuses on the land use or property identification of areas with high ecologically significant opportunity. LUCIS has the unique ability to easily identify ecologically significant areas that are either in conflict or have no conflicts or have a special opportunity to coexist with other major land-use categories because of the greenfield conflict values in the LUCIS greenfield

CEM. Within the greenfield CEM, LUCIS has a variable GFCONFLICT that identifies land-use conflict among three major land-use categories: (1) agriculture, (2) conservation or ecological significance, and (3) urban. This section emphasizes conservation. Conservation planning looks at large areas of space with significant ecological preference. Query 7.1 selects lands that possess a high or moderate LUCIS ecological significance preference with no conflict identified. In the LUCIS GFCONFLICT preference values, the middle digit is the largest value of the three digits: 121, 131, 132, 231, and 232. No-conflict areas are identified by a GFCONFLICT value in which one digit contains the highest preference value and its value is not the same as either of the other two digits. Figure 7.3 shows the areas that query 7.1 selects.

Query 7.1.
```
("GFCONFLICT" = 121 OR "GFCONFLICT" = 131 OR "GFCONFLICT" = 132 OR "GFCONFLICT" = 231 OR "GFCONFLICT" = 232) AND "ACRES" > 5.
```

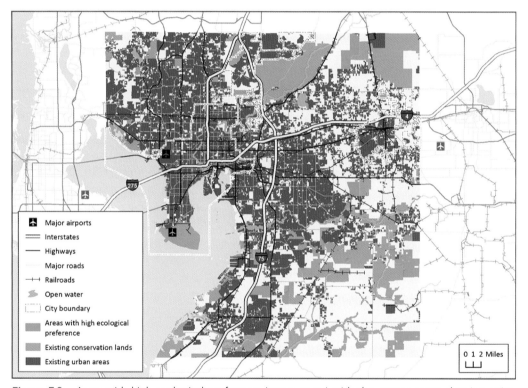

Figure 7.3. Areas with high ecological preference (green areas) with the orange areas showing existing conservation lands and the dark-gray areas showing existing urban/suburban areas. These areas were selected using query 7.1 and selecting GFCONFLICT values in which the middle preference value was the highest among all stakeholders and in which there was no conflict.

Figure from Paul D. Zwick and the Arizona Board of Regents on behalf of the University of Arizona. Airports from the Federal Aviation Administration and the Research and Innovative Technology Administration Bureau of Transportation Statistics National Transportation Atlas Databases (NTAD) 2012; water bodies from the US Geological Survey; state boundary from US Census Bureau TIGER/Line Files; interstates, highways, and major roads from the Florida Department of Transportation; railroads from the Federal Railroad Administration; parcels from the Florida Department of Revenue.

Figure 7.3 illustrates that many locations are next to existing conservation areas. Selecting adjacent areas that are larger than five acres could significantly add to the conservation areas within the county. The selection finds 14,750 acres of land with no LUCIS conflict for conservation use. If the landowners are willing, the county could use the previously discussed fee simple acquisition or conservation easements to add to the conservation land base. If the county could use conservation easements to add the areas selected by query 7.1 to its conservation lands, the lands would remain on the tax rolls. The landowners could continue to use the lands with limited or restricted development use. Because the new areas would typically add to the existing conservation lands, it could improve the opportunity for connecting greenway corridors to benefit both humans and wildlife.

The second query used to identify ecologically significant areas that might add to existing conservation looks for areas of high or moderate ecological significance that are in conflict, either with the agriculture or urban land-use categories.

Query 7.2.
```
"GFCONFLICT" = 221 OR "GFCONFLICT" = 122 OR "GFCONFLICT" = 222 OR
"GFCONFLICT" = 233 OR "GFCONFLICT" = 332 OR "GFCONFLICT" = 331 OR
"GFCONFLICT" = 133 OR "GFCONFLICT" = 333".
```

Query 7.2 finds urban and agricultural lands that have moderate or high preference values in conflict with ecologically significant areas. These values hold significant meaning for conservation efforts. When the middle digit equals the third digit, a conflict exists between the ecological significance of the area and its urban preference. If a governmental organization or nongovernmental organization (NGO) wanted to protect these areas, the value of the land would likely be much greater because of the development pressure present. That same development pressure also indicates an urgency for conservation acquisition or easement. Both would likely be more difficult to obtain if the landowner was swayed by the opportunity for a high sales price or much higher development profit. Conversely, when the first digit and middle digit are equal and represent moderate or high preference values, there is an opportunity for protecting agriculture because the land is already in agricultural production with moderate to high ecological significance. One example might be an agricultural property with high connectivity values for green infrastructure. Perhaps the property contains a hydrologic connection from a river tributary or stream.

Table 7.1 shows that the LUCIS process has selected 20,425.24 acres of potential conservation lands within 462 specific areas (figure 7.4). Within the 20,000-plus acres, 17,539.76 acres have high preference for both agriculture and conservation. They represent 85.9 percent of the identified acres. Therefore, the conservation easement could provide working landscapes juxtaposed with protected conservation areas. These lands will provide a buffer between future development and many locations on the edge of conservation areas. Ultimately, the lands identified might also provide working landscapes within a connected conservation corridor.

Table 7.1. Conservation conflict values with corresponding number of acres and locations

Conflict Description	LUCIS Conflict	Area Count	Acres
Minor Conflict; Moderate Preference Conservation–Urban	122	4	93.04
Major Conflict; Moderate Preference All	222	11	239.00
Minor Conflict; High Preference Conservation–Urban	233	10	129.36
Minor Conflict; High Preference Conservation–Agriculture	331	4	358.25
Minor Conflict; High Preference Conservation–Agriculture	332	310	17,181.51
Major Conflict; High Preference All	333	123	2,424.08
Totals		462	20,425.24

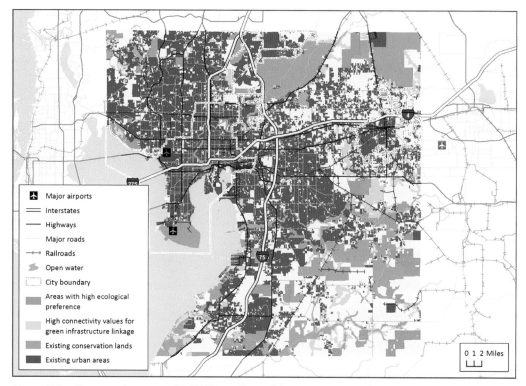

Figure 7.4. Conservation areas with LUCIS conflict. Table 7.1 summarizes the conflict descriptions for 462 areas of conservation opportunities illustrated here within Hillsborough County.

Figure from Paul D. Zwick and the Arizona Board of Regents on behalf of the University of Arizona. Airports from the Federal Aviation Administration and the Research and Innovative Technology Administration Bureau of Transportation Statistics National Transportation Atlas Databases (NTAD) 2012; water bodies from the US Geological Survey; state boundary from US Census Bureau TIGER/Line Files; interstates, highways, and major roads from the Florida Department of Transportation; railroads from the Federal Railroad Administration; parcels from the Florida Department of Revenue.

Agriculture preservation and land-use planning

A joint project of the American Farmland Trust and the Agricultural Issues Center of the University of California, Davis, supported by the Farm Foundation, analyzed five tests of effectiveness for preserving or protecting agricultural farmland from urban influences. The five tests (Sokolow et al. 2006, 4–6) and related measures include:

1. *Whether acres put under easement add up to a substantial portion of a community's total farmland base and significantly advance preservation goals.* Numerical achievements certainly are a sign of positive program impacts. Putting more parcels in a community under easement expands the farmland base that is off-limits to development. Especially if this leads to large blocks of protected land, it increases the probability that individual easement-covered farms will be buffered from incompatible land uses.
2. *Whether easements help assure that land will be retained in agriculture, as measured by resale of protected properties and related land market trends.* This test specifically asks about the affordability of agriculture for encumbered farms put up for sale, the characteristics of the purchasers of such properties (whether farmers or others), and how the properties are then used.
3. *Whether easements help sustain local agricultural economies.* Related to but broader than land market trends, this test deals with the measures of a community's agricultural prosperity. It initially considers the stability of the underlying economic infrastructure; the range of support provided by businesses that service individual farms and commodity trends, which changes over time; and a number of other measures of the economic health of local agriculture.
4. *Whether easement programs positively influence urban land-use patterns.* Moving from the focus on continued agricultural production of easement-covered properties, this fourth test is a more proactive one that concentrates on the sources of the threat to farmland. It asks about the capacity of easements to control or influence the pressures of urbanization, residential, and other nonagricultural demands for farmland. Can easements—either alone or in conjunction with local government planning and land-use regulations—reduce the negative effects of these pressures on agriculture? Can easements help redirect growth; block its expansion; or change its direction, rate, or efficiency?
5. *Whether the short history of the agricultural easement technique to date suggests that the promise of long-term, if not perpetual, preservation of farmland is a credible scenario.* Considering the difficulty of predicting the future, this final test is the least definitive of all five. But there are clues in how the sample programs in the study are prepared or not prepared for the long term, in what managers and others say about program strengths and weaknesses, and, more importantly, in the attention given by programs to postacquisition stewardship work.

The findings that emerge from this study are not uniformly definitive among the five tests, as they apply to the easement programs in the national research sample. Generally, the results of tests 1 and 2 are more conclusive than the results of tests 3, 4, and especially 5. There are two interrelated reasons for these differences. One is the inherent difficulty in isolating the specific impacts of easement programs from other influences on farmland and farming. For example, the prosperity of local agricultural economies (test 3) is affected by powerful forces beyond the control of public conservation efforts. These include global market trends for agricultural commodities, changing economies of scale, and generational patterns in farm families. The second reason for variation among test results is inherent in the limits of the project's scope and research methods. The research is based primarily on the perceptual information generated by phone interviews with program managers and others (Sokolow et al. 2006, 4–6).

Why agriculture preservation?

Many reasons exist to support the preservation of agricultural lands:

- Farmland and ranchland are a finite resource.
- Areas with prime agricultural soils are limited.
- Agriculture is an important part of many state, regional, and local economies.
- Agricultural lands possess substantial market value.
- Growing food locally helps meet local sustainability goals.
- Agriculture is part of the rural lifestyle for many Americans, and is part of America's culture.
- Agriculture may also (when properly managed) offer the potential for groundwater recharge.
- Agriculture also offers environmental benefits as wildlife habitat.
- Agriculture provides jobs, including ancillary jobs in food processing and related industries.
- Agricultural lands have scenic value.
- Agricultural lands pay for the cost of the community services that they require, unlike many urban land uses.
- The protection of agricultural lands promotes sound growth management and sustainable smart growth.

The Farmland Information Center provides valuable statistics about US farmland for the period 1982 to 2007. Table 7.2 shows that 23.2 million acres of land have been lost to development since 1982, representing 3.4 percent of the farmland present in 1982. In 1982, there were 286.4 million acres of prime agricultural land. By 2007, the United States had lost 8.7 million acres of prime agricultural farmland.

Table 7.2. Natural resource inventory for US farmland (from 1982 to 2007)

Category	1997–2002	2002–7	1982–2007
Agricultural land converted to developed land (acres)	5,135,500	4,080,300	23,163,500
Agricultural land at the beginning of the reporting period (acres)	936,151,700	925,082,100	968,342,700
Prime agricultural land converted to developed land (acres)	1,815,700	1,431,100	8,716,600
Prime agricultural land at the beginning of the reporting period (acres)	275,953,800	271,452,200	286,426,200
Rural land converted to developed land (acres)	9,705,500	7,491,300	41,324,800
Rural land at the beginning of the reporting period (acres)	1,391,496,800	1,381,269,800	1,418,965,400
Prime rural land converted to developed land (acres)	2,555,500	2,036,300	11,573,800
Prime rural land available at the beginning of the reporting period (acres)	331,149,200	327,946,900	339,432,200
Total surface area (acres)	1,937,664,200	1,937,664,200	1,937,664,200

Source: Farmland Information Center.

Table 7.3. Purchase of agricultural conservation easements (from 2010 to 2012)

State-Level Program Activity Totals	2010	2011	2012
Easements or restrictions acquired	11,899	12,415	12,970
Acres protected	2,023,230	2,185,996	2,284,005
Program funds spent to date	$3,058,480,491	$3,243,684,871	$3,416,498,572

Source: Farmland Information Center.

Table 7.3 shows that during the period 2010 to 2012, approximately $3.4 billion was spent on the protection of US farmland. It include the acquisition of 12,970 easements and the protection of 2.3 million acres.

Table 7.4 shows US farmland data from the Census of Agriculture. In 2007, there were 2.2 million farms in the United States. These farms occupy approximately 1 billion acres, or 41 percent of the total acres. Interestingly, 57 percent of the farms in 2007 were managed by people age 55 and older. This age bracket indicates that these lands may change ownership within the next 10 to 15 years if the operators retire between the ages of 65 and 70. The market value of farmland in 2007 was one-third of a trillion dollars, and 48 percent of the production was from cropland.

Table 7.4. Census of Agriculture (from 1997 to 2007)

Category	1997	2002	2007
Farms	2,215,876	2,128,982	2,204,792
Land in farms (acres)	954,752,502	938,279,056	922,095,840
Total land area (acres)	N/A	2,263,960,501	2,260,994,361
Full-time operators	1,044,388	1,224,246	993,881
Part-time operators	1,171,488	904,736	1,210,911
Percentage of operators 55 and older	N/A	50	57
Land managed by operators 55 and older (acres)	N/A	472,956,653	527,405,083
Market value of agricultural products sold (× 1,000)	$201,379,812	$200,646,355	$297,220,491
Percentage from crop production		50	47	48

Source: Farmland Information Center.

Clearly, US farmland and green infrastructure represent a significant economic and land-use resource. Farmland and green infrastructure are not only important for food security and wildlife protection, but also for their value to the national economy, and specifically to local rural economies. Traditional urban structure argues that the agricultural and natural hinterlands in the country are as important to an urbanizing nation as any other aspect of urban life—and it is because they constitute the fundamental support system for urban life. Therefore, it is important for the US government and local communities to protect these vital resources. That said, agriculture preservation and the creation of green infrastructure are an integral community land-use issue.

Other resources for information regarding agricultural lands include, but are not limited to, (1) the Bureau of Economic Analysis, (2) the Bureau of Labor Statistics, (3) Foreign Agricultural Trade of the United States data, (4) the Statistical Abstract of the United States, and (5) the US Census Bureau.

LUCIS agriculture goals

Following a process for agriculture or farmland protection described by T. Daniels and K. Daniels (2003) in *The Environmental Planning Handbook for Sustainable Communities and Regions*, LUCIS uses defined goals and objectives to identify, analyze, and form an action strategy for protection. The LUCIS agriculture modeling process has five identification goals for agricultural suitability:

- Lands suitable for row crops
- Lands suitable for livestock (including intensive and nonintensive uses)
- Lands suitable for specialty farms (including orchards)

- Lands suitable for nurseries
- Lands suitable for timber

Each major goal has objectives and subobjectives. For example, the row crop objectives and subobjectives include (1) physical objectives for production rates for 15 crop soils and the identification of prime farmland, (2) proximal objective for accessibility to local and regional markets and transportation accessibility, and (3) land values. Figures 7.5 and 7.6 show the final combined agriculture suitability and collapsed preference, respectively. The highest agriculture preferences are in eastern Hillsborough County. There is a large area of moderate preference in the southeastern corner of the county.

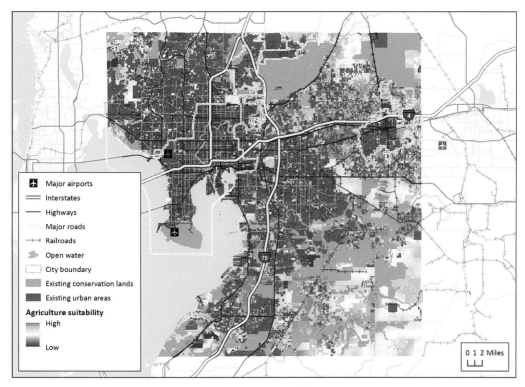

Figure 7.5. Agriculture suitability. Suitability values range from 1 to 9, with 9 (green) being the highest suitability. The orange areas are the existing conservation lands, and the dark-gray areas are existing urban and suburban areas. The highest agriculturally suitable locations are located in the eastern and southeastern areas of Hillsborough County.

Figure from Paul D. Zwick and the Arizona Board of Regents on behalf of the University of Arizona. Airports from the Federal Aviation Administration and the Research and Innovative Technology Administration Bureau of Transportation Statistics National Transportation Atlas Databases (NTAD) 2012; water bodies from the US Geological Survey; state boundary from US Census Bureau TIGER/Line Files; interstates, highways, and major roads from the Florida Department of Transportation; railroads from the Federal Railroad Administration; parcels from the Florida Department of Revenue.

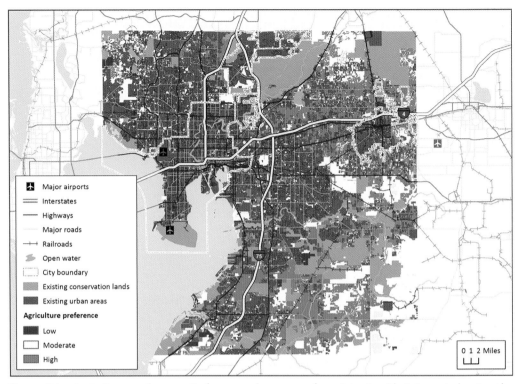

Figure 7.6. Agriculture preference. Preference values range from 1 to 3, with 3 (green) showing the highest preference areas. The orange areas are the existing conservation lands, and the dark-gray areas are existing urban and suburban areas.

Figure from Paul D. Zwick and the Arizona Board of Regents on behalf of the University of Arizona. Airports from the Federal Aviation Administration and the Research and Innovative Technology Administration Bureau of Transportation Statistics National Transportation Atlas Databases (NTAD) 2012; water bodies from the US Geological Survey; state boundary from US Census Bureau TIGER/Line Files; interstates, highways, and major roads from the Florida Department of Transportation; railroads from the Federal Railroad Administration; parcels from the Florida Department of Revenue.

Using LUCIS to identify agriculture lands for the preservation or protection of agricultural productivity and scenic views

In the first half of this chapter, LUCIS identifies lands with high or moderate conservation preference. This section locates lands with high or moderate agriculture preference. Some of these lands are in conflict with either urban or conservation interests, or both. Query 7.3 selects agriculture lands that have high or moderate preference and no conflict. Selected lands are larger than five acres. The requirement of being larger than five acres identifies areas of significance but that are still small enough to include family farms.

Query 7.3.
`("GFCONFLICT" = 321 OR "GFCONFLICT" = 322) AND "ACRES" > 5`.

Query 7.3 selects 8,970.32 acres of agricultural lands with a potential for preservation. The average size is 42.7 acres. The smallest area contains 5.1 acres, and the largest contains 405 acres. Figure 7.7 shows the distribution of the selected lands. Most of these lands are near the Interstate 4 corridor between Tampa and Plant City. The remaining properties are in the southern portion of the county.

Query 7.4 selects agricultural lands with high or moderate preference and some LUCIS conflict with urban high or moderate preference. The lands are larger than five acres. Query 7.4 identifies agricultural properties with low or moderate conservation preference and high urban development potential. Depending on owner priority, protection easements could be used to prioritize protection of the agricultural lands in figure 7.8 based on their development potential.

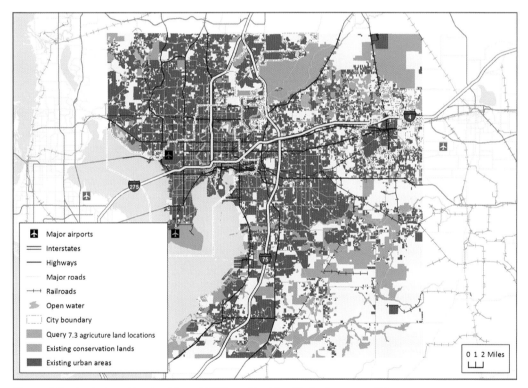

Figure 7.7. Agricultural lands with no LUCIS conflict identified using query 7.3.

Figure from Paul D. Zwick and the Arizona Board of Regents on behalf of the University of Arizona. Airports from the Federal Aviation Administration and the Research and Innovative Technology Administration Bureau of Transportation Statistics National Transportation Atlas Databases (NTAD) 2012; water bodies from the US Geological Survey; state boundary from US Census Bureau TIGER/Line Files; interstates, highways, and major roads from the Florida Department of Transportation; railroads from the Federal Railroad Administration; parcels from the Florida Department of Revenue.

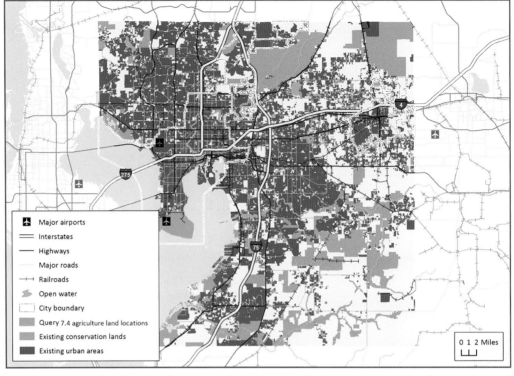

Figure 7.8. Agricultural lands with LUCIS conflicts. The agriculture opportunities identified may be used for agriculture protection, using land-use policies described at the beginning of this chapter.

Figure from Paul D. Zwick and the Arizona Board of Regents on behalf of the University of Arizona. Airports from the Federal Aviation Administration and the Research and Innovative Technology Administration Bureau of Transportation Statistics National Transportation Atlas Databases (NTAD) 2012; water bodies from the US Geological Survey; state boundary from US Census Bureau TIGER/Line Files; interstates, highways, and major roads from the Florida Department of Transportation; railroads from the Federal Railroad Administration; parcels from the Florida Department of Revenue.

Query 7.4.
`("GFCONFLICT" = 313 OR "GFCONFLICT" = 323) AND "ACRES" > 5`.

Table 7.5 indicates the LUCIS greenfield conflict (GFCONFLICT) values, easement priority (agricultural acquisition priority), and acreage for the agricultural lands identified using queries 7.3 and 7.4. The table also contains statistics for the areas previously identified by query 7.2 as having major LUCIS conflict (that is, `GFCONFLICT = 333 OR GFCONFLICT = 222`). Although possessing high or moderate conservation preference, these lands also have high or moderate agriculture and urban preference and are also under urban development pressure.

Table 7.5. Agriculture preference statistics

LUCIS Conflict Category	GFCONFLICT Values	Agricultural Acquisition Priority	Properties	Acres
Agriculture–Urban Conflict High Preference	313	High	17	295.6
Agriculture–Urban Conflict High Preference; Moderate Conservation Preference	323	High	222	4,098.4
High Agriculture Preference	322	Moderate	210	8,970.3
Agriculture–Conservation Conflict High Preference; Low Urban Preference	331	Low	4	358.3
Agriculture–Conservation Conflict High Preference; Moderate Urban Preference	332	Moderate	310	17,181.5
All Conflict High Preference	333	High	123	2,424.1
All Conflict Moderate Preference	222	Low	11	239.0

High-hazard flood areas, connectivity, agriculture, and conservation protection

Figure 7.9 shows the county's high-hazard flood areas for both coastal and inland flooding. The 100-year floodplain areas are excellent areas for conservation connectivity, primarily because they follow the natural hydrological surface water links. If there are connection opportunities using the high-hazard flood areas within the greenfield CEM, these areas might also represent excellent conservation opportunity. The flood areas could also help planners decide between conservation easements and agriculture or farmland preservation easements to protect areas of ecological significance.

In Florida, wetland areas—not floodplain areas—are restricted to development, but both areas can be cleared for agricultural use. The result is little protection for wetlands and floodplains if the land-use classification is for agriculture use. In fact, wetlands are often cleared or filled for agriculture use before the land is sold for development. Using a permanent easement to protect these areas is therefore of significant interest to environmental and agricultural NGOs.

A linkage based on flood protection within lands identified for agriculture preservation opportunity would clearly represent an area in need of high conservation protection. It would protect areas subject to natural floods; provide connectivity for species movement through the landscape; and if acquired, represent a buffer between agriculture and natural waters. If the areas are managed properly, the floodplain would help reduce the pollution of natural waters by agriculture activities.

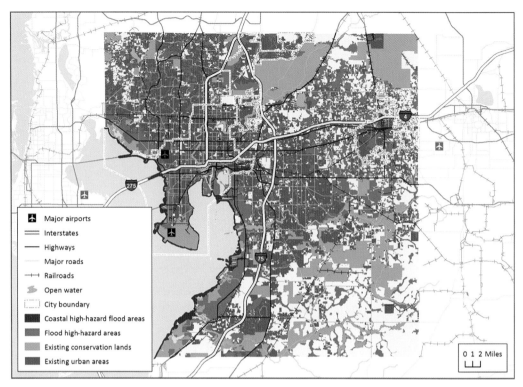

Figure 7.9. Coastal and inland high-hazard flood areas. The mauve areas are the inland high-hazard flood areas, and the red areas are coastal high-hazard flood areas.

Figure from Paul D. Zwick and the Arizona Board of Regents on behalf of the University of Arizona. Airports from the Federal Aviation Administration and the Research and Innovative Technology Administration Bureau of Transportation Statistics National Transportation Atlas Databases (NTAD) 2012; water bodies from the US Geological Survey; state boundary from US Census Bureau TIGER/Line Files; interstates, highways, and major roads from the Florida Department of Transportation; railroads from the Federal Railroad Administration; parcels from the Florida Department of Revenue.

Property in agricultural use with significant floodplain areas would also support an increased priority for agricultural protection. Such protection might include easement restrictions on the use of pesticides and fertilizer within a 100- to 400-foot buffer around the floodplain area. The buffer would provide even better protection of the natural waters within the floodplain. The value of the conservation/agriculture combined easement to the property owner would be more substantial because of the conservation opportunity. It would still provide agricultural use of the areas not in the floodplain or floodplain buffer, but with restrictions. The conservation/agriculture combined easement value would be split, and increase depending on the percentage of the area to be restricted for agricultural uses by the conservation easement portion of the contract. Figure 7.9 shows the coastal and inland high-hazard flood areas within the county. The red areas in figure 7.9 are coastal high-hazard flood areas, and the mauve areas are inland high-hazard flood areas.

Table 7.6 displays the acreage and percentage of parcel acreage within the two categories of flood high risk, coastal and inland.

Table 7.6. Flood high-risk hazard acres

Risk Category	Parcel Acres	Flood Acres	Percent of Flood	Frequency
Agriculture				
Coastal Flood High-Risk Area	1,219.57	80.2	6.57	19
Inland Flood High-Risk Area	392,146.33	46,165.35	11.73	6,234
Totals	393,365.90	46,245.55	11.76	6,253
Urban				
Coastal Flood High-Risk Area	13,499.81	935.63	6.93	1,244
Inland Flood High-Risk Area	605,303.83	94,834.12	15.67	112,930
Totals	618,803.64	95,769.75	15.48	114,174

Figure 7.10 shows the inland flood hazard areas that intersect with the greenfield CEM. The agriculture section of table 7.6 describes the agriculture area subject to flood high risk. The table also shows significantly more urban areas subject to flood high risk than agriculture areas. Table 7.7 shows the full parcel acres associated with agriculture preservation opportunity areas. The percentage of agricultural land within the greenfield CEM needed for agriculture or conservation protection is listed in the column Preservation Percentage. All acres listed in this column do not necessarily need easements. In fact, the cropland and orchards may only need conservation easements for the actual flood hazard areas within the parcels. These same parcels will also need agriculture easements for protection if the urban development pressure is high. Table 7.5 identifies only 6,800 acres in high-priority need of protection from urban development pressures. Table 7.5 also identifies approximately 26,000 acres of agricultural lands with moderate priority for easement contracts.

Conservation and agriculture preservation conclusions

Figure 7.11 shows the final agriculture and conservation priorities for land acquisition. Any or all of the acquisition methods described at the beginning of the chapter may be used, depending on owner preference and available acquisition funding in the county. The orange areas in figure 7.11 are the existing conservation lands; the dark-green areas are priority areas for conservation or agriculture preservation; and the light-green areas are lands with agriculture preservation opportunity, depending on funding availability and owner interest.

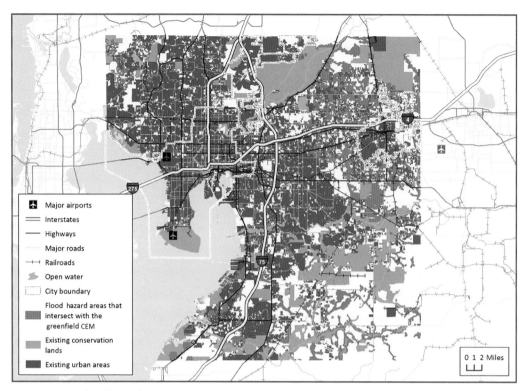

Figure 7.10. Flood hazard areas that intersect with the greenfield CEM.

Figure from Paul D. Zwick and the Arizona Board of Regents on behalf of the University of Arizona. Airports from the Federal Aviation Administration and the Research and Innovative Technology Administration Bureau of Transportation Statistics National Transportation Atlas Databases (NTAD) 2012; water bodies from the US Geological Survey; state boundary from US Census Bureau TIGER/Line Files; interstates, highways, and major roads from the Florida Department of Transportation; railroads from the Federal Railroad Administration; parcels from the Florida Department of Revenue.

Table 7.7. Agriculture parcel acres containing agriculture preservation opportunity

Agriculture Land Use	Parcel Acres	Greenfield Acres	Preservation Percentage	Frequency
Cropland	26,826.12	63,045.11	42.55	1,322
Orchards	11,000.79	18,909.56	58.17	512
Pasture	86,249.10	244,989.25	35.20	4,180
Plant Nurseries	3,290.55	7,731.27	42.56	418
Timber	4,806.47	23,609.52	20.35	144
Totals	**132,173.03**	**358,284.71**	**36.89**	**6,576**

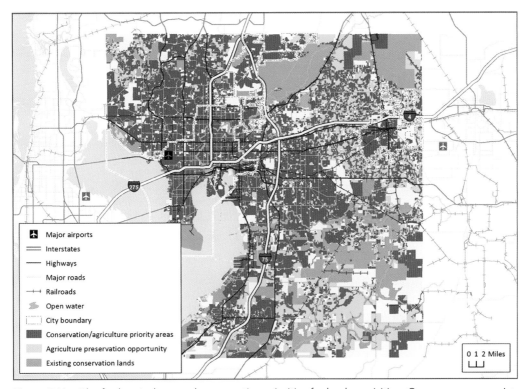

Figure 7.11. The final agriculture and conservation priorities for land acquisition. Orange areas are existing conservation, dark-green areas are conservation/agriculture priority areas, and light-green areas are agriculture preservation opportunity.

Figure from Paul D. Zwick and the Arizona Board of Regents on behalf of the University of Arizona. Airports from the Federal Aviation Administration and the Research and Innovative Technology Administration Bureau of Transportation Statistics National Transportation Atlas Databases (NTAD) 2012; water bodies from the US Geological Survey; state boundary from US Census Bureau TIGER/Line Files; interstates, highways, and major roads from the Florida Department of Transportation; railroads from the Federal Railroad Administration; parcels from the Florida Department of Revenue.

The final question is, How does the protection of conservation impact the county's development? Using allocation scenario 5 from chapter 6, table 7.7 displays the impact acreage. Because the conservation and agriculture preservation opportunities were selected from the greenfield CEM, any impacts to the scenario 5 allocation could occur only in greenfield development.

Table 7.8 shows that conservation/agriculture priority acquisition impacted 17.3 acres of greenfield scenario 5 institutional employment developments. The agriculture preservation or conservation opportunity acres impacted 1,200 acres of institutional employment and 1,627 acres of residential greenfield development. In terms of jobs and residential population, the conservation and agriculture preservation and protection opportunities would remove 4,969 institutional employment locations. They would also result in relocating 45,727 residents to alternative locations, most likely in multifamily mixed-use opportunity.

Table 7.8. Development impacts for conservation and agriculture preservation or protection opportunity

Category	Conservation/ Agriculture Priority Acres	Agriculture Preservation or Conservation Opportunity Acres	Employment	Population
Industrial	0.0	0.0	0	0
Commercial	0.0	0.0	0	0
Retail	0.0	0.0	0	0
Service	0.0	0.0	0	0
Institutional	17.3	1,200.1	4,969	0
Residential	0.0	1,627.0	0	45,727
Totals	17.3	2,827.1	4,969	45,727

Chapter summary

The following points are discussed in this chapter:

- Ultimately, LUCIS shows the location and helps identify the priority level of easement needs. But fieldwork and experienced conservation acquisition management may identify a need for preserving or protecting even more of the 132,173 available acres of opportunity.
- Finally, the LUCIS process described in this chapter can identify the actual property parcels and names of owners for properties of interest. Using the greenfield CEM and a list of property owners and parcels, you can identify preservation and protection opportunities for leaseback agreements or other methods of easement based on willing landowners' participation.

References

Benedict, M., and J. Drohan 2004. "Maryland's Green Infrastructure Assessment and GreenPrint Program." Green Infrastructure—Linking Lands for Nature and People Case Study Series. Arlington, VA: The Conservation Fund. http://www.conservationfund.org/images/programs/files/Marylands_Green_Infrastructure_Assessment_and_Greenprint_Program.pdf.

Carr, Margaret H., and Paul D. Zwick. 2007. *Smart Land-Use Analysis: The LUCIS Model*. Redlands, CA: Esri Press.

Craighead, F. L., and C. L. Convis. 2013. *Conservation Planning: Shaping the Future*. Redlands, CA: Esri Press.

Daniels, T., and K. Daniels. 2003. *The Environmental Planning Handbook for Sustainable Communities and Regions*. Chicago, IL: Planning Press American Planning Association.

EPA (US Environmental Protection Agency). 2014. Green Infrastructure. Green and Gray Infrastructure Research. http://www.epa.gov/nrmrl/wswrd/wq/stormwater/green.html#gray.

Farmland Information Center. 2010. 2010 National Resources Inventory. http://www.farmlandinfo.org/statistics#National%20Resources%20Inventory.

Florida Department of Environmental Protection. 2013. *Florida Greenways & Trails System Plan 2013–2017*. Tallahassee, FL: Florida Department of Environmental Protection. http://www.dep.state.fl.us/gwt/FGTS_Plan/PDF/FGTS_Plan_2013–17_publication.pdf.

Hoctor, T., M. Carr, and P. D. Zwick. 2000. "Identifying a Linked Reserve System Using a Regional Landscape Approach: The Florida Ecological Network." *Conservation Biology* 14:984–1000.

Hoctor, T., and J. Wood. 2002. *Identification of Critical Linkages within the Florida Ecological Greenways Network*. Gainesville, FL: University of Florida GeoPlan Center and Florida Department of Environmental Protection Office of Greenways and Trails.

Maryland DNR (Department of Natural Resources). 2014. "Maryland's GreenPrint Program: Preserving Our Green Infrastructure and Safeguarding Maryland's Most Valuable Ecological Lands." http://www.dnr.state.md.us/greenways/greenprint.html.

Maryland Greenways Commission. 2000. *Maryland Atlas of Greenways, Water Trails, and Green Infrastructure*. Annapolis, MD: Maryland Greenways Commission. http://www.dnr.state.md.us/greenways/introduction.html.

Oetting, Jon, Tom Hoctor, and Beth Stys. 2012. *Critical Lands and Waters Identification Project (CLIP) Version 2.0 Technical Report*. Tallahassee, FL: Florida Natural Areas Inventory.

Randolph, J. 2004. *Environmental Land-Use Planning and Management*. Washington DC: Island Press.

Sokolow, A. D., J. Speka, K. Richter, and M. Cotromanes. 2006. *A National View of Agricultural Easement Programs: Measuring Success in Protecting Farmland—Report 4*. American Farmland Trust and Agricultural Issues Center.

Wood, J., Hoctor, T., and M. Benedict 2004. "Florida's Ecological Network." Green Infrastructure—Linking Lands for Nature and People Case Study Series. Arlington, VA: The Conservation Fund. http://www.conservationfund.org/images/programs/files/Floridas_Ecological_Network.pdf.

Online resources

Florida Geographic Data Library, at http://www.fgdl.org.
Florida Natural Areas Inventory, at http://www.fnai.org/clip.cfm.

Chapter 8

Summarizing LUCIS land-use results

Paul D. Zwick, Iris E. Patten, and Abdulnaser Arafat

What this chapter covers

Often, elected officials, planning management, or clients want to know the local results of LUCIS regional scenarios. For example, what are the local population allocations for a city, a neighborhood, or a special planning district within the region? Perhaps, the local school district is interested in how many new schools must be located to support the new population growth. What type of school (primary, middle, or high school) must be located in the local area, and specifically, how many new students will the new school support? These local summaries depend on the specific regional scenario selected. Local residents often have one, and only one, concern: How will each of the regional alternatives impact my local community or neighborhood? Understanding how specific regional scenario allocations impact a local community helps residents within the community make decisions about which regional scenario best supports their local goals.

A number of methods are useful when summarizing spatial data within raster layers. This chapter focuses on three primary methods: (1) summarizing variables directly within the value attribute table (VAT) with the table open, (2) using the Summary Statistics tool in ArcGIS Analysis Tools, and (3) using the Zonal Statistics tools in ArcGIS Spatial Analyst Tools. Other useful tools for summarizing raster layer data are in the Extraction toolset, also in Spatial Analyst Tools.

Criteria used in selection queries

The following list of criteria is used for land-use selections structured in ArcGIS query language (see allocation scenario 5 in chapter 6):

- *S5COMYR*: Allocation Year Commercial Employment for Scenario 5. Item used to hold the allocation year when the cell was allocated for commercial use in allocation scenario 5 in chapter 6. For example, when this value equals 0, the cell has not been allocated. When the value equals 2045, the cell has been allocated for commercial use by 2045.
- *S5ALLYR*: Allocation Year Residential Population for Scenario 5. Item used to hold the allocation year when the cell was allocated for residential use during the scenario 5 allocations in chapter 6. For example, when this value equals 0, the cell has not been allocated. When the value equals 2045, the cell has been allocated for residential use by 2045.
- *S5COMEMP*: Available Commercial Employment for Scenario 5. Item used to hold the allocation employment when the cell was allocated for commercial use during the scenario 5 allocations in chapter 6.
- *S5INDYR*: Allocation Year Industrial Employment for Scenario 5. Item used to hold the allocation year when the cell was allocated for industrial use during the scenario 5 allocations in chapter 6. For example, when this value equals 0, the cell has not been allocated. When the value equals 2045, the cell has been allocated for industrial use by 2045.
- *S5SERYR*: Allocation Year Service Employment for Scenario 5. Item used to hold the allocation year when the cell was allocated for service use during the scenario 5 allocations in chapter 6. For example, when this value equals 0, the cell has not been allocated. When the item equals 2045, the cell has been allocated for service use by 2045.
- *S5INSYR*: Allocation Year Institutional Employment for Scenario 5. Item used to hold the allocation year when the cell was allocated for institutional use during the scenario 5 allocations in chapter 6. For example, when this value equals 0, the cell has not been allocated. When the item equals 2045, the cell has been allocated for institutional use by 2045.
- *S5RETYR*: Allocation Year Retail Employment for Scenario 5. Item used to hold the allocation year when the cell was allocated for retail use during the scenario 5 allocations in chapter 6. For example, when this value equals 0, the cell has not been allocated. When the item equals 2045, the cell has been allocated for retail use by 2045.
- *ACRES*: Available Allocation Acres. Item that contains the number of acres represented by the raster zone. A raster zone is a record within an integer raster that contains all the cells in a raster that have the same value (for example, all the cells that represent single-family residential land use).
- *CITYNAME*: City Name. Item representing the city the cell is located within: (1) Tampa, (2) Temple Terrace, (3) Plant City, or (0) in the county.

Summarizing LUCIS suitability and conflict

Often, data summary is a simple geoprocessing task, and you must organize a sequence of tools to get the required information. The following sections of this chapter present the tools and provide examples of the geoprocessing structure used to summarize or synthesize Land-Use Conflict Identification Strategy (LUCIS) scenario allocations.

Summarizing data example 1: Extracting raster data and summarizing information from the extracted raster

Summarizing LUCIS raster data may require summarizing information for local areas that are not represented in a specific LUCIS^{plus} criteria evaluation matrix (CEM). For example, in the Hillsborough County, Florida, study region, there are areas known as developments of regional impact (DRIs). DRI is a development that, because of its characteristics, has been determined by Florida regulations to have a regional impact. Figure 8.1 shows some example DRIs within Hillsborough County.

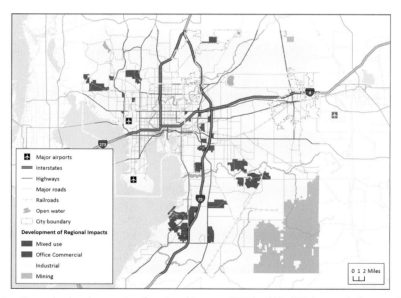

Figure 8.1. Example developments of regional impact (DRIs) within Hillsborough County. DRIs substantially impact the health, safety, and welfare of citizens of more than one county.

Figure from Paul D. Zwick and the Arizona Board of Regents on behalf of the University of Arizona. Airports from the Federal Aviation Administration and the Research and Innovative Technology Administration Bureau of Transportation Statistics National Transportation Atlas Databases (NTAD) 2012; water bodies from the US Geological Survey; state boundary from US Census Bureau TIGER/Line Files; interstates, highways, and major roads from the Florida Department of Transportation; railroads from the Federal Railroad Administration; parcels from the Florida Department of Revenue; developments of regional impact from the State of Florida Department of Community Affairs, subject to Florida Statutes Section 380.06(1), and the University of Florida GeoPlan Center.

The DRI color schema for the figure displays office commercial in purple, industrial in yellow, mixed use in red, and mining in light brown. To determine how many acres of office commercial DRI areas were allocated from the outcome of allocation scenario 5 in chapter 6 (mixed-use allocation for infill), you could use the Extract by Mask tool and the infill CEM. The result would illustrate cells from the infill CEM that were coincident to DRIs categorized as office commercial (figure 8.2).

Figure 8.2 indicates some shared areas that hold opportunity for commercial infill development. The Extract by Mask tool can help determine how much commercial office infill occurred within this type of DRI.

Figure 8.3 shows the Extract by Mask dialog box populated with the appropriate inputs to select the cells from the infill CEM that are shared with the office development of regional impact areas. The tool uses a mask to extract cells from an existing raster layer. For this example, the existing raster layer from which cells are to be extracted is named InfillCEM, and the mask layer is a polygon feature class named Office_DRI. The output raster layer is named InfOfficeCEM. It will contain only cells from InfillCEM that are contained within

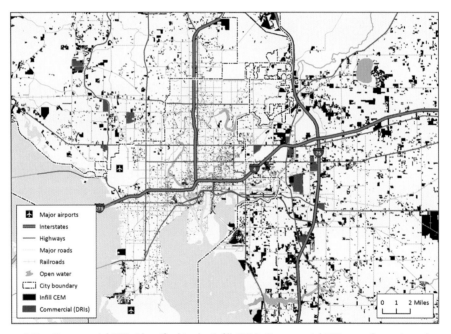

Figure 8.2. Commercial DRIs identified in the infill CEM. Commercial DRIs are shown in red overlaid with the infill CEM displayed in black. Visual inspection of the figure indicates some overlay, although not significant, of the two layers.

Figure from Paul D. Zwick and the Arizona Board of Regents on behalf of the University of Arizona. Airports from the Federal Aviation Administration and the Research and Innovative Technology Administration Bureau of Transportation Statistics National Transportation Atlas Databases (NTAD) 2012; water bodies from the US Geological Survey; state boundary from US Census Bureau TIGER/Line Files; interstates, highways, and major roads from the Florida Department of Transportation; railroads from the Federal Railroad Administration; parcels from the Florida Department of Revenue; developments of regional impact from the State of Florida and the University of Florida GeoPlan Center.

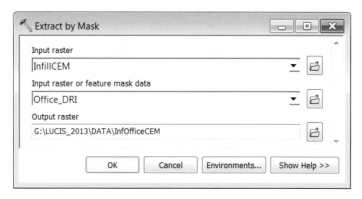

Figure 8.3. Extract by Mask tool dialog box. Cells are extracted from InfillCEM that are within, or masked by, areas within the Office_DRI dataset. The cells that satisfy this criteria are included in the InfOfficeCEM raster.

Figure from the Arizona Board of Regents on behalf of the University of Arizona.

the polygons from the Office_DRI feature class. Once this extraction is complete, you can summarize data regarding how many of the cells within InfillCEM were allocated for scenario 5.

Figure 8.4 displays the overlay of commercial DRIs with the cells extracted by the tool. Now the analyst can summarize the data within the new InfOfficeCEM raster layer to determine how scenario 5 allocated infill commercial development within the commercial DRIs.

You can summarize the data within the InfOfficeCEM raster layer in a number of ways. The simplest and fastest method is to open the raster layer VAT and summarize data for a specific field. InfOfficeCEM contains three fields of interest for this summary: s5comyr (S5COMYR), s5comemp (S5COMEMP), and Acres (ACRES). (The names of variables can vary in ArcGIS.) The summary starts with the selection of cells allocated for commercial employment. Query 8.1 selects the cells allocated.

Query 8.1.
`"S5COMYR" = 2045`.

Figure 8.5 shows that query 8.1 selects 1,222 of the 1,905 records in the InfOfficeCEM table. The figure also shows that you can use the summary method available within the table to do the summary.

Figure 8.6 displays the summary process for the two variables of interest, (a) S5COMEMP and (b) Acres, within InfOfficeCEM. Table 8.1 displays the result of the InfOfficeCEM summary.

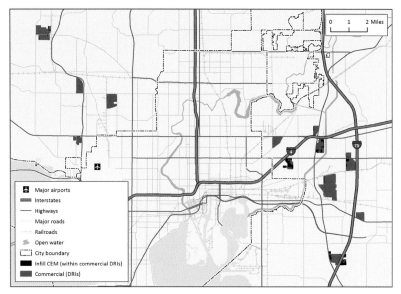

Figure 8.4. Overlay displaying cells from InfillCEM within commercial DRIs.

Figure from Paul D. Zwick and the Arizona Board of Regents on behalf of the University of Arizona. Airports from the Federal Aviation Administration and the Research and Innovative Technology Administration Bureau of Transportation Statistics National Transportation Atlas Databases (NTAD) 2012; water bodies from the US Geological Survey; state boundary from US Census Bureau TIGER/Line Files; interstates, highways, and major roads from the Florida Department of Transportation; railroads from the Federal Railroad Administration; parcels from the Florida Department of Revenue; developments of regional impact from the State of Florida and the University of Florida GeoPlan Center.

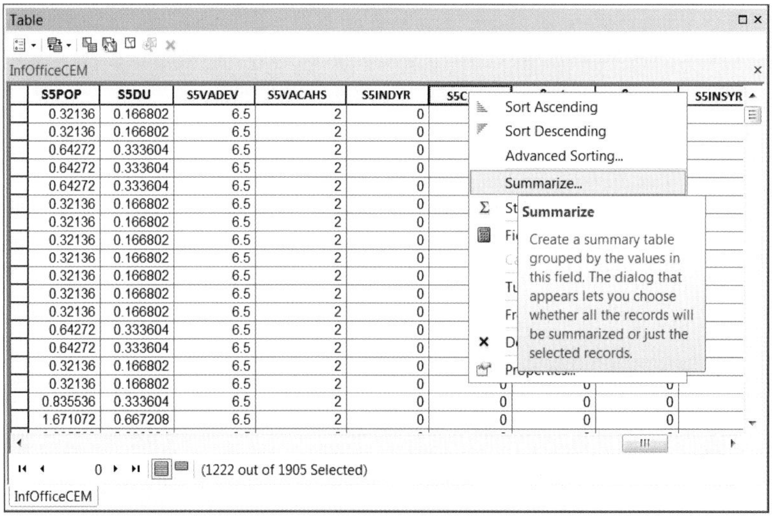

Figure 8.5. The InfOfficeCEM VAT shows how query 8.1 selects 1,222 records from the 1,905 records within InfOfficeCEM. The results can be summarized using the Summarize option and right-clicking each respective attribute heading.

Figure from the Arizona Board of Regents on behalf of the University of Arizona.

 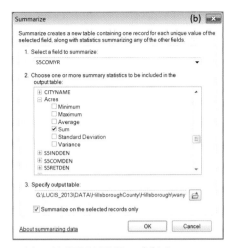

Figure 8.6. Table summary field identification. Two variables, (a) S5COMEMP and (b) Acres, are summarized for the selected records (representing allocated jobs and commercial acres, respectively) for allocation scenario 5.

Figure from the Arizona Board of Regents on behalf of the University of Arizona.

Table 8.1. Table summary method for infill commercial employment allocations within the commercial DRIs for allocation scenario 5

Table Record Row Count	Acres	Employment
1,222	67.07	715

If you want to extract only the cells from InfillCEM that have been allocated (the 1,222 records within InfOfficeCEM), the Extract by Attributes tool will extract the cells based on a query (figure 8.7).

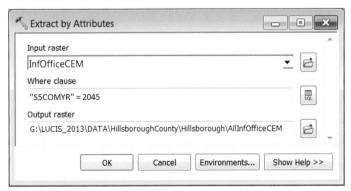

Figure 8.7. Extract by Attributes tool dialog box. Cells in the InfOfficeCEM raster that were allocated for commercial use by 2045 in scenario 5 are extracted. Only these allocated commercial cells are included in the AllInfOfficeCEM raster.

Figure from the Arizona Board of Regents on behalf of the University of Arizona.

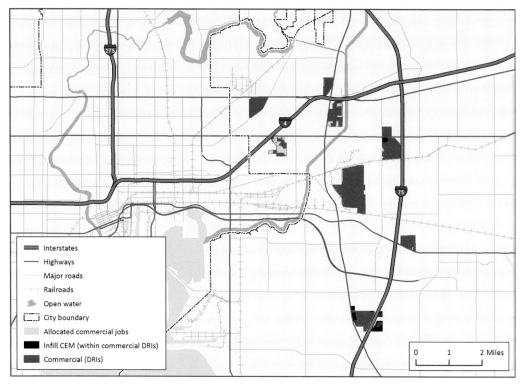

Figure 8.8. The result selection for the Extract by Attributes tool shown in figure 8.7.

Figure from Paul D. Zwick and the Arizona Board of Regents on behalf of the University of Arizona. Airports from the Federal Aviation Administration and the Research and Innovative Technology Administration Bureau of Transportation Statistics National Transportation Atlas Databases (NTAD) 2012; water bodies from the US Geological Survey; state boundary from US Census Bureau TIGER/Line Files; interstates, highways, and major roads from the Florida Department of Transportation; railroads from the Federal Railroad Administration; parcels from the Florida Department of Revenue; developments of regional impact from the State of Florida and the University of Florida GeoPlan Center.

Figure 8.8 displays the results of the Extract by Attributes tool for commercial DRI infill development. The commercial DRIs are shown in red, the infill CEM values within the DRIs are in black, and the areas allocated for commercial jobs are in aqua.

Summarizing data example 2: Summarizing information from a LUCIS raster layer

Another method of summarizing LUCIS raster data is using the Summary Statistics tool in the Statistics toolset in ArcGIS Analysis Tools. Suppose that after summarizing the LUCIS commercial infill associated with the commercial DRIs in the previous example, you wanted to know how much mixed-use retail was redeveloped within those commercial DRIs by 2045. How would you do this? It is simple using the Summary Statistics tool. Figure 8.9

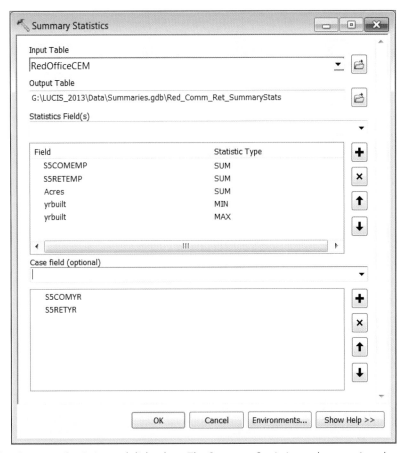

Figure 8.9. Summary Statistics tool dialog box. The Summary Statistics tool summarizes the total number of mixed-use acres, total number of commercial and retail jobs, and the minimum and maximum year built for existing structures within the commercial DRIs.

Figure from the Arizona Board of Regents on behalf of the University of Arizona.

shows the tool dialog box with the appropriate selections completed for summarizing the statistics.

The input raster layer is RedOfficeCEM, which was created in the same manner as InfOfficeCEM, using the Extract by Mask tool. The fields used to summarize statistical information include:

- S5COMEMP—total commercial employment job number for DRIs
- S5RETEMP—total retail employment job number for DRIs
- Acres—total acres of mixed-use commercial-retail employment for DRIs
- yrbuilt—building year-built data with both the minimum and maximum values collected

Table 8.2 displays the summary data collected using the Summary Statistics tool.

Table 8.2. Summary statistics for redevelopment commercial-retail mixed use in commercial DRIs

Table Record Count	Commercial Employment	Retail Employment	Minimum Year Built	Maximum Year Built	Acres
51	1,690	1,130	1970	2009	303.29

The statistics indicate that the minimum year built for all structures in the commercial DRIs used for 2045 mixed-use redevelopment is 1970, and the maximum year built is 2009. The year-built data indicates that by 2045, all structures to be either removed or rebuilt would be at least 36 years old. Redevelopment does not mean that all buildings are removed. They may also be adapted for the new mixed-use development. The total number of mixed-use jobs is 2,820, with 1,690 commercial jobs and 1,130 retail jobs. The total number of acres used is 303.29, producing a mean employment density of 9.3 employees per acre.

Figure 8.10 displays the physical area containing the acreage described in table 8.2. The green acres are the mixed-use commercial-retail areas, and the combination of green, red, and black areas are the total acres (1,283 acres) of redevelopment opportunity within the commercial DRIs. The 303.29 acres represent about one-quarter of the redevelopment opportunity (23.64 percent) of the total 1,283 acres within the commercial DRIs.

Summarizing data example 3: Summarizing information from a LUCIS raster layer to traffic analysis zones

Transportation planning and traffic simulation often rely on the summary of population and employment data within defined areas known as traffic analysis zones (TAZs). Each TAZ has a unique ID. The areas often coincide with US Census Block Group boundaries, but not always. By knowing the number of residents, average household size, number of employees, number of schoolchildren, and number of automobiles, among other data, transportation planners can analyze the traffic demand and daily peak- and off-peak-hour traffic activity.

Figure 8.11 shows the TAZ boundaries for part of downtown Tampa. A portion of Hillsborough County's TAZ boundaries is also shown in the map. These locations are important for the land-use transportation interaction across the city's boundary. Typically, the more dense an area's development, especially dense employment centers and areas with high numbers of residential units, the more automobile traffic. Again typically, the more traffic, the more congestion. Congestion is especially likely when a roadway supports more traffic than the roadway's capacity, measured in trips per hour or day. Considering the impact of population and employment growth on traffic roadway capacity, you can summarize new population and employment growth to existing TAZ polygons.

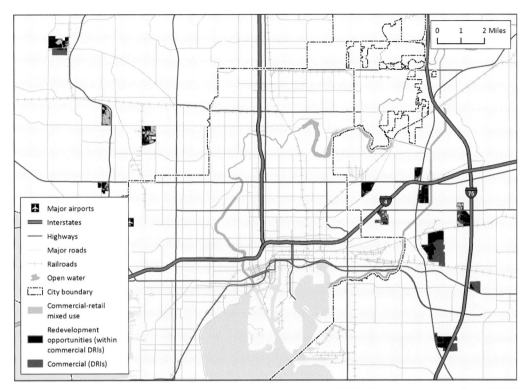

Figure 8.10. Redevelopment opportunity areas within commercial DRIs. The green areas reflect 303.29 acres of redevelopment commercial-retail mixed use in commercial DRIs. These areas are about one-quarter of the redevelopment opportunity within the commercial DRIs.

Figure from Paul D. Zwick and the Arizona Board of Regents on behalf of the University of Arizona. Airports from the Federal Aviation Administration and the Research and Innovative Technology Administration Bureau of Transportation Statistics National Transportation Atlas Databases (NTAD) 2012; water bodies from the US Geological Survey; state boundary from US Census Bureau TIGER/Line Files; interstates, highways, and major roads from the Florida Department of Transportation; railroads from the Federal Railroad Administration; parcels from the Florida Department of Revenue; developments of regional impact from the State of Florida and the University of Florida GeoPlan Center.

Using the Summary Statistics tool is the best method to summarize LUCIS data from all three CEMs. The only requirement necessary for the summary is to include the TAZ unique identification (ID) number from CEM. To do this, you must create a raster layer using the TAZ ID number that is in the Value field in VAT. The normal LUCIS process for creating a CEM includes the TAZ ID number.

To summarize redevelopment commercial-retail acres and employment job numbers for each TAZ polygon, the process is simple. First, use query 8.2 to select the records included in the redevelopment CEM (RedevCEM) that have been allocated for commercial-retail mixed-use development. Next, use the Summary Statistics tool to collect the data from the selected records within CEM (figure 8.12). Finally, join or relate the summary statistics table to the original TAZ feature class or shapefile.

Query 8.2.
```
"S5COMYR" = 2045 AND "S5RETYR" = 2045.
```

Figure 8.13 graphically represents the TAZ summaries for the raw redevelopment commercial employment allocation. The raw employment numbers range from eight to 1,134 jobs. The highest raw job numbers are in the red areas, and the lowest raw job numbers are in dark blue. The black areas show the physical location of redevelopment commercial jobs selected using the LUCIS^{plus} process for allocation scenario 5. The graphic represents a portion of the county, specifically the Interstate 4 corridor, because the commercial employment that the scenario allocated was in those areas. The concept for mixed-use allocations in scenario 5 is to investigate the confinement of development to proximity zones 9 to 6. The proximity zones 9 to 6 are used to allocate residential development and employment in closer proximity to the center of employment activities. For the scenario, the center of employment activities was determined to be near the center of Tampa, the county seat.

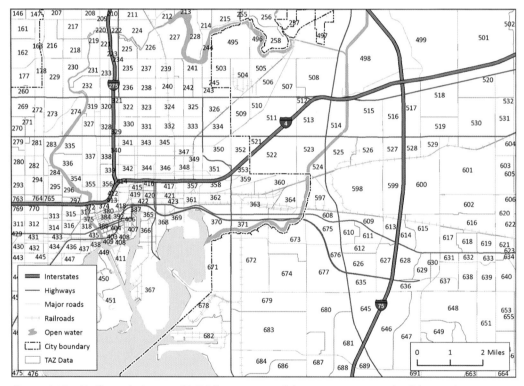

Figure 8.11. Traffic analysis zones (TAZs) for a portion of downtown Tampa and Hillsborough County are shown. Each TAZ has a unique identification number. Unique identification numbers allow for table joins and links to display or summarize land-use data.

Figure from Paul D. Zwick and the Arizona Board of Regents on behalf of the University of Arizona. Water bodies from the US Geological Survey; state boundary from US Census Bureau TIGER/Line Files; interstates, highways, and major roads from the Florida Department of Transportation; railroads from the Federal Railroad Administration; traffic analysis zones from the US Census Bureau.

Figure 8.12. Summary Statistics tool for the summary of commercial and retail redevelopment employment and the employment acres for these categories for the county's TAZ polygons. Query 8.2 selects the allocated commercial and retail employment acres for allocation scenario 5.

Figure from the Arizona Board of Regents on behalf of the University of Arizona.

Figure 8.14 displays the new locations for retail redevelopment employment by the year 2045. Clearly, the employment allocations are in mixed use with commercial employment opportunity. The number of retail jobs ranges from a low of seven to a maximum allocation of 994. Table 8.3 presents the total commercial-retail mixed-use employment as summarized.

Table 8.3. Summary statistics for redevelopment commercial-retail mixed use by TAZ

Number of TAZ	Total Commercial Employment	Total Retail Employment	Total Mixed-Use Employment	Average Employment Density	Total Acres Developed
396	39,096	34,277	73,373	11.4	6,440.41

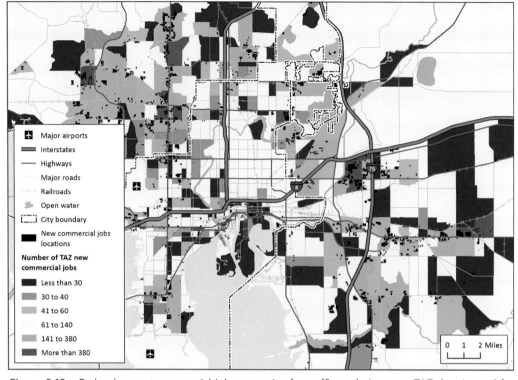

Figure 8.13. Redevelopment commercial job summaries for traffic analysis zones. TAZ data is used for transportation system modeling.

Figure from Paul D. Zwick and the Arizona Board of Regents on behalf of the University of Arizona. Airports from the Federal Aviation Administration and the Research and Innovative Technology Administration Bureau of Transportation Statistics National Transportation Atlas Databases (NTAD) 2012; water bodies from the US Geological Survey; state boundary from US Census Bureau TIGER/Line Files; interstates, highways, and major roads from the Florida Department of Transportation; railroads from the Federal Railroad Administration; parcels from the Florida Department of Revenue; traffic analysis zones from the US Census Bureau.

Summarizing data example 4: Summarizing LUCIS allocation information by city boundary

Even a cursory inspection of allocation scenario 5 clearly indicates that Plant City is outside the chosen proximity requirements for the scenario. Plant City lies within the I-4 corridor and has a number of significant opportunities for urban/suburban growth. Growth opportunities include transport accessibility, a growing urban fabric (although not nearly that of Tampa's), and its location halfway between Orlando and Tampa along the primary connector between these two rapidly growing cities. Data indicates that Plant City has an opportunity for growth as an alternative to development within Tampa and Orlando because of lower land values and that it fits the bedroom community category for both of these major cities.

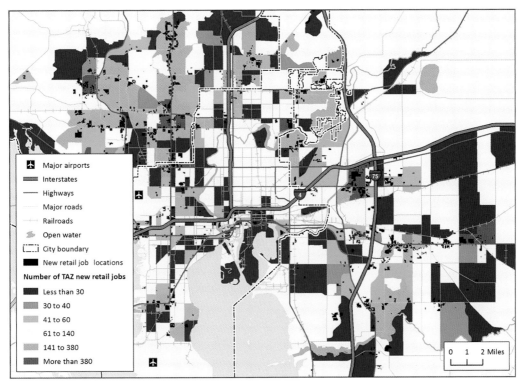

Figure 8.14. Redevelopment retail job summaries for traffic analysis zones. TAZ data is used for transportation system modeling.

Figure from Paul D. Zwick and the Arizona Board of Regents on behalf of the University of Arizona. Airports from the Federal Aviation Administration and the Research and Innovative Technology Administration Bureau of Transportation Statistics National Transportation Atlas Databases (NTAD) 2012; water bodies from the US Geological Survey; state boundary from US Census Bureau TIGER/Line Files; interstates, highways, and major roads from the Florida Department of Transportation; railroads from the Federal Railroad Administration; parcels from the Florida Department of Revenue; traffic analysis zones from the US Census Bureau.

Hypothetically, the mayor of Plant City might challenge the development results from scenario 5, concluding that Plant City should receive more residential and employment growth because of the city's commuter-friendly characteristics. Therefore, the mayor requests a summary of the residential and employment growth proposed by the scenario that is specifically within Plant City.

The first step in summarizing the data is to determine whether all three CEMs have the boundary of Plant City as a variable within each of the CEMs. Inspection shows that all three CEMs (RedevCEM, InfillCEM, and the greenfield CEM, GFDevCEM) have a variable of "cities," which contains a value of 1 representing the cells within the boundary of Plant City. Having this variable within the three CEMs greatly simplifies the summary process. You do not have to use an extraction tool to extract any cells. You can summarize the data either within the tables or by using the Summary Statistics tool.

The second step in the process is done within a single CEM VAT. The query "CITIES" = 1 spatially selects the cells within Plant City. For greenfield development in Plant City, the summary starts by determining whether any cells have already been allocated for either residential development or employment. The summary uses the Summary Statistics tool, as shown in figure 8.15. The fields holding allocation for individual land use are S5INSYR, S5ALLYR, S5COMYR, S5RETYR, S5SERYR, and S5INDYR. If any of these fields contain a value greater than 0, that type of land use was allocated within greenfield development in Plant City. As table 8.4 shows, no allocations within Plant City for either residential development or employment were made in greenfield areas.

Summarizing redevelopment and infill development is slightly more complicated but follows the same process. The complication arises because of the multiple allocations that can occur including mixed-use development.

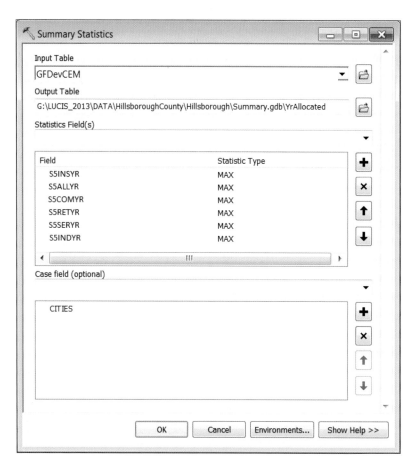

Figure 8.15. The Summary Statistics tool used to summarize GFDevCEM for scenario 5 within Plant City. The attributes listed in the Field input contain the allocations for individual land uses.

Figure from the Arizona Board of Regents on behalf of the University of Arizona.

Table 8.4 shows that if scenario 5 is selected, the new development in Plant City is in residential redevelopment and industrial redevelopment. A total of 160 acres of greenfield development is used to allocate 2,843 residents. The redevelopment is a mixture of multifamily and single-family residential development for an average of approximately seven dwelling units per acre. An additional 174 acres of infill residential multifamily development support 4,656 people. The average density is 12 dwelling units per acre. The total job creation allocated is in industrial land use, with 482 industrial jobs on 1,594 acres.

The summary of 2045 residential and employment allocations for Plant City supports the mayor's concerns. A developing bedroom community with excellent transportation access and ample greenfield opportunity would be expected to support a significant amount of single-family residential development. The amount of multifamily residential development is appropriate. However, there is a low allocation of commercial, retail, service, and institutional jobs. In fact, the concept of allocation scenario 5, which restricted residential development to proximity to an urban center, has impeded what any elected official, planning director, or community developer would expect to be a more robust land-use allocation within the city limits of Plant City.

The mayor of Plant City would clearly be justified in challenging the land-use allocations proposed for scenario 5. Scenario 5 has unintended consequences that are detrimental to Plant City. A simple summary of LUCIS allocations supports the mayor's concerns. The mayor would be justified in expecting that Plant City should be offered a greater role in an alternative scenario based on the results generated by scenario 5. A scenario that allocates employment closer to the three major city centers, and not just the center of employment activity in Tampa, could be completed as an alternative to scenario 5.

Table 8.4. Plant City development summary

CEM and Land-Use Category	Acres	Residential	Employment				
			Commercial	Retail	Service	Institutional	Industrial
Greenfield	0	0	0	0	0	0	0
Residential Redevelopment	160	2,843	0	0	0	0	0
Industrial Redevelopment	1,594	0	0	0	0	0	482
Residential Infill	174	4,656	0	0	0	0	0

Chapter summary

The following points are discussed in this chapter:

- Using LUCIS^{plus} CEMs and the ArcGIS Extraction toolset, including Extract by Mask, Extract by Attributes, and other tools, you can summarize the results of LUCIS^{plus} allocations based on constraints such as specific development areas (for example, DRIs).
- In many land-use planning processes, indicators such as proposed residential unit density and the acreage proposed for residential development are used to evaluate the performance of land-use scenarios. The ability to summarize LUCIS^{plus} results serves a similar purpose. You can use the summary to evaluate the spatial performance or allocation of employment and residential development. The summary can help determine whether policy decisions are justified or should be changed or altered based on LUCIS^{plus} results.

ArcGIS tools referenced in this chapter

Tool Name	Version 10.2 Toolbox/Toolset
Summary Statistics	Analysis Tools/Statistics
Zonal Statistics	Spatial Analyst Tools/Zonal
Extract by Mask	Spatial Analyst Tools/Extraction
Extract by Attributes	Spatial Analyst Tools/Extraction

Part III

Advanced allocation techniques using LUCIS^plus

Chapter 9

Analyzing and mapping land use for natural disasters: Hurricane storm surge and sea level rise

Paul D. Zwick

What this chapter covers

Flooding is the most common result of storm surge. Weather events that cause storm surge are typically organized cyclonic systems that include high winds and low-level circulation. In the United States, depending on maximum sustained winds, many of these systems are called *tropical depressions*, *tropical storms*, or *hurricanes*. Even storms that do not meet the wind thresholds of these named systems can pose a threat of flooding and storm surge. The Multi-Hazard Loss Estimation Methodology (Hazus-MH) Flood Model "produces loss estimates for vulnerability assessments and plans for flood risk mitigation, emergency preparedness, and response recovery" (DHS FEMA 2013c, xv). Hazus-MH is a standardized method created by the Federal Emergency Management Administration (FEMA) "to estimate potential losses from earthquakes, hurricane winds, and floods" (Hazus 2013).

Hazus-MH models can estimate damage from hurricane winds and storm surge flooding, riverine flooding, and earthquakes. This chapter uses the Hazus-MH hurricane storm surge or coastal flooding model to analyze the impact of a baseline 100-year storm surge for three scenarios. The three scenarios are (1) the base year 2010 for Hillsborough County, Florida; (2) the year 2045 with new LUCIS development from allocation scenario 5 (see chapters 6 and 8); and (3) the change in the base year 2010 as a result of a one-meter sea level rise (SLR). The goal is to analyze future impacts of storm surges on increased development and understand the implications of an increase in the mean sea level as a result of SLR.

A few basics about hurricanes

Hurricanes are large, powerful storms that cause billions of dollars in damage along the Gulf of Mexico and East Coast of the United States. The structure of a hurricane is a simple concept though it depends on many factors. Hurricane winds circulate in a counterclockwise direction around a low-pressure center, called the *eye of the hurricane*. Typically, the lower the central pressure, the higher the hurricane wind speeds. The highest wind speeds are located along the *eye wall*. Taking into account the direction of the storm's movement, the overall wind speeds are greater in the front right quadrant of the storm than in the other three quadrants. Figure 9.1 shows the relationship between wind direction and storm surge. The *storm surge* is the abnormal rise of water generated by a storm, over and above the astronomical tide. For the Atlantic Ocean, hurricane season starts June 1 and ends November 30. For the Pacific Ocean, the season starts May 1 and also ends November 30.

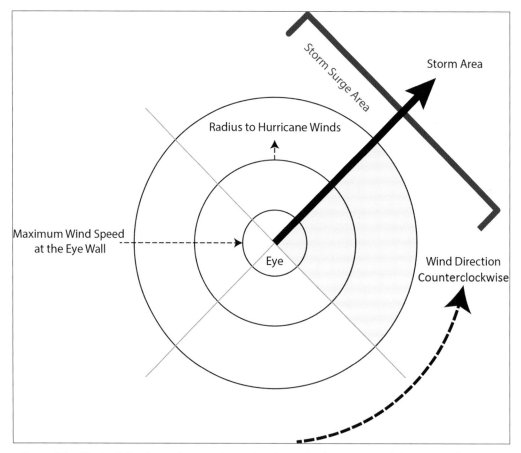

Figure 9.1. The basic hurricane diagram shows the relationship between wind direction and storm surge. Overall winds are greater in the front right quadrant (blue area).

Figure from Paul Zwick and the Arizona Board of Regents on behalf of the University of Arizona.

Table 9.1 shows the Saffir-Simpson Hurricane Wind Scale (National Hurricane Center 2013). The Saffir-Simpson scale is divided into five categories, in which Category 1 is the lowest wind speed category and Category 5 the highest. This scale also helps estimate potential property damage. According to the National Hurricane Center, hurricanes reaching Category 3 and higher are considered major hurricanes because of their potential for significant loss of life and damage. All hurricanes are dangerous, including Category 1 and 2 storms.

Table 9.1. The National Hurricane Center Saffir-Simpson Hurricane Wind Scale

Category	Sustained Winds	Types of Damage Due to Hurricane Winds
1	74–95 mph 119–153 km/h	**Very dangerous winds will produce some damage:** Well-constructed frame homes could have damage to roof, shingles, vinyl siding, and gutters. Large branches of trees will snap and shallowly rooted trees may be toppled. Extensive damage to power lines and poles likely will result in power outages that could last a few to several days.
2	96–110 mph 154–177 km/h	**Extremely dangerous winds will cause extensive damage:** Well-constructed frame homes could sustain major roof and siding damage. Many shallowly rooted trees will be snapped or uprooted and block numerous roads. Near-total power loss is expected with outages that could last from several days to weeks.
3 (major)	111–129 mph 178–208 km/h	**Devastating damage will occur:** Well-built framed homes may incur major damage or removal of roof decking and gable ends. Many trees will be snapped or uprooted, blocking numerous roads. Electricity and water will be unavailable for several days to weeks after the storm passes.
4 (major)	130–156 mph 209–251 km/h	**Catastrophic damage will occur:** Well-built framed homes can sustain severe damage with loss of most of the roof structure and/or some exterior walls. Most trees will be snapped or uprooted and power poles downed. Fallen trees and power poles will isolate residential areas. Power outages will last weeks to possibly months. Most of the area will be uninhabitable for weeks or months.
5 (major)	157 mph or higher 252 km/h or higher	**Catastrophic damage will occur:** A high percentage of framed homes will be destroyed, with total roof failure and wall collapse. Fallen trees and power poles will isolate residential areas. Power outages will last for weeks to possibly months. Most of the area will be uninhabitable for weeks or months.

Source: National Hurricane Center.
Note: Hurricane Andrew in 1992 was the most powerful hurricane to hit Florida and was the most costly hurricane until Hurricane Katrina hit the Gulf Coast and New Orleans in 2005.

Hurricane Andrew made a direct hit on Homestead, Florida, and nearby Florida City (both not too far from Miami). The storm destroyed 28,006 homes; 82,000 businesses were destroyed or damaged; 107,380 homes were damaged; 250,000 people were left homeless; 700,000 people were evacuated; 1.4 million homes were without electricity; Homestead US Air Force Base was totally destroyed; and the storm spared only nine mobile homes, damaging 1,167. The total cost of the storm was between $26.5 billion and $30 billion in 1992 dollars. The National Hurricane Center says a total of 44 people died from Andrew (15 directly killed by the hurricane, 29 indirect deaths). The Centers for Disease Control puts the death toll at 32 (14 directly killed by the hurricane and just 18 indirect deaths). (StormFacts.net 2011)

Table 9.2 shows the hurricanes by category and location that have hit the US Gulf and Atlantic coastlines from Texas to Maine. The percentage of all hurricanes from 1851 to 2012 that have hit Florida is 39.3 percent of all recorded storms, or 114 out of 290. Of all the major US hurricanes on the Saffir-Simpson Hurricane Wind Scale, 37 out of 97, or 38.14 percent, hit Florida. The frequency of hurricanes hitting Florida is one storm every 1.4 years, and one major hurricane every 4.35 years.

The basic storm surge

As previously stated, the storm surge is the abnormal rise of water generated by a storm, over and above the astronomical tide. The maximum water level during a storm is the sum of the astronomical tide level and the storm surge, combined with the wave crests. A storm surge is calculated by subtracting the astronomical tide level from the maximum water level with wave height. Additionally, figure 9.1 shows that a hurricane storm surge is pushed by the winds in the front quadrants of the storm. As a result, the direction and size of the storm are important in predicting the possible storm surge area. Storm surge is responsible for more deaths related to hurricanes than winds (National Hurricane Center 2014). The general misconception in public thinking seems to be that a storm surge is linked to a storm's maximum wind speed. However, other factors also contribute to the height of the storm surge. Table 9.3 shows selected storms and their relationship between the Saffir-Simpson Hurricane Wind Scale and the storm surge at landfall. Clearly, the table indicates that wind speed is not the only factor that impacts the height of a storm surge.

Table 9.2. Hurricanes that have hit the US coastline from Texas to Maine (from 1851 to 2012)

AREA	CATEGORY						Major Hurricanes, Categories 3, 4, and 5
	1	2	3	4	5	Totals	
US Coastline (Texas to Maine)	117	76	76	18	3	290	97
Texas	25	19	12	7	0	63	19
—North	13	8	3	4	0	28	7
—Central	7	5	2	2	0	16	4
—South	10	5	7	1	0	23	8
Louisiana	19	15	15	4	1	54	20
Mississippi	2	5	8	0	1	16	9
Alabama	12	5	6	0	0	23	6
Florida	44	33	29	6	2	114	37
—Northwest	27	16	12	0	0	55	12
—Northeast	13	8	1	0	0	22	1
—Southwest	16	8	7	4	1	36	12
—Southeast	13	13	11	3	1	41	15
Georgia	12	5	2	1	0	20	3
South Carolina	19	6	4	2	0	31	6
North Carolina	24	13	11	1	0	49	12
Virginia	9	2	1	0	0	12	1
Maryland	1	1	0	0	0	2	0
Delaware	2	0	0	0	0	2	0
New Jersey	2	0	0	0	0	2	0
Pennsylvania	1	0	0	0	0	1	0
New York	6	1	5	0	0	12	5
Connecticut	4	3	3	0	0	10	3
Rhode Island	3	2	4	0	0	9	4
Massachusetts	5	2	3	0	0	10	3
New Hampshire	1	1	0	0	0	2	0
Maine	5	1	0	0	0	6	0

Source: National Oceanic and Atmospheric Administration Hurricane Research Division Atlantic Oceanographic and Meteorological Laboratory.

According to the National Weather Service (2010), factors determining storm surge height include (1) storm central pressure, (2) storm intensity, (3) forward wind speed, (4) size of the storm, (5) angle of the storm's approach to the coastline, (6) the shape of the coastline, (7) width and slope of the ocean bottom, and (8) local features on or along the coastline. The eight storm surge factors are as follows:

1. *"The central pressure of the storm.* The lower the storm's central pressure, the higher the storm surge because the ocean water level rises as a result of decreased atmospheric pressure on the water. However, NOAA (National Oceanic Atmospheric Administration) indicates that the impact of central pressure on the storm surge height is minimal.
2. *"The storm's intensity,* other factors held constant, increases the height of the storm surge. Generally, the larger storms result in increased storm surge because a larger storm affects the ocean's water surface over a larger area. For example, Katrina was a much larger storm at landfall than Charley.
3. *"Faster storms,* making landfall on open coastline, will produce a higher surge. However, a higher surge is produced in bays, sounds, and other enclosed bodies of water with a slower storm.
4. *"The size of the storm.* The physical size of the storm usually measured by its diameter. Generally the larger the storm, the higher the storm surge.
5. *"The angle of the storm.* As the storm approaches the coastline, the angle of the storm with the coastline has an impact on the storm surge height. Storms that form a perpendicular angle with the coastline produce a higher storm surge than a storm that approaches parallel to the coastline.
6. *"The shape of the coastline* has an impact on the height of the storm surge. A coastline that curves inward produces a higher storm surge than a coastline that curves outward.
7. *"The width and shape of the ocean bottom* impacts the height of the storm surge. Higher storm surge occurs with wide, gently sloping continental shelves, while lower storm surge occurs with narrow, steeply sloping shelves. Areas along the Gulf Coast, especially Louisiana and Mississippi, are particularly vulnerable to storm surge because the ocean floor gradually deepens offshore. Conversely, areas such as the east coast of Florida have a steeper shelf, and storm surge is not as high.
8. *"The local features of the coastline* highly influence storm surge. Local features influence the flow of water with the storm surge and the distance the surge propagates inland. Features such as man-made barriers, dunes, and coastal islands significantly impact storm surge." (National Weather Service 2010)

Finally, the amount of structural damage clearly is impacted by the amount of development, including infrastructure, along or within close proximity to the coastline. For example, Hurricane Sandy in 2012 was a large storm that struck the coastline of New York and New Jersey almost perpendicular to the path of the storm. It created a

Table 9.3. Storm surge and the Saffir-Simpson Hurricane Wind Scale

Name	Year	Category at Landfall	Coastal Landfall Location	Storm Surge
Irene	2011	1	North Carolina	8–11 feet
Ike	2008	2	Texas	20 feet
Charley	2004	4	Florida	6–8 feet
Katrina	2005	3	Louisiana	28 feet

Source: National Weather Service 2010.

storm surge along the coastline from three to nine feet high with a total water level of 12 to 13 feet (National Weather Service 2014). The storm was a high to moderate Category 1 storm, but it generated billions of dollars in damage. The close proximity of urban structures, including buildings, subways, highways, and other infrastructure, resulted in increased urban exposure to the storm surge, and therefore to increased flood damage. The *Huffington Post*, in "Superstorm Sandy: 3 Months Later, Losses Mount" (January 29, 2013), reported that Sandy damaged or destroyed 305,000 housing units and disrupted more than 265,000 businesses in New York. In New Jersey, 346,000 housing units were destroyed or damaged, and 190,000 businesses were affected. The article also reported that in December 2013, state governments reported a total of $62 billion in damage and other losses as a result of Hurricane Sandy.

The Hazus-MH coastal flood model

[The Hazus-MH methodology] deals with nearly all aspects of the built environment, and a wide range of losses. The user can evaluate losses from a single flood event, or for a range of flood events allowing for annualized estimates of damages. Using the extensive national databases that are embedded in Hazus, users can make general loss estimates for a region. These databases contain information such as demographic aspects of the population in a study region, square footage for different occupancies of buildings, and numbers and locations of bridges. The Hazus methodology and software are flexible enough so that locally developed inventories and other data that more accurately reflect the local environment can be substituted, resulting in improved loss estimates. (DHS FEMA 2013c, xv)

The Hazus Flood Model analyzes both riverine and coastal flood hazards, where flood hazard is defined as a relationship between depth of flooding and the annual chance of inundation to that depth. In different contexts, flood hazard may have different meanings. ... Hazard can mean risk in some contexts, and it can mean a source of danger in others. The hazard may be that an area is inundated about once every 10 years (risk), or it may be that an area is subject to flood depths ranging from 5 to 10 feet (source of danger). Flood frequency studies combine these ideas and define flood hazard in terms of the chance that a certain magnitude of flooding is exceeded in any given year. Flood magnitude is usually measured as a discharge value, flood elevation, or depth. For example, one may refer to the 100-year flood elevation. It is the elevation, at the point of interest, that has a one percent (1%) annual chance of being exceeded by floodwater. Using the flood frequency convention, flood hazard is defined by a relation between depth of flooding and the annual chance. (DHS FEMA 2013c, 1–3)

Hazus-MH modeling can be done at three levels. Level 1 modeling does not require the user to be experienced or have extensive technical knowledge. Level 1 modeling is completed with data provided with the software from national databases. It uses default damage estimate methodologies and Hazus-MH coastal flood parameters. Levels 2 and 3 require increased experience, and level 3 modeling requires extensive experience and engineering expertise.

Hazus-MH coastal flood definitions

This section includes some important definitions for coastal flood modeling using the Hazus-MH Flood Model. The following definitions are taken directly from the *Hazus-MH Flood Model Technical Manual* (DHS FEMA 2013a, section 4: 83–91). The manual can be downloaded from the FEMA website (see "Online resources" at the end of this chapter). Figure 9.2 graphically displays the structure of storm surge used in the Hazus-MH Flood Model. The stillwater elevation depth plus wave setup depth includes the mean sea level, tide, pressure setup depth, and wind setup depth. The wave crest elevation is set to one-half the stillwater elevation depth plus wave setup. The wave crest elevation decreases as the ground depth increases until the wave crest elevation is zero when stillwater elevation is the same as ground elevation.

Coast: The Coastal Flood Model distinguishes between four coasts: Atlantic, Gulf of Mexico, Pacific, and Great Lakes. Certain attributes of each coast (such as *reference elevation*, default *flood elevation ratios, wave regeneration* factors, and *dune reservoirs*) are contained in lookup tables used in the Coastal Flood Model.

Reference elevation: This is the water level used in the computation of *n*-year stillwater elevations. For Atlantic, Gulf of Mexico, and Pacific shorelines, the reference elevation is taken to be 0.0 National Geodetic Vertical Datum of 1929 (NGVD29) or North American Vertical Datum of 1988 (NAVD88), which is approximately mean sea level.

Shoreline: This is the intersection of the land with the sea, bay, or lake under normal conditions. The Coastal Flood Model uses two shorelines: (1) a smoothed shoreline, based on TIGER shorelines and from which *transects* are drawn (this is the shoreline displayed to the user); and (2) a digital elevation model (DEM) shoreline, which is the intersection of the terrain along a transect with the reference elevation.

100-year stillwater elevation: The 100-year stillwater elevation is a required user input. The parameter represents the water surface elevation as a result of tides or storm surge. It does not include the effects of wave height, wave run-up, or wave setup. The value for this parameter can be obtained from the "Summary of Stillwater Elevations" table contained in a county's Flood Insurance Study (FIS) report. Figure 9.3 is the stillwater elevations (SWEL) from the Hillsborough County FIS.

Wave setup: Wave setup is a local rise in the stillwater level due to the presence of breaking waves in the nearshore region (see figure 4.57 in *The Hazus Multi-Hazard Loss Estimation Methodology Flood Model User Manual* [DHS FEMA 2013c]). The Coastal Flood Model assumes that the wave setup reaches a maximum near the dune toe (profile intersection with 10-year SWEL), and decays 10 percent per 100 feet inland from that point. If the terrain rises above the stillwater flood level and wave effects diminish to zero, the Coastal Flood Model terminates the wave setup at the same point.

If the FIS is unclear whether wave setup is included in the 100-year SWEL, the 10-year, 50-year, 100-year, and 500-year stillwater values given in the table can be graphed on semilog graph paper. If the 100-year SWEL does not contain wave setup, it should fall on a line drawn through the other SWELs.

Wave height: This is the vertical distance between the trough and crest of a wave. Because most waves are irregular (height, length, and period are not constant at a given location and point in time), the Coastal Flood Model adopts the "significant" wave height concept: the *significant wave height*, H_s, is approximately the average of the highest one-third of wave heights at a site.

Wave run-up: This is the height above the stillwater level that waves rush up a slope after breaking. Because incoming waves vary, wave run-up also varies at a given site. The Coastal Flood Model uses the average wave run-up height, or rave, to determine flood hazards associated with wave run-up. *This feature is not currently available in the flood model.*

Transect: This is an imaginary line drawn perpendicular to the shoreline, across which dune erosion and wave effect calculations are made. The Coastal Flood Model automatically draws transects at approximately 1,000-foot intervals along the shoreline. The user cannot add, delete, or edit transects.

Wave setup: This is a local increase in the stillwater level due to the presence of breaking waves. The Coastal Flood Model assumes that the wave setup component is at maximum near the shoreline (at the dune toe), and decays in the inland direction at a rate of 10 percent per 100 feet. Wave setup is not calculated for interior fetches.

Figure 9.2 shows the basic concept that Hazus-MH uses for coastal flooding caused by a storm. The base of the storm surge is caused by three factors: (1) winds piling water above the mean sea level plus tide level, equaling approximately 85 percent of the storm surge level; (2) waves pushing water inland faster than the water can drain off the land (wave setup), approximately 5–10 percent of the storm surge level; and (3) the low-pressure effect of the storm increasing the water higher near the eye of the storm, approximately 5–10 percent of the storm surge level.

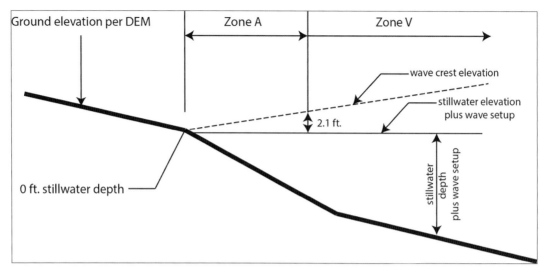

Figure 9.2. The Hazus-MH Wave Height Model shows the basic structure used for coastal flood hazard modeling. The model shows the relationship between wave crest, stillwater flood depth, and wave setup.

Source: Department of Homeland Security Federal Emergency Management Agency (DHS FEMA 2013c, section 4: 98). Figure reproduced by the Arizona Board of Regents on behalf of the University of Arizona.

The base model sets the mean sea level for the coastline DEM equal to zero elevation, which is the mean value of sea level at a location on the coastline. The components of the storm surge, including the tide level above or below mean sea level, the wind setup, pressure setup, and sea bottom configuration, are all included in the model. These variables may be changed if necessary. The tide level fluctuates with a number of variables, but tidal elevations change primarily as a result of the moon's gravitational interaction with the earth. As the moon revolves around the earth, the tide level increases as the moon's gravitational effect increases because of the moon's location in relation to the earth. The inverse barometer level, or pressure setup, increases the storm surge as a result of the low pressure within the storm. The lower the storm central pressure, the larger the storm surge elevation increases because of the inverse barometric pressure setup. The wind setup elevation is the height of the water pushed in front of the storm by wind and does not include wave effects.

SWEL is determined by engineers and scientists as part of an FIS. Most counties have an FIS, and coastal counties have a coastal flooding portion of the report. The coastal flooding portion of the FIS report includes the "Summary of Coastal Stillwater Elevations" (figure 9.3), which provides SWELs for 10-, 50-, 100-, and 500-year storms that are calculated for specific sections along the county's shoreline.

Figure 9.4 shows the Hillsborough County DEM. DEM is an important component of coastal (storm surge) or riverine flood modeling. Hazus-MH uses a DEM as one part of its model to determine the extent and depth of coastal flooding commonly caused by a storm surge. A DEM is a digital model of land surface for the county or specified area on the earth's surface.

SUMMARY OF COASTAL STILLWATER ELEVATIONS

FLOODING SOURCE AND LOCATION	PEAK ELEVATION (feet NAVD 88)*			
	10-PERCENT	2-PERCENT	1-PERCENT	0.2-PERCENT
GULF OF MEXICO				
OLD TAMPA BAY				
Near mouth of Boat Bayou	4.6	7.8	9.2	12.3
Near mouth of Rocky Creek	4.4	7.7	9.3	12.5
HILLSBOROUGH BAY				
Near mouth of Delaney Creek	4.8	8.4	10.1	13.3
Near mouth of Alafia Creek	5.1	8.4	10.2	13.4
TAMPA BAY				
Near Apollo Beach Road	4.4	7.9	9.4	12.6
Near mouth of Little Manatee River	4.1	8.3	8.6	11.7
Near Mill Bayou on the Little Manatee River	4.1	7.2	8.5	11.5
Near intersection of Cockroach Bay Road and Gulf City Road	4.1	7.1	8.5	11.4
Near mouth of Piney Point Creek	3.9	6.8	8.1	10.7

*North American Vertical Datum of 1988

Figure 9.3. Summary of coastal stillwater elevations for the 10-year (10 percent), 50-year (2 percent), 100-year (1 percent), and 500-year (0.2 percent) storms. FEMA classifies the 100-year storm as the base storm.

Source: Federal Emergency Management Agency 2008.

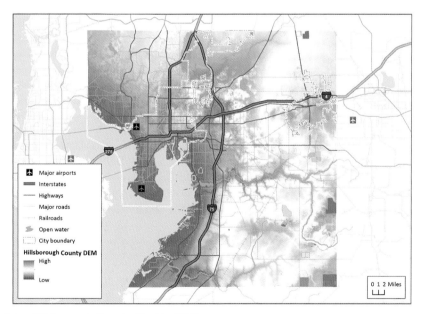

Figure 9.4. Hillsborough County, Florida, DEM.

Figure from Paul D. Zwick and the Arizona Board of Regents on behalf of the University of Arizona. Airports from the Federal Aviation Administration and the Research and Innovative Technology Administration Bureau of Transportation Statistics National Transportation Atlas Databases (NTAD) 2012; water bodies from the US Geological Survey; state boundary from US Census Bureau TIGER/Line Files; interstates, highways, and major roads from the Florida Department of Transportation; railroads from the Federal Railroad Administration; digital elevation model from the US Geological Survey.

Once the SWEL value is set in the Hazus-MH Flood Model, the model calculates the wave setup and wave crest elevations using a DEM and Hazus-MH-generated transects. The transects progress inland from the shoreline until SWEL and the wave setup plus wave crest value equal zero. Figure 9.5 shows Hazus-MH transects generated for the base 100-year storm surge.

Most planning analysts probably do not need to use a model like Hazus-MH to analyze storm surge for land-use planning. To become a professional user of Hazus-MH requires attending and passing FEMA training on the model. Unless planners are part of an organization for disaster or emergency management, they probably would not need to model coastal flooding. The Flood Insurance Rate Maps (FIRM) or storm surge polygons provided by county or state emergency management offices should suffice for land-use analysis of storm surge impacts. However, you can use the Hazus-MH Flood Model to investigate the impact of SLR and an accompanying storm surge in combination with SLR to estimate the increased area and depth of coastal inundation. Storm surge increase as a result of SLR is, or should be, an important factor in developing future land-use plans for coastal counties in states along the Gulf and the East Coast. This chapter uses the Hazus-MH Flood Model because (1) ArcGIS software supports the application; (2) FEMA recognizes it as a tool for disaster cost estimations; and (3) the model provides storm surge damage estimates by census boundaries, which are useful for planners and county officials.

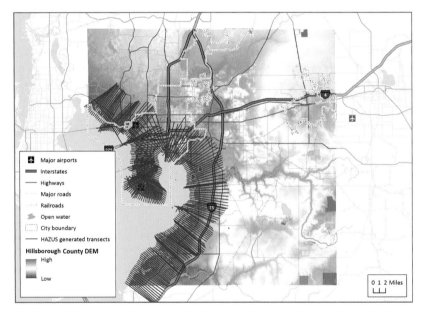

Figure 9.5. Hazus-MH-generated transects. Each transect is created perpendicular to the coastline at the location of each transect. Transects are used to sample DEM and determine differences between SWEL and a DEM necessary to Hazus-MH modeling of storm surge (for example, for wave height and wave crest height).

Figure from Paul D. Zwick and the Arizona Board of Regents on behalf of the University of Arizona. Airports from the Federal Aviation Administration and the Research and Innovative Technology Administration Bureau of Transportation Statistics National Transportation Atlas Databases (NTAD) 2012; water bodies from the US Geological Survey; state boundary from US Census Bureau TIGER/Line Files; interstates, highways, and major roads from the Florida Department of Transportation; railroads from the Federal Railroad Administration; digital elevation model from the US Geological Survey.

Sea level rise—the problem

According to a National Academy of Sciences report (2010, 244–45), *Advancing the Science of Climate Change: America's Climate Choices*:

> Research on current and potential future rates of SLR has advanced considerably since the IPCC Fourth Assessment Report, which was based on data published in 2005 or earlier. Some research conducted during the past several years suggests that SLR during the 21st century could be several times the [Intergovernmental Panel on Climate Change] IPCC estimates. … Empirical techniques (e.g., Grinsted et al. 2009; Rahmstorf 2007; Vermeer and Rahmstorf 2009) that relate sea level to historical average temperatures suggest that an SLR of up to nearly 5 feet (1.4 meters) is possible by 2100. By incorporating this empirical effect into models, Horton et al. (2008) estimate an SLR of 2 to 2.6 feet (0.62 to 0.88 meters) by 2100. In other work, Rohling et al. (2008) find that a rise rate of up to 5 feet (1.6 meters) per century is possible, based on paleoclimatic evidence from past interglacial periods (including the most recent interglacial period, 110,000 years ago, when global temperatures were 3.6° F [2°C] higher than today and sea levels were 13 to 20 feet [4 to 6 meters] higher). Kopp et al. (2009) estimate that sea level peaked at 22 to 31 feet (6.6 to 9.4 meters) higher than today during the last interglacial period and had a 1,000-year average rise rate between 1.8 and 3 feet (0.56 to 0.92 meters) per century. Pfeffer et al. (2008) used geophysical constraints of ice loss to suggest that a 2.5-foot (0.8-meter) SLR is more likely, with a 6.5-foot (2-meter) rise the maximum to be expected by 2100. Others (Siddall et al. 2009) suggest that a 2.5-foot (0.8-meter) rise is the most we could experience by 2100, based on a model that is fit to data only since the last glacial maximum.

It is also logical to assume that SLR will greatly impact the losses from storm surge. The logic is simple. If an increased tide above mean sea level increases the storm surge elevation, an increase in mean sea level will also increase storm surge elevation. The impact of that increased storm surge will still depend on the other variables contributing to or decreasing the storm surge elevation. Following that assumption, the following scenarios provide examples for using Hazus-MH to investigate the impacts of storm surge with and without increased sea level elevation caused by climate change. The scenarios investigate the 100-year base storm surge without an increase in SLR and then the 100-year storm surge with SLR, of both a half meter and one meter.

Level 1 Hazus-MH analysis of a 100-year base storm, with and without SLR impacts

The storm surge of a 100-year base storm, with and without SLR impacts, can have wide-ranging implications for land use. You can explore these impacts using a level 1 Hazus-MH analysis.

To develop a base 100-year storm surge without SLR for a level 1 Hazus-MH analysis, use the base US Census data provided with the application and get SWEL from the FIS. To develop a base 100-year storm surge with SLR for a level 1 analysis (figure 9.6), collect the same data, using the base US Census data provided with the application. In fact, using Hazus-MH analysis to compare the effects of storm surge with and without SLR allows you to analyze the relative difference in expected exposure and damage values. However, for SLR level 1 estimates, make one slight modification to the Hazus-MH base 100-year storm surge data. The Hazus-MH Flood Model application uses SWEL to calculate the wave height and wave crest elevations (see figure 9.2). This requires calculating how many feet are in a half meter. If the scenario uses a half-meter SLR, add 1.65483 feet to SWEL for the calculation, in equation (9.1).

(39.5 inches per meter / 12 inches per foot) * 0.5 meter SLR = 1.65483 feet.   (9.1)

So equation (9.2) would then be as follows:

SWEL depth + SLR in feet + wave depth setup.   (9.2)

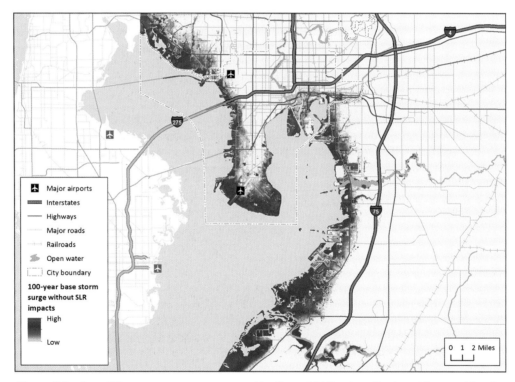

Figure 9.6. Base 100-year storm surge generated by Hazus-MH flood application with coastal flood selected.

Figure from Paul D. Zwick and the Arizona Board of Regents on behalf of the University of Arizona. Airports from the Federal Aviation Administration and the Research and Innovative Technology Administration Bureau of Transportation Statistics National Transportation Atlas Databases (NTAD) 2012; water bodies from the US Geological Survey; state boundary from US Census Bureau TIGER/Line Files; interstates, highways, and major roads from the Florida Department of Transportation; railroads from the Federal Railroad Administration.

To compare land-use implications of storm surge, the first two periods that are important are the current time and scenario end time. The following sequence of tables presents a level 1 Hazus-MH base 100-year storm surge with no SLR, base 100-year storm surge with a half-meter SLR, and base 100-year storm surge with one-meter SLR analysis using the technique described in this chapter. Because the version of Hazus-MH software available at the time this chapter was written uses the 2006 updated census data, the base year for the Hazus-MH SLR modeling of Hillsborough County is 2006. Tables 9.4 to 9.9 present selected information generated by the Hazus-MH Flood Model application.

Table 9.4 presents the 2006 census data basic demographics for Hillsborough County. The percentage values for residential land uses within the county show that residential use accounts for 70.8 percent of building structural value and 90.69 percent of the total structures. The building structural value in 2006 was approximately $79 billion for 405,461 structures.

Table 9.4. Hazus-MH summary data for Hillsborough County (2006)

Area in Sq. Mi.	Census Blocks	Number of Households	Population	Number of Structures	Value in Dollars (× 1,000)	Percent Value Residential	Percent Structures Residential
1,500	17,009	391,000	998,948	405,461	78,949,485	70.8%	90.69%

Source: Department of Homeland Security Federal Emergency Management Agency (DHS FEMA 2013b).

Table 9.5 shows Hillsborough County's 2006 base 100-year storm surge exposed building value by occupancy for the base storm surge and the two SLR scenarios. The values presented in this table are the maximum exposures generated by the maximum envelope of water. The maximum envelope of water is the total area of 100-year storm surge exposure from all possible 100-year storms, not from an individual 100-year storm. Essentially, the area is the total area exposed in the county to the probabilistic set of 100-year storms. For land-use planning, the area and exposure is the maximum exposure in the county. The shaded rows in table 9.5 indicate that residential and commercial buildings are greater than 91 percent of the structures in Hillsborough County (between $15 billion and $18 billion) potentially impacted by all three scenarios, from a base 100-year storm surge to a 100-year storm surge with one-meter SLR. As SLR increases from no SLR to a one-meter SLR, the structural value exposed increases by approximately $3.3 billion in 2006 dollars ($20,047,383,000–$16,783,568,000). Interestingly, the percentage of the exposed structural value remains consistent with 67 percent residential, 24 percent commercial, 4 percent industrial, and the remaining 5 percent distributed between four other categories (agriculture, religion, government, and education). The maximum flood area of the base 100-year (figure 9.6) storm surge without SLR is 57,454 acres and the area of the maximum base 100-year storm surge with one-meter SLR is 73,214 acres, representing 15,760 more acres. Because a square mile is 640 acres, the increased storm surge area represents an additional 24.625 square miles of flood area.

Many land-use planners in coastal counties are interested in the impacts resulting from a base storm surge. With climate change, they are also worried about how a future base storm surge will change with increased sea level. Intuitively, they should expect a significant increase in the number of impacted structures. Urban/suburban developments located in lower elevations close to the coastline are especially vulnerable. In fact, the additional 24.625 square miles in Hillsborough County represent 2.3 percent of the county's total land area. Tables 9.6 (occupancy type) and 9.7 (building type) present expected building damage for the base 100-year storm surge, again with and without SLR.

Table 9.5. Hillsborough County building exposure by occupancy type (2006)

Occupancy Type	Base 100-Year Storm Surge		Scenario with Base 100-Year Storm Surge with Half-Meter SLR		Scenario with Base 100-Year Storm Surge with One-Meter SLR	
	Exposure ($1,000)	Percent of Total	Exposure ($1,000)	Percent of Total	Exposure ($1,000)	Percent of Total
Residential	11,140,485	66.38%	12,072,069	66.97%	13,541,128	67.54%
Commercial	4,154,776	24.76%	4,371,485	24.26%	4,798,186	23.93%
Industrial	708,135	4.22%	770,254	4.27%	832,010	4.15%
Agriculture	71,409	0.42%	76,327	0.42%	82,129	0.41%
Religion	247,188	1.47%	264,081	1.47%	297,698	1.50%
Government	264,162	1.58%	264,469	1.47%	269,675	1.35%
Education	197,413	1.17%	205,060	1.14%	226,557	1.12%
Totals	16,783,568	100.00%	18,023,745	100.00%	20,047,383	100.00%

Note: For base 100-year storm surge, with and without SLR.

Table 9.6 shows expected building damage by occupancy type. The table indicates that 4,169 structures (4,166 residential) for the base storm surge with no SLR are substantially damaged. Substantial damage means greater than 50 percent of the structure is damaged. The number of substantially damaged structures increases to 9,445 (9,397 residential) as SLR increases to one meter. The most structure damage by occupancy type is 10,345 residential buildings in the category in which 21–30 percent of the building is damaged during a base storm surge with no SLR. It is 17,083 residential structures in the 41–50 percent category for the one-meter SLR scenario. The most obvious trend is a shift in the most structures damaged, which leads to greater damage as the number of damaged structures increases. With a half-meter SLR, the greatest number of residential structures (11,600) is in the category of 41–50 percent damaged. With a one-meter SLR, the greatest number of total structures damaged (17,109) is also in the 41–50 percent damaged category. A simple conclusion regarding damage from SLR is that the trend in expected structural damage will increase for most of these residential structures. Structures damaged in the base storm surge near the coastline will experience greater wave and storm surge depths as the sea level rises and will experience greater structural damage. Conceptually, the data in table 9.6 shows that the increased area of the storm surge because of SLR exposes

Table 9.6. Expected building damage by occupancy type (2006)

Base 100-Year Storm Surge						
Occupancy Type	1%–10%	11%–20%	21%–30%	31%–40%	41%–50%	Substantial Damage
Agriculture	1	1	1	0	0	0
Commercial	102	381	40	48	11	1
Education	11	1	0	0	0	0
Government	10	18	2	1	0	0
Industrial	0	16	8	5	4	2
Religion	0	30	0	0	0	0
Residential	29	3,139	10,345	3,287	4,561	4,166
Totals	153	3,586	10,396	3,341	4,576	4,169
Base 100-Year Storm Surge with Half-Meter SLR						
Occupancy Type	1%–10%	11%–20%	21%–30%	31%–40%	41%–50%	Substantial Damage
Agriculture	1	7	3	1	0	0
Commercial	53	830	86	47	27	6
Education	6	3	0	0	0	0
Government	20	29	3	1	0	1
Industrial	2	11	22	5	17	2
Religion	1	51	0	0	0	0
Residential	14	2,575	11,513	4,592	11,600	5,993
Totals	97	3,506	11,627	4,646	11,644	6,002
Base 100-Year Storm Surge with One-Meter SLR						
Occupancy Type	1%–10%	11%–20%	21%–30%	31%–40%	41%–50%	Substantial Damage
Agriculture	0	6	4	3	3	0
Commercial	22	965	243	64	7	33
Education	6	5	0	0	0	0
Government	21	49	3	2	1	2
Industrial	1	19	20	24	15	13
Religion	1	43	0	0	0	0
Residential	64	2,685	10,259	5,292	17,083	9,397
Totals	115	3,772	10,529	5,385	17,109	9,445

Note: For base 100-year storm surge, with and without SLR.

Table 9.7. Expected building damage by building type (2006)

Building Type	1%–10%	11%–20%	21%–30%	31%–40%	41%–50%	Substantial Damage
Base 100-Year Storm Surge						
Concrete	7	100	346	80	143	3
Manufactured Housing	0	8	25	0	29	4,006
Masonry	48	2,360	7,256	2,335	3,157	126
Steel	67	219	26	22	11	2
Wood	19	878	2,798	909	1,247	33
Totals	141	3,565	10,451	3,346	4,587	4,170
Base 100-Year Storm Surge with Half-Meter SLR						
Concrete	8	129	374	112	441	16
Manufactured Housing	0	4	32	0	44	5,172
Masonry	29	2,113	8,065	3,288	8,022	607
Steel	44	426	67	15	33	7
Wood	11	793	3,127	1,224	3,152	207
Totals	92	3,465	11,665	4,639	11,692	6,009
Base 100-Year Storm Surge with One-Meter SLR						
Concrete	5	176	358	128	638	55
Manufactured Housing	0	4	27	0	60	6,694
Masonry	52	2,207	7,285	3,773	11,862	1,924
Steel	32	505	140	45	35	30
Wood	18	845	2,765	1,432	4,627	738
Totals	107	3,737	10,575	5,378	17,222	9,441

Note: For base 100-year storm surge with and without SLR.

a greater number of structures to damage. Moreover, the expected damage may increase significantly because the structures impacted by a base storm surge with no SLR are exposed to greater storm surge impacts because of an increase in sea level. The expected damage for individual areas could vary greatly, depending on surface elevations near the coastline. Some areas could experience little to no increase in storm surge area. However, the damage to structures impacted by a base storm surge with no SLR might clearly be exposed to greater damage with SLR.

For occupancy types other than residential, the most damage is to commercial structures, ranging from 381 structures with 11–20 percent damage for the base 100-year storm with no SLR to 965 structures with 11–20 percent damage for the 100-year storm surge with one-meter SLR scenario. The number of commercial structures also expected to be significantly damaged increases from one to 33 for the same one-meter SLR scenario.

Table 9.7 shows the expected building damage by building type. Again, the substantial-damage category is dominated by residential damage. But most of the residential damage is to manufactured housing units. They account for 96.07 percent of the expected damage in the substantial-damage category. Also, 98.48 percent of the damage to manufactured housing is in the substantial-damage category. The category with the most structural damage by type is masonry structures in the 21–30 percent damaged range.

Table 9.8 shows the expected damage to essential facilities, which include fire stations, police stations, hospitals, and schools. The category of most interest is "Loss of Use." It shows that 28 essential structures, including 24 schools, are expected to be damaged enough to be closed or removed from use for a while. According to the Hazus-MH analysis, the model also estimated a loss of three hospitals, representing a loss of 852 beds, or a 20 percent reduction in bed capacity.

The following section in this chapter, "2006 property parcel-based level 1 Hazus-MH analysis, with and without SLR impacts," compares the number of parcels with structures built before 2007 to the expected loss presented in the Hazus-MH 100-year storm surge report for a 2006 storm. The comparison is not meant to check the reliability of the Hazus-MH expected losses. Rather, it is to see how closely the land-use alternatives in the property tax data for Hillsborough County support or vary from the 2006 census data that Hazus-MH uses. Hazus-MH uses nationally available data, whereas the property parcel data for Hillsborough County is locally available data. The locally available data could provide extra information for the base storm surge comparisons of expected damage in the LUCIS 2045 land-use scenario numbers.

Table 9.9 shows building value lost by category: (1) residential, (2) commercial, (3) industrial, (4) other, and (5) total structural loss.

Table 9.8. Expected damage to essential structures (2006)

Classification	Total Facilities	Number of Facilities Damaged		
		Moderate	Substantial	Loss of Use
Fire Stations	20	1	0	1
Hospitals	18	3	0	3
Police Stations	20	0	0	0
Schools	382	28	0	24

Table 9.9. Building value lost (expected loss in thousands of 2006 dollars)

Scenario	Residential	Commercial	Industrial	Other	Totals
Base	1,836,260	519,600	102,040	47,330	**2,505,230**
Half-Meter	2,789,350	770,450	146,010	72,120	**3,777,930**
One-Meter	3,570,400	1,095,990	193,840	101,180	**4,961,410**

Note: For base 100-year storm surge, with and without SLR.

2006 property parcel-based level 1 Hazus-MH analysis of a 100-year base storm, with and without SLR impacts

The figures in this section are for land-use implications associated with a 100-year base storm surge, with and without SLR impacts. The level 1 Hazus-MH analysis uses 2006 property parcel data. Figure 9.7 shows an area including much of the Tampa peninsula and the surrounding area. The multicolored areas (red, blue, and aqua) are commercial parcels impacted by the base 100-year storm surge with no SLR. The red parcels are in the significantly flooded, Hazus-MH 50+ Percent Parcel Flooded category. Table 9.10 presents Hazus-MH data describing the number of commercial structures exposed in "percent flooded" categories.

As shown in the Hazus-MH base 100-year storm surge census data summaries (tables 9.4 to 9.9), the largest amount of damage is in the significant-damage category. Of the 920 parcels totaling 1,525.5 acres impacted in the 50+ Percent Parcel Flooded category, 91.37 percent of the parcel is flooded. Some 1,159 commercial structures on 1,393.46 acres are flooded. The structural value of the 1,159 commercial buildings is approximately $147.7 million. The parcel analysis does not depreciate the building value based on the year built nor does it calculate the damage to the building structure or estimate the building content damage, as the Hazus-MH model does. The intent of reviewing this data is neither to verify nor replicate the Hazus-MH analysis, but to examine the building market values on the significantly flooded parcels. It is reasonable to assume that commercial parcels with 92 percent of their area flooded and $147.7 million worth of structures will require extensive repairs. These repairs could take months, and therefore clearly impact the workforce employment provided by these commercial employers.

The impact on these same commercial parcels located within base 100-year storm surge areas is greater if SLR is added to the mix. Trying to estimate the value of commercial structural damage for a 100-year storm occurring in or after 2045 when the impacts of SLR will begin to be felt, but using 2006 parcel data as the base value, is not a constructive analysis. However, LUCIS offers a viable method to estimate the impacts of a base 100-year storm surge with a half-meter SLR on the new 2045 development. Building on allocation scenario 5 (see chapter 6), this method is presented in the next section of this chapter, "Using LUCIS to assess future land-use implications, with and without SLR impacts."

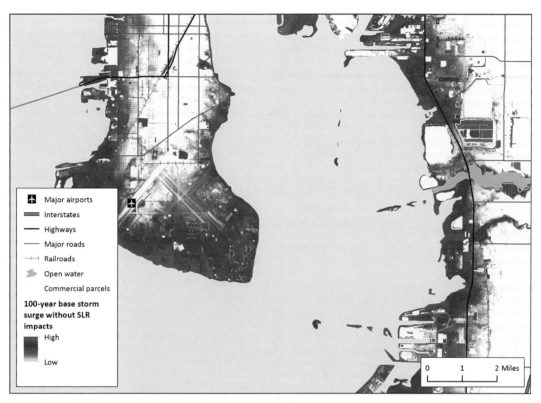

Figure 9.7. Commercial parcels expected to flood during a base storm surge 100-year flood with no SLR. The plane runways on the peninsula tip of Tampa are on MacDill Air Force Base.

Figure from Paul D. Zwick and the Arizona Board of Regents on behalf of the University of Arizona. Airports from the Federal Aviation Administration and the Research and Innovative Technology Administration Bureau of Transportation Statistics National Transportation Atlas Databases (NTAD) 2012; water bodies from the US Geological Survey; state boundary from US Census Bureau TIGER/Line Files; interstates, highways, and major roads from the Florida Department of Transportation; railroads from the Federal Railroad Administration; parcels from the Florida Department of Revenue and the Hillsborough County property appraiser.

Table 9.10. Commercial parcels flooded (with structures built before 2007)

Category	Parcels	Number of Buildings	Percent of Total Structures	Parcel Acres	Flood Acres	Percent Parcel Flooded
1%–10% Flood	85	148	9.09	426.4	18.32	4.30
11%–20% Flood	45	68	4.17	244.1	38.32	15.69
21%–30% Flood	47	70	4.30	172.4	43.04	24.97
31%–40% Flood	29	39	2.39	113.0	37.66	33.34
41%–50% Flood	54	145	8.90	1,847.9	866.93	46.92
50+% Flood	920	1,159	71.15	1,525.5	1,393.46	91.37

Note: Within the base 100-year storm surge with no SLR.

Figure 9.8 shows the total parcel base for 2006 with the shaded areas indicating the percentage of the parcel flooded. In the red areas, greater than 50 percent of the parcel's acreage is flooded by the storm surge. The figure includes all categories of land use, not only commercial properties. The parcels with greater than 50 percent of the area flooded clearly show the significant impact of storm surge on the coastal areas of the county. The total area flooded within the category of "greater than 50 percent" contains 41,000 parcels on approximately 33,029 acres. This damage accounts for 85 percent of the total flood area.

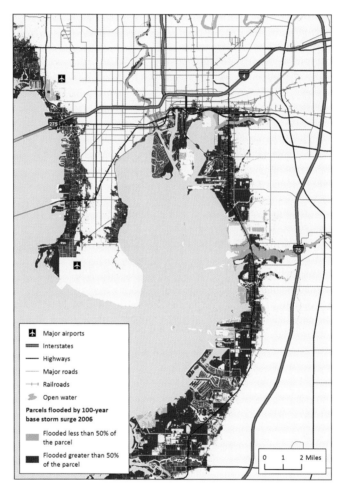

Figure 9.8. Parcels for 2006 flooded by the base 100-year storm surge with no SLR. The red areas include parcels flooded at greater than 50 percent.

Figure from Paul D. Zwick and the Arizona Board of Regents on behalf of the University of Arizona. Airports from the Federal Aviation Administration and the Research and Innovative Technology Administration Bureau of Transportation Statistics National Transportation Atlas Databases (NTAD) 2012; water bodies from the US Geological Survey; state boundary from US Census Bureau TIGER/Line Files; interstates, highways, and major roads from the Florida Department of Transportation; railroads from the Federal Railroad Administration; parcels from the Florida Department of Revenue and the Hillsborough County property appraiser.

Using LUCIS to assess future land-use implications, with and without SLR impacts

This section uses LUCIS to assess future land-use implications associated with a 100-year base storm surge, with and without SLR impacts. Figure 9.9 shows the advancing 100-year storm surge, first with no SLR and then as SLR increases from a half-meter to a one-meter increase in sea level. The area of the storm surge increases from a base 100-year storm surge with no SLR of 57,454 acres (dark-blue area) to 66,135 acres for a half-meter SLR (dark-blue and lighter blue areas) to 73,214 acres for a one-meter SLR (all blue areas combined).

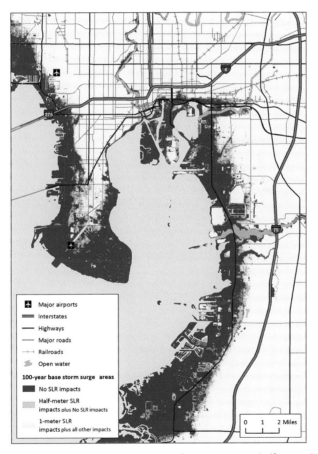

Figure 9.9. Increasing 100-year base storm surge areas from no SLR to a half-meter SLR to a one-meter SLR.

Figure from Paul D. Zwick and the Arizona Board of Regents on behalf of the University of Arizona. Airports from the Federal Aviation Administration and the Research and Innovative Technology Administration Bureau of Transportation Statistics National Transportation Atlas Databases (NTAD) 2012; water bodies from the US Geological Survey; state boundary from US Census Bureau TIGER/Line Files; interstates, highways, and major roads from the Florida Department of Transportation; railroads from the Federal Railroad Administration; parcels from the Florida Department of Revenue and the Hillsborough County property appraiser.

Tables 9.11, 9.12, and 9.13 present the LUCIS scenario 5 allocations for 2045 impacted by the base 100-year storm surge and the 100-year storm surge with half-meter SLR. Table 9.11 shows the population, residential acres, dwelling units, mean residential unit density, employment acres, jobs, and mean employee density for the allocated acres. The summary information shown in the final row of the table indicates that for the selected LUCIS 2045 scenario, 92,790 new residents are residing in 35,420 new dwelling units exposed to the base 100-year storm surge with no SLR. Also, 4,778 employment acres with 42,236 employees are

Table 9.11. New 2045 urban/suburban development (population and employment)

LUCIS Type	Population	Residential Acres	Dwelling Units	Mean Residential Unit Density	Employment Acres	Jobs	Mean Employee Density
Infill							
Commercial-Retail	0	0.00	0	0.00	401.68	4,282	10.66
Residential-Commercial	1,152	81.80	387	4.73	81.80	872	10.66
Residential	25,056	1,646.13	8,167	4.96	0.00	0	0.00
Institutional	0	0.00	0	0.00	219.27	1,171	5.34
Institutional-Service	0	0.00	0	0.00	390.47	7,480	19.16
Industrial	0	0.00	0	0.00	609.00	3,998	6.56
Service	0	0.00	0	0.00	226.95	3,542	15.61
Redevelopment							
MU Residential-Commercial-Retail	56,320	776.00	23,116	29.79	776.00	6,902	8.89
Residential	3,820	213.00	1,542	7.24	0.00	0	0.00
Industrial	0	0.00	0	0.00	502.00	1,882	3.75
Service	0	0.00	0	0.00	970.00	9,357	9.65
Greenfield							
Residential	6,442	225.00	2,208	9.81	0.00	0	0.00
Industrial	0	0.00	0	0.00	163.00	783	4.80
Institutional	0	0.00	0	0.00	438.00	1,967	4.49
Summary	92,790	2,941.93	35,420	12.04	4,778.17	42,236	8.84

Note: For a base 100-year storm surge with no SLR.

exposed to the same storm surge. The table indicates that significant numbers of 2045 new residential dwellings and jobs are exposed to a present-day sea-level base 100-year storm surge. As stated earlier, the expected probability of a base 100-year storm surge for any given year in 100 years is 1 percent.

Finally, calculating the expected building losses, economic loss of jobs, and internal residential and business building losses for furniture and other real property is difficult, since the hypothetical storm occurs decades from now, in 2045. However, considering the expected losses from the Hazus-MH base storm surge using 2006 US Census data, you can crudely estimate the losses using the mean building loss value for residential structures of $71,934 (table 9.6 residential structures damaged for the base storm surge with no SLR of 25,527 and table 9.9 residential base building loss of nearly $1.8 billion). Table 9.11 indicates that for the 35,420 new 2045 residential structures, the expected building damage could be nearly $2.6 billion in 2006 dollars. Although the dollar figure is substantial, the more significant statistic is the number of additional new residential structures and residents exposed to the storm surge.

Table 9.12 shows the same information. However, in this case, the property parcels are impacted by a 100-year storm surge with a half-meter SLR. The summary information shown in the final row of the table indicates that for the LUCIS 2045 residential development, 104,905 people, residing in 39,812 dwelling units on 3,391.02 acres, are exposed to the 100-year storm surge. Some 5,560.38 employment acres supporting 48,456 employees are also exposed to the storm surge. Again, calculating the building losses, economic loss of jobs, and internal residential and business building losses for furniture and other real property is difficult. However, considering the expected losses from the Hazus-MH base storm surge using updated 2006 US Census data, you can crudely estimate the losses using the mean building loss value for residential structures of $71,934. Considering that there are 39,812 new 2045 residential structures, the expected building damage for residential structures alone could be nearly $2.9 billion in 2006 dollars. Again, the dollar losses could be significant, but the chance that the storm surge with a half-meter SLR could expose an additional 12,115 residents (104,905 – 92,970) and 4,392 residential dwellings (39,812 – 35,420) is of much greater concern.

Table 9.13 presents the change (in most cases, increases) in property exposed to the storm surge with a half-meter SLR. There are 12,115 more people exposed to the half-meter 100-year base storm surge compared with the base 100-year storm surge. An increase of 4,392 dwelling units is exposed to the surge, and an extra 449 acres of residential units are exposed to storm surge impacts. The business economic losses include 6,220 jobs exposed and 781 acres of businesses impacted, including the associated business structures. Because some of the business structures could be significantly damaged, the affected jobs could be lost for a lengthy period, or even eliminated in the recovery.

Figure 9.10 displays the major land-use categories for LUCIS 2045 allocations. The yellow locations are residential uses with no distinction between redevelopment, infill, or greenfield residential development. The red areas are commercial, purple are mixed-use development (multifamily, commercial, and retail or service), black are industrial, and

Table 9.12. New 2045 urban/suburban development

LUCIS Type	Population	Residential Acres	Dwelling Units	Mean Residential Unit Density	Employment Acres	Jobs	Mean Employment Density
Infill							
Commercial-Retail	0	0.00	0	0.00	448.00	4,777	10.66
Residential-Commercial	1,383	91.00	469	5.15	91.00	970	10.66
Residential	29,554	1,884.86	9,567	5.08	0.00	0	0.00
Industrial	0	0.00	0	0.00	753.29	4,758	6.32
Institutional	0	0.00	0	0.00	278.47	1,477	5.30
MU Institutional-Service	0	0.00	0	0.00	419.22	8,029	19.15
Service	0	0.00	0	0.00	246.29	3,818	15.50
Redevelopment							
MU Residential-Commercial-Retail	61,410	917.01	25,256	27.54	917.01	8,159	8.90
Residential	4,467	240.87	1,762	7.32	0.00	0	0.00
Industrial	0	0.00	0	0.00	502.00	1,882	3.75
Service	0	0.00	0	0.00	1,146.95	10,990	9.58
Greenfield							
Residential	8,091	257.28	2,758	10.72	0.00	0	0.00
Industrial	0	0.00	0	0.00	191.77	965	5.03
Institutional	0	0	0	0	566.38	2,631	4.65
Summary	104,905	3,391.02	39,812	11.74	5,560.38	48,456	8.71

Note: For a base 100-year storm surge with a half-meter SLR.

orange are institutional. The residential acreage and number of dwelling units, along with the number of jobs and employment acres, are shown in table 9.12. You can get statistics for individual parcels using the ArcGIS Zonal Statistics as Table tool. These statistics include the mean depth, minimum depth, maximum depth, and standard deviation of the depth for each unique parcel exposed to the storm surge.

Table 9.13. Change from base 100-year storm surge with no SLR to a half-meter SLR

LUCIS Type	Population Change	Residential Acre Change	Residential Dwelling Unit Change	Employment Acre Change	Job Change
Infill					
Commercial-Retail	0	0	0	46	495
Residential-Commercial	231	9	82	9	98
Residential	4,498	239	1,400	0	0
Industrial	0	0	0	144	760
Institutional	0	0	0	59	306
Institutional-Service	0	0	0	29	549
Service	0	0	0	19	276
Redevelopment					
Residential-Commercial-Retail	5,090	141	2,140	141	1,257
Residential	647	28	220	0	0
Industrial	0	0	0	0	0
Service	0	0	0	177	1,633
Greenfield					
Residential	1,649	32	550	0	0
Industrial	0	0	0	29	182
Institutional	0	0	0	128	664
Summary	**12,115**	**449**	**4,392**	**781**	**6,220**

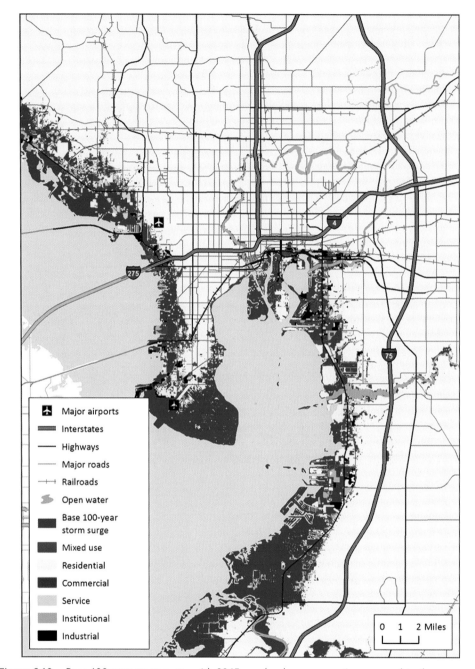

Figure 9.10. Base 100-year storm surge with 2045 new land-use properties exposed to the surge.

Figure from Paul D. Zwick and the Arizona Board of Regents on behalf of the University of Arizona. Airports from the Federal Aviation Administration and the Research and Innovative Technology Administration Bureau of Transportation Statistics National Transportation Atlas Databases (NTAD) 2012; water bodies from the US Geological Survey; state boundary from US Census Bureau TIGER/Line Files; interstates, highways, and major roads from the Florida Department of Transportation; railroads from the Federal Railroad Administration; parcels from the Florida Department of Revenue and the Hillsborough County property appraiser.

Chapter summary

The following points are discussed in this chapter:

- Most scientists have declared that climate change, and global warming, exists. As a result, the issues related to climate change, global warming, and SLR are, or should be, of interest to land-use planners, elected officials, developers, and the general public. For planners grappling with the issue of SLR, perhaps the adage to "plan for the worst and hope for the best" is a wise suggestion. Not surprisingly, the issue of SLR is not without significant, and even confrontational, political debate, especially for groups that do not, or will not, accept that global warming and SLR are real.
- This chapter demonstrates that base 100-year storm surge events have significant impacts, even without SLR as a component of the storm surge. As the sea level rises, a storm surge can greatly increase the physical impacts to buildings and other properties. The SLR areas studied in this chapter expand approximately 22 percent in size above the base 100-year storm surge for the half-meter SLR and 34 percent for the one-meter SLR. The use of LUCIS allocation scenario 5 clearly indicates that redevelopment of urban areas that are close to Florida's coastline will produce increased structural damage and loss of structures that support existing and future jobs in the event of a 100-year flood because of storm surge.
- An answer to this dilemma might be to remove significant redevelopment from areas that will experience structural damage, both from future storm surge and storm surges that increase with SLR included. Another option might be to change the design of buildings to accommodate storm surge flooding more readily. For example, architects could use the lower floors in multistory buildings for parking garages rather than residential. The loss of vehicles parked on the lower floors would be a lesser value than the loss of residential units on the lower floors of the structure. A precedent for change might be found in the structural design of buildings in areas that are subject to earthquakes. A property tax solution might collect increased property taxes based on the probability of individual structures being impacted by a storm surge. Local government could use the funds collected to pay for services necessary to evacuate residents or repair public infrastructure or for the removal of storm surge debris. A regulatory remedy might be to place restrictions on the development of parcels subject to the percentage of the parcel that is flooded by a 100-year storm surge. Clearly, it is reasonable to assume that the damage to property is impacted by the percent of parcel flooded. Or all the previous alternatives could be combined into a strategy that uses the percentage of a parcel expected to be flooded, the structure's design, and the land elevation upon which the structure is built to place reasonable restrictions on development. Governments could require new

(continued)

(continued)

> structures to be placed at the highest elevations of the property, which would presumably reduce the damage caused by a storm surge.
> - Many of the previous alternatives could be eliminated by removing or reducing future development in areas subject to significant storm surge impacts. Whatever solutions are implemented, the necessity to retrofit existing urban areas to accommodate SLR and the accompanying storm surge impacts will be significant. The costs of retrofitting might well outweigh the depreciated value of individual structures within the storm surge area when the 100-year storm hits.

ArcGIS tools referenced in this chapter

Tool Name	Version 10.2 Toolbox/Toolset
Zonal Statistics as Table	Spatial Analyst Tools/Zonal

References

DHS FEMA (Department of Homeland Security Federal Emergency Management Agency). 2008. *Hillsborough County and Incorporated Areas Florida Flood Insurance Report* Vol. 1 of 5. Washington, DC: FEMA.
———. 2013a. *Flood Model Hazus-MH Technical Manual*. Washington, DC: Federal Emergency Management Agency.
———. 2013b. Hazus-MH software.
———. 2013c. *Hazus Multi-Hazard Loss Estimation Methodology, Flood Model User Manual*. Washington, DC: FEMA. www.fema.goc/plan/prevent/hazus.
Hazus. 2013. https://www.fema.gov/hazus.
National Academy of Sciences. 2010. *Advancing the Science of Climate Change: America's Climate Choices*. Panel on Advancing the Science of Climate Change. Washington, DC: The National Academies Press.
National Hurricane Center. 2010. National Weather Service. http://www.nhc.noaa.gov/surge.
National Hurricane Center. 2012. Saffir-Simpson Hurricane Wind Scale. http://www.nhc.noaa.gov/aboutsshws.php.
National Weather Service. 2010. *Introduction to Storm Surge*. Miami, FL: National Hurricane Center. http://www.nws.noaa.gov/om/hurricane/resources/surge_intro.pdf.
National Weather Service. 2012. Hurricane Sandy. http://www.weather.gov/okx/HurricaneSandy.
NOAA Digital Coast. 2013. Hazards US Multi-Hazard (Hazus-MH). http://www.csc.noaa.gov/digitalcoast/tools/Hazus-MH.
StormFacts.net. 2011. Hurricane Andrew Facts Page. http://stormfacts.net/handrew.htm.

Online resources

Federal Emergency Management Agency (FEMA), www.fema.gov.

Chapter 10

Providing LUCIS maps and information to users, clients, and the general public

Danny Downing and Paul D. Zwick

What this chapter covers

Up to this point, the focus of this book has been on learning the methods to apply the LUCIS model in various planning scenarios. Using LUCIS for land-use analysis provides powerful data for making tough planning decisions. This chapter is slightly different. It focuses on how to develop a simple web application for sharing data and models with both clients and the public. The LUCIS web application allows readers to view selected data from chapters 5 through 10 in this book. This chapter discusses the process used to create this application and then allows you to experiment with the application on the sample web page (see "Online resources" at the end of this chapter). ArcGIS Viewer for Flex software can display web-enabled services in an easily configurable and accessible viewer. A couple of assumptions are as follows:

- There is a web server running on the machine in which the viewer is installed.
- The user has read/write access to the folders containing the viewer application.
- There is access to ArcGIS for Server service.

ArcGIS Viewer for Flex Application Builder is a lightweight, XML-based viewer. It requires no coding knowledge to start using the application builder. You can download ArcGIS Viewer for Flex at http://resources.arcgis.com/en/communities/flex-viewer.

To download ArcGIS Viewer for Flex, you need an ArcGIS public account. You can create a public account at www.arcgis.com/home/createaccount.html.

Before you can install ArcGIS Viewer for Flex Application Builder, Adobe AIR must be installed on the server. Adobe AIR is available at http://get.adobe.com/air.

When you initialize the application for the first time, ArcGIS Viewer for Flex Application Builder will verify the web server base folder and web server base folder URL. When the base folder and URL are both correct, everything is ready to start building an application. Before using the viewer, it is a good idea to prepare and create the ArcGIS for Server services that will be used in the application.

Creating ArcGIS for Server services

Again, we used ArcGIS for Server to share the data that was created in chapters 5 through 10 of this book, using LUCIS. Creating services is a relatively simple process using the ArcGIS ArcMap application. For this example, we created a single service from some example suitability raster layers used in this book.

To serve GIS data, start by adding data layers to the Table Of Contents window in ArcMap (figure 10.1). Make sure that these data layers reside in a folder that ArcGIS for Server has the proper permission to access. If the permission access is unknown, you can change it later when the service is analyzed.

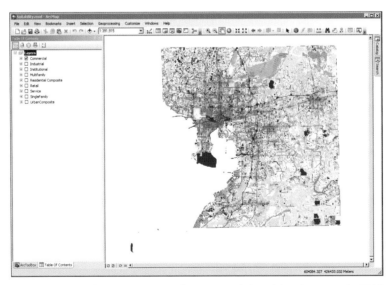

Figure 10.1. Creating ArcGIS for Server services begins simply by adding layers to the Table Of Contents window in ArcMap.

Figure from Danny Downing. Data derived from Paul D. Zwick and parcels from the Florida Department of Revenue and the Hillsborough County property appraiser.

The appearance of the layers in the map document (MXD) is how they will also appear in the viewer application, so any changes made there will carry over to the published services. Set the symbology of the layers, as well as labels, scale-dependent rendering, layer names, and layer visibility, in the map document. All data to be shared can be in one map document and published to a single service in your application. If the data lends itself to logical groupings that need to be maintained, you can create multiple services. For example, in the sample LUCIS application discussed in this chapter, there are separate services for some chapters of the book as well as a service for basemap data. (No maps or data are available with this book. The Flex Viewer application is available on the sample web page. See the "Using the viewer application" section later in this chapter.)

The layers in the map document do not have to be turned on in the table of contents. It may be preferable to have them turned off. This will allow users of the application to choose which layers to display on their own. It is okay if your map is blank. The data is still there and can be turned on from the viewer interface.

When the map document has been completed, it is ready to publish. Make sure to save the map document as it will be used in the future to make edits to the service.

The following steps cover taking a map document and publishing it to ArcGIS for Server as a web map service (WMS). By following the steps for this example, you can add your own resulting service to ArcGIS Viewer for Flex for easy sharing across the web. Flex Viewer configuration is covered in the next section of this chapter, "Creating the Flex Viewer application."

Publishing a WMS

On the File menu, click Share As > Service (figure 10.2).

Figure 10.2.

Figure from Danny Downing.

In the Share as Service window, click Publish a service (figure 10.3).

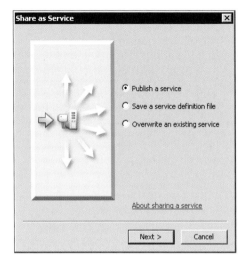

Figure 10.3.

Figure from Danny Downing.

In the Publish a Service window (figure 10.4), click the connection to the ArcGIS for Server instance listed in the drop-down menu. If this connection has not already been made, create a new connection. A prompt will ask you to add the connection information for the server. It will add the ArcGIS for Server instance to the Choose a connection drop-down list. Click the newly added server, keep or create a name for the service, and click Next.

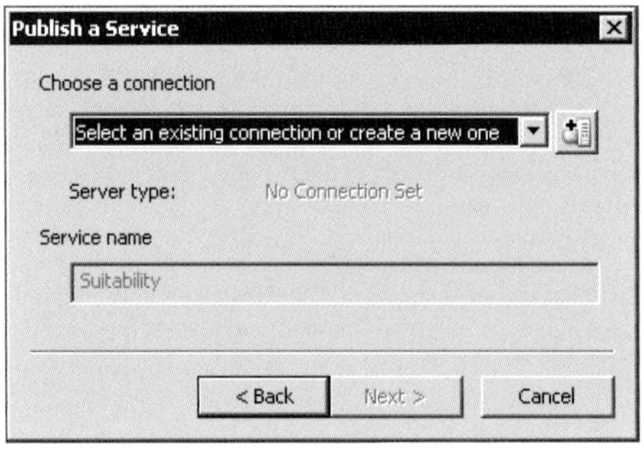

Figure 10.4.

Figure from Danny Downing.

In the subsequent Publish a Service window (figure 10.5), choose where to publish the service. Choose either an existing folder on the server or create a new one.

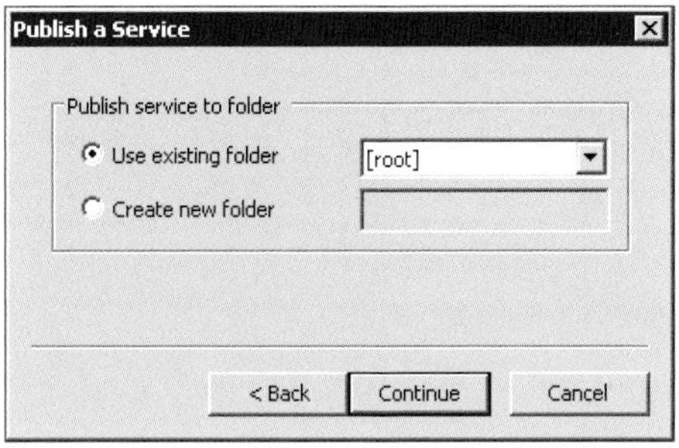

Chapter 10 Providing LUCIS maps and information to users, clients, and the general public

Figure 10.5.

Figure from Danny Downing.

The final window is the Service Editor (figure 10.6). Here you can edit the service properties.

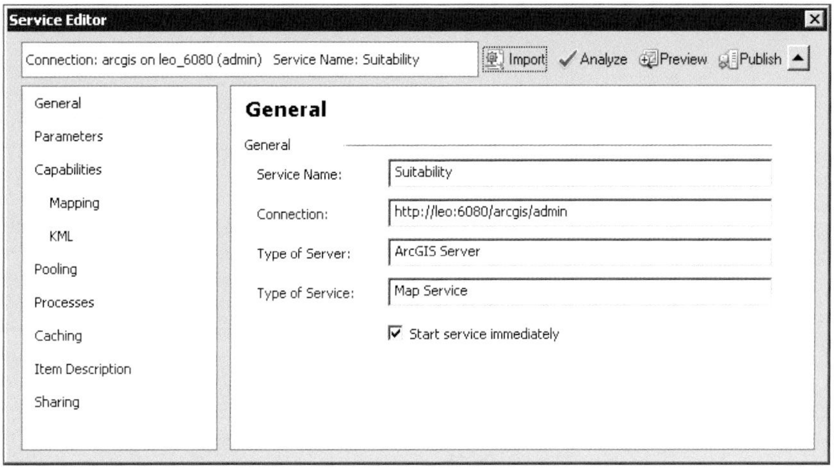

Figure 10.6.

Figure from Danny Downing.

For this example, the default settings were used whenever possible. Exploring the many options available will show the ability to add metadata (Item Description), create a cached data service instead of a dynamic service (Caching), configure pooling and processes, and add or remove service capabilities (Web Coverage Service [WCS], WMS, Feature Access, Schematics, Mobile Data Access, Network Analysis, KML, Web Feature Service [WFS]).

Before publishing the service, it is important to run the Analyze command at the top of the Service Editor. It will inspect the map document for common errors that can keep a service from being published or slow down performance. When completed, the analyzer will provide feedback on how to improve the service or whether any errors were encountered (figure 10.7).

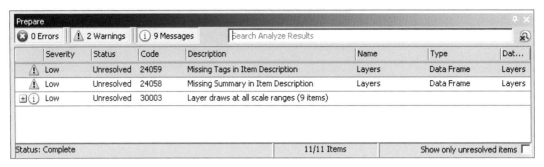

Figure 10.7.

Figure from Danny Downing.

Inspecting these messages and resolving any errors or warning messages can help in the creation of well-documented services that function more reliably and perform more smoothly.

Once you have resolved all the issues found in the Flex Viewer analysis, click the Publish button in the upper-right corner of the Service Editor. The service will be published to the server. The services created with ArcGIS for Server can now be added to Flex Viewer, providing a powerful platform to display and share data.

Creating the Flex Viewer application

The next step is to use the application builder to create a viewer for the service created in the previous section, "Creating ArcGIS for Server services."

Create an application

To build the LUCIS example application, we started by clicking the Create a New application button (figure 10.8). In the Create a New application dialog box, provide a name for the application. After the application name is chosen, the viewer application URL below it will be updated. It will show where the application will be viewable on the web. Click the Create button to continue.

Chapter 10 Providing LUCIS maps and information to users, clients, and the general public

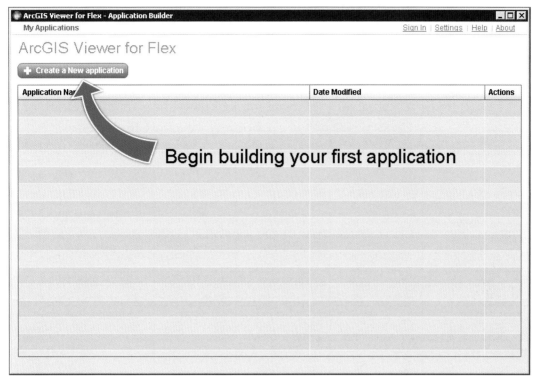

Figure 10.8.

Figure from Danny Downing.

Maps

You can create the new application using the Web Maps or Basemaps & Operational Layers tabs. The Web Maps option allows the use of existing maps from ArcGIS Online. This example uses Basemaps & Operational Layers to highlight user-created data from the ArcGIS for Server service created earlier in this chapter.

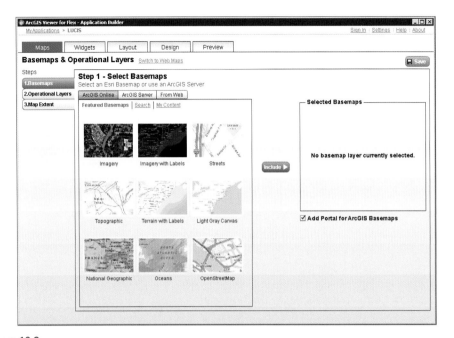

Figure 10.9.

Figure from Danny Downing.

The Basemaps & Operational Layers page (figure 10.9) has three tabs on the left, under Steps:

- *Basemaps.* For this application, the featured basemaps from ArcGIS Online are used as the base layer. All the available featured basemaps can be added, or limited to only the ones that work with the data in the hosted services. The first basemap added will be the default basemap for the application, or the basemap that is shown when the viewer first opens. If multiple basemap options are selected, they will be available to the user through the basemap switching widget.
- *Operational Layers.* The Operational Layers tab allows you to add the services created using ArcGIS for Server to the application. Click the ArcGIS Server tab. The REST Services URL of the ArcGIS for Server site should populate the address bar. Click Browse and a list of services should populate the Operational Layers window. Click the services to be displayed and then click Include. The Operational Layers can be configured by clicking the edit pencil next to the layer entry in the selected Operational Layers box. When editing the selected operational layers, you can change the layer label, visibility, transparency, and add information pop-up windows to layers.
- *Map Extent.* The third and last tab allows you to set the map extent. This is the area of the map that will be displayed when it is first opened. The sample LUCIS site we created is interested in only the area of the map associated with the provided operational layers. By clicking the Zoom to operational layers extent, the map will

automatically center to the provided layers. Alternatively, you can pan and zoom the map to any extent that is appropriate.

Widgets

The Widgets page (Esri 2013) in figure 10.10 contains a catalog of available built-in widgets. These include the following:

- *Bookmark widget:* spatial bookmarks for quick navigation.
- *Chart widget:* display charts for selected features.
- *Data Extract widget:* extracts data (for export).
- *Draw widget:* quick drawing (such as redlining).
- *Edit widget:* allows you to edit any editable feature layers in the operational layers.
- *Geoprocessing widget:* allows you to work with any geoprocessing task.
- *GeoRSS widget:* creates layers based on a GeoRSS feed.
- *Layer List widget:* displays the operational data in the Flex Viewer application.
- *Legend widget:* displays a legend for the layers you choose.
- *Locator widget:* allows address geocoding.
- *Print widget:* advanced or simple printing of the current map.
- *Query widget:* displays clickable features based on a query to a layer from an ArcGIS for Server map service or feature service.
- *Search widget:* allows users to select a clickable feature to display on the map. Similar to the Query widget, except the Query widget shows all the features by default.
- *Time widget:* enables time animation of time-aware layers.

Figure 10.10.

Figure from Danny Downing.

Third-party widgets are also available for download from the Esri ArcGIS Viewer for Flex Resources page, at http://resources.arcgis.com/en/communities/flex-viewer.

To include a widget in an application, select it and click the Include button. After adding widgets to the application, you can configure them by clicking the edit pencil below the widget icons. The configuration page for each widget is different. However, there are some items that are common to all widgets that can be configured (for example, the widget label, widget icon, and whether the widget is opened or closed by default when the application opens). Explore the configuration tab of each widget to determine the full range of settings available.

Adding a third-party widget to Application Builder

Third-party Flex Viewer widgets are available for download in the gallery on the Esri ArcGIS Viewer for Flex Resources page. The gallery allows you to browse and download widgets created by the developer community. To add a third-party widget to the Application Builder interface, make sure the versions match (Flex Viewer application version and widget version), and download the ZIP File that contains Application Builder integration. On the Application Builder Settings page, click Manage Custom Widgets. It opens the Manage Custom Widgets window (figure 10.11).

Figure 10.11.

Figure from Danny Downing.

Click the Add button and navigate to the widget ZIP File you downloaded. The third-party widget will now appear in the widget catalog and can be added to the application. For the LUCIS sample site being created here (figure 10.12), two third-party widgets have been added. The Enhanced Layer List widget is a table of contents widget that will be used to display and organize ArcGIS for Server services. The Swipe Spotlight widget will be used to quickly swipe between operational layers and the basemap below them.

Chapter 10   Providing LUCIS maps and information to users, clients, and the general public

Figure 10.12.

Figure from Danny Downing.

Layout

The Layout page (figure 10.13) allows for configuration of the elements that will be displayed by default on the viewer application.

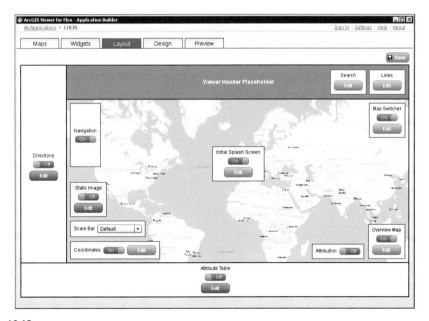

Figure 10.13.

Figure from Danny Downing.

Most elements can be turned on and off, and some can be configured further using the included Edit button.

Layout elements include the following:

- *Directions:* driving directions.
- *Navigation:* pan and zoom around the map.
- *Static Image:* displays an image on the map.
- *Scale Bar:* displays a scale bar in your choice of units.
- *Coordinates:* displays map coordinates while panning and zooming.
- *Initial Splash Screen:* an information window that greets users upon opening the viewer.
- *Attribution:* displays map attribution.
- *Overview Map:* displays an overview map.
- *Map Switcher:* display that allows switching basemaps and access to included services.
- *Search:* allows address geocoding.
- *Links:* allows you to add external links.
- *Attribute Table:* provides an interface for viewing the attribute table of map data.

For the LUCIS sample site, Directions, Static Image, Attribute Table, and Attribution will be turned off. Some additional configuration using the Edit button will be done to the Initial Splash Screen, Map Switcher, and Links elements. In the Initial Splash Screen edit window (figure 10.14), HTML will be added to share information from the example chapters of this book. All the code needed is already provided in the application builder.

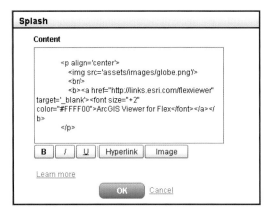

Figure 10.14.

Figure from Danny Downing.

To change the image on the splash screen, replace the image path (assets/images/globe.png) with a path to a new image. For this example, the LUCIS image has been changed to (assets/images/LUCIS.gif). To change the web address, replace the provided code (http://links.esri.com/flexviewer) with a direct link to a different web page or file (in this case, a link to a PDF chapter of this book). The text associated with the link can be changed by modifying the text (ArcGIS Viewer for Flex) to whatever is appropriate for the link. For this example, "LUCIS Pdf" was chosen. The edited splash content for the sample site is shown in figure 10.15.

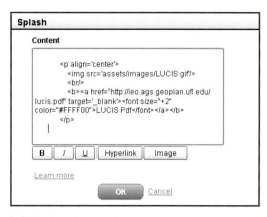

Figure 10.15. The Splash dialog box displays the code indicating the content that will appear on the splash screen when users open the website viewer. Using basic HTML code, you can customize color, images, and reference documents.

Figure from Danny Downing.

The HTML modifications produce the resulting splash screen for LUCIS^{plus} (figure 10.16).

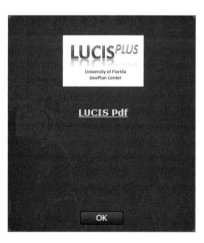

Figure 10.16.

Figure from Danny Downing.

In the Map Switcher edit window (figure 10.17), the layer list will be turned off so that the third-party widget added earlier will perform the function of giving access to operational layers.

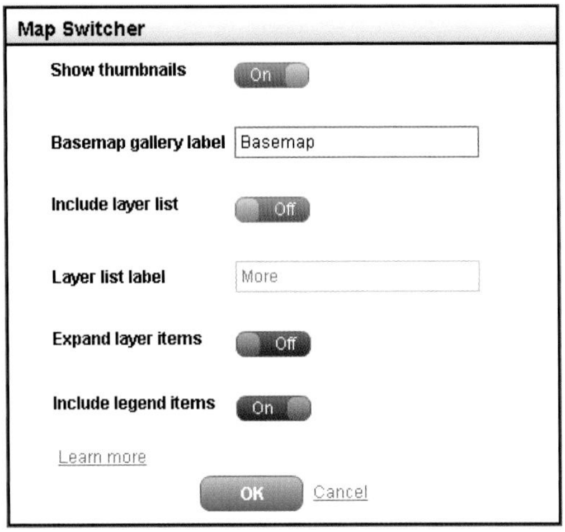

Figure 10.17.

Figure from Danny Downing.

Finally, the Links edit window (figure 10.18) will be set up to display the same information that is available in the initial splash screen.

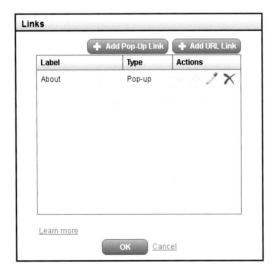

Figure 10.18.

Figure from Danny Downing.

To add material for the About link, click the edit pencil in the Actions column. HTML is presented that is identical to the original splash screen. Copy the code from the splash screen to this column to provide the same information in both the splash screen and the Links pop-up window.

Chapter 10 Providing LUCIS maps and information to users, clients, and the general public   273

Design

The Design page (figure 10.19) has three tabs and a preview window. The preview window shows the viewer application and updates as edits are made.

- *Logo.* The viewer displays an image logo (PNG, JPEG, GIF) in the upper-left corner of the site. This tab allows you to change the default logo (blue world cube). Drag a new image into the image drop area or browse to the file to change the image logo.
- *Title & Fonts.* This tab allows you to change the viewer title and add a subtitle. Drop-down lists allow you to change the fonts used throughout the viewer application. Selections include title font, subtitle font, and text font.
- *Color Scheme.* This tab allows you to set the color scheme of the viewer application. Choose from a number of predefined styles or build your own. Some elements of the viewer are built to be transparent. The level of transparency can be set here as well.

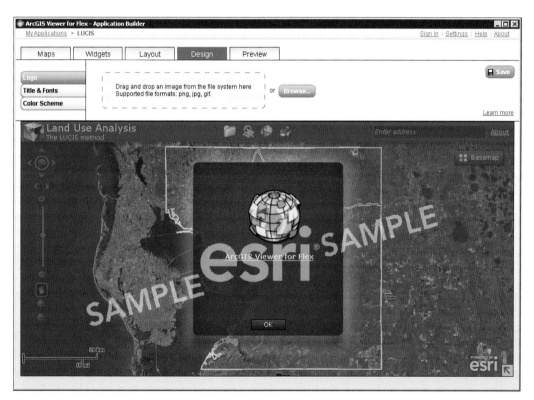

Figure 10.19.

Figure from Danny Downing.

Preview

The Preview page (figure 10.20) shows an interactive preview of the viewer application in its current state. Any visual changes or manual panning and zooming on this page will not be saved and will not affect the application in any way. Use this page to decide whether any viewer elements should be added, removed, or altered. Click the Done button to save the completed work and exit the application builder. All projects are listed on the My Applications page and can be accessed by clicking the edit pencil in the Actions column for the application name.

Figure 10.20.

Figure from Danny Downing.

Using the viewer application

Now that the example viewer application is complete, it is accessible on the web using any modern web browser. Project data and analysis results can be shared easily with both clients and the public. The completed viewer application is shown in figure 10.21. The viewer created here is available at http://lucisgeodesign.geoplan.ufl.edu. Now that you have read chapters 5 through 10, you can interact with the map while reading this chapter to experience the true capabilities of the software.

Figure 10.21. Completed viewer application for land-use analysis of Hillsborough County, Florida. The ArcGIS Viewer for Flex site that you see here contains all the data generated from the land-use analysis in chapters 5–10 of this book.

Figure from Danny Downing.

In the LUCIS example viewer, the Hillsborough County, Florida, boundary and interstates are turned on by default in the basemap layers group in the Layer List widget. These layers are turned on to help define the study area and give a sense of location within the study area.

From here, you can use Flex Viewer to view analysis results from select chapters of this book. Take, for example, the industrial suitability figure from chapter 5 (figure 5.6, shown again in figure 10.22).

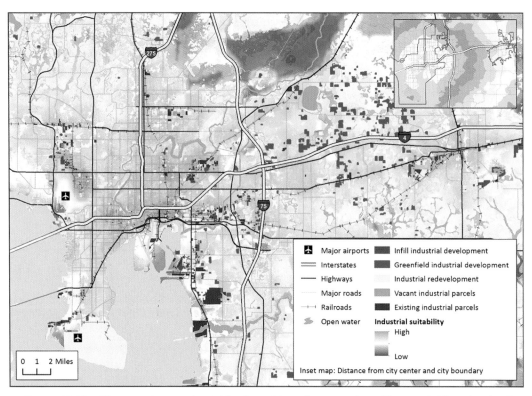

Figure 10.22. The map shows industrial land-use 2045 allocations for infill, greenfield, and redevelopment. The dark-gray areas are existing 2010 industrial parcels, and the lavender areas are the vacant industrial property allocations. Using the viewer application, you can take a more detailed look at this data. You can focus on specific areas of interest at whatever scale is needed, from block level to countywide.

Figure from Paul D. Zwick and Wanyi Song. Airports from the Federal Aviation Administration and the Research and Innovative Technology Administration Bureau of Transportation Statistics National Transportation Atlas Databases (NTAD) 2012; water bodies from the US Geological Survey; state boundary from US Census Bureau TIGER/Line Files; interstates, highways, and major roads from the Florida Department of Transportation; railroads from the Federal Railroad Administration; parcels from the Florida Department of Revenue.

Open the Chapter 5 folder in the Layer List widget and turn on the data layers that are visible in figure 10.23.

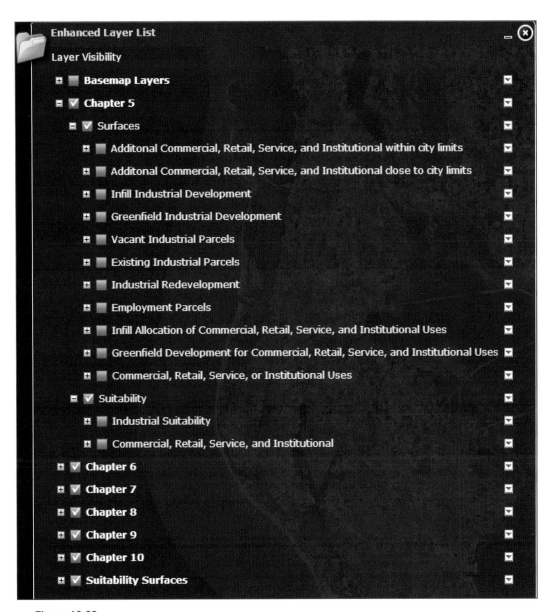

Figure 10.23.

Figure from Danny Downing.

The lower-right corner of all widgets can be used to resize the window view. The Layer List widget in the figure has been expanded to show the full layer names in chapter 5. Layers in the sample viewer application have been set to draw with a slight transparency so that the underlying imagery can be seen. This may make the symbology look slightly different from the book graphic, but the data is the same. The initial transparency can be adjusted in the Maps section of the Application Builder as described earlier and in the viewer application by clicking the down arrow of a parent group (in this case, chapter 5) and clicking Transparency.

Figure 10.24 shows figure 10.22 re-created at the county scale.

Figure 10.24. The industrial land-use 2045 allocations re-created at the county scale in the Flex Viewer application.

Figure from Danny Downing.

For map imagery with place labels, use the Basemap widget. Hover over the basemap icon, and the basemap options will expand. The basemaps shown (figure 10.25) were set from the Basemaps tab on the Basemaps & Operational Layers page.

Chapter 10 Providing LUCIS maps and information to users, clients, and the general public

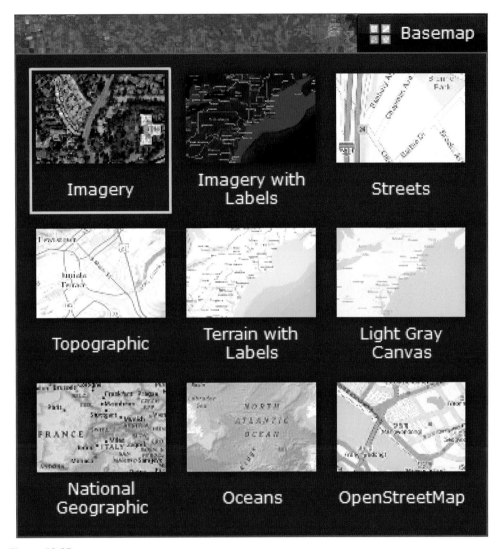

Figure 10.25.

Figure from Danny Downing.

Figure 10.26 shows the same data from figure 10.22 in the viewer application focused on the East Bay area of Hillsborough County.

Figure 10.26. The industrial land-use 2045 allocation zoomed in to the East Bay area of Hillsborough County.

Figure from Danny Downing.

Here, the map shows the areas that the LUCIS analysis has allocated as industrial land use on top of the industrial suitability raster layer. Use the Swipe Spotlight widget to swipe over allocations and examine the suitability or the imagery below them to see what the actual parcels look like. Quickly switch between suitability raster layers to see what locations are highly suitable for more than one use. Use the Draw widget to highlight areas of the map on the fly. Add and remove relevant data to better see the story your analysis is telling. Do the allocations look correct? Why are the target areas unsuitable? Why are competing parcels more suitable? Many of these questions require consensus building, multiorganization collaboration, and public input. Often, all these things happen at great distances from each other, both literally and figuratively. The Flex Viewer application is a powerful tool to expand the reach and explanatory power of your data and analysis.

Chapter summary

The following points are discussed in this chapter:

- Tools for creating simple web applications are within reach, even for the nonweb developer.
- The sample site described maximizes the default options and uses only a small sampling of the available widgets.
- With a little time and experience, even the simplest land-use analysis can be expanded and modified to accommodate any project or research.

ArcGIS applications referenced in this chapter

Tool Name	Version 10.3 Application
Flex Viewer Application Builder	ArcGIS Viewer for Flex

References

Esri. 2013. ArcGIS Viewer for Flex Help. http://resources.arcgis.com/en/communities/flex-viewer.

Online resources

Sample viewer application at the University of Florida GeoPlan Center, at http://lucisgeodesign.geoplan.ufl.edu.

Chapter 11
Applications of LUCIS on tribal lands

Iris E. Patten

What this chapter covers

Sound land-use planning is not solely reserved for traditional jurisdictions. Using geographic information system (GIS) techniques on tribal lands encourages a collaborative decision-making process. Land-use and scenario planning in nontribal jurisdictions is typically challenging. But developing that process is further challenged in an environment in which the preservation of tradition and the landscape must ensure that the needs of an entire people are met for another 500 years. It is a process that draws on public participation, data collection, modeling, and information distribution.

The Aldrich Nation is a federally recognized Indian tribe in Kentucky. (The name and location of the tribe have been changed for legal reasons.) Although the main reservation is substantial in size (1,979,002 acres), providing quality critical resources such as water, adequate transportation networks, and education has always been a challenge to the tribe. The Land-Use Conflict Identification Strategy (LUCIS) method was used as part of the development of a modern long-range plan (LRP) for the Aldrich Nation. LUCIS uses GIS tools to consider the intrinsic value of land and help evaluate complex problems. This method identifies land-use opportunities, minimizing the impact to Aldrich culture. Further, this strategy shifts the paradigm of planning practice by facilitating an informed decision-making structure that is truly by "the people."

The planning process

In 1979, a private consultant created the Aldrich Nation's first LRP. Although the document was composed using a typewriter and most charts were drawn by hand, the maps and methods reflected an appreciation for suitability modeling as an approach to inform land-use planning. When the tribe commissioned an LRP in 2012, improvements in GIS and mapping technology ensured a more detailed spatial planning process. But as with any GIS project, the result is only as good as the inputs. During this time, other scenario planning processes were being done in the surrounding region. But participation by the Aldrich, as well as other Native Americans, was low. It was a discouraging state of affairs, considering the large land area that the nation occupies in the region. One Aldrich tribal member suggested that it might be because the other scenario-planning processes did not lend themselves to preserving or including cultural and historical values that shape the tribe's decision-making ethos. Subsequently, a national planning agency was chosen to complete the Aldrich LRP. The LUCIS method was used because of its flexibility in customizing criteria for modeling and its capacity for including human values in land-use evaluations.

At the outset of creating the Aldrich LRP, many challenges existed. These obstacles included unofficial and often conflicting population figures, lack of current (or any) spatial data that is typically used in the base LUCIS goals and objectives matrix, uneasiness over the inclusion of data pertaining to sensitive tribal assets, and reluctance by tribal members to participate in a process so heavily influenced by technology. Taking these concerns into account, the planning project team developed a custom tribal land-use process, driven by collaborative planning and continual education of all stakeholders.

Step 1: Data gathering

The Aldrich Nation is divided into seven districts, which vary in size and landownership structure. Although all districts generally employ an individual landownership under trust, or allotment, model, each district exercises discretion in the amount of land distributed in each allotment. Additionally, the nation has about 10,000 residents that live on the reservation, and most homesteads are not surveyed. The first step in the tribal LUCIS process was gathering data. Clearly, understanding land management practices within each district was the key to determining how to effectively illustrate and aggregate the LUCIS result. This phase also allowed the project team to begin conceptualizing modeling challenges early in the process.

The LUCIS process relies on a core set of goals and objectives related to land-use stakeholders. The tribal LUCIS model included the typical agriculture and conservation stakeholders. But considering the low density of development and modest on-reservation population, identifying the third stakeholder as "urban" did not seem appropriate. During the first phase, tribal leadership, district leadership, and district community members were given a presentation on the LUCIS process. It concluded with a discussion about whom the process should plan for. The consensus was that the modeling strategy and LRP process should result in a vision that was not just for projected annual population growth, but that it should also embrace an ethic that "land is life." Through an identification of *opportunities* for future land use instead of an *assignment* of future land use, the tribal LRP would be a living, working document. Thus, the final stakeholder was labeled *development*. District-level land-use and leadership committees would make the final decision on land-use assignments as development prospects presented themselves. Also, the project team would provide insight into the impacts on the land if all enrolled tribal members were to return to the reservation.

Land-use discussions typically involve an explanation of possible activities within different land uses. The benefit of this discussion is to support a landscape that provides stakeholders with the services and resources necessary to minimize unsafe environments and create a balance of compatible land uses. However, on many tribal lands, including the Aldrich Nation's, there is no formal declaration of land use, except in cases of range management or agriculture. This situation is not the tribe's fault because, in most cases, it reflects tribal values dating back to the delineation of formal reservation boundaries. The challenge facing tribes today is finding a balance of preserving valued lands and planning for a future that includes development-based uses. In regard to the Aldrich tribe, an understanding of the landownership structure provides only part of the picture. To fully use the LUCIS^{plus} (planning land-use scenarios) method, a more detailed stratification of land uses is needed. Fundamentally, the LUCIS strategy is enough to identify opportunities for future land use. Use of the advanced strategy LUCIS^{plus} can more quickly and accurately evaluate options for development in a context that is sensitive to both culture and the preservation of landscape. Using core concepts of the land-based classification system (LBCS), the project team developed a tribal LBCS system that provides for activity, function, ownership, site, and structure for each unit of land.

Commissioned by the tribe to study environmental issues in conjunction with the project planning team, Ellis University developed a process for collecting the tribe's GIS data. The university created a GIS database of Aldrich Nation assets to facilitate tribal management of resources. This database served as the base inventory for the LUCIS process. Further examination revealed problems regarding the quality of the data. In most cases, no metadata was available so it became a challenge to figure out attribute information. In other cases, data was not available for all districts, and this same data, more often than not, was the key to evaluating the basic LUCIS criteria. Existing Aldrich Nation literature and feedback from tribal leadership indicated that the project should embrace informal, especially verbal, data collection methods. During the first phase of public engagement, tribal members were used as the QA/QC team. The hope was that by using community mapping techniques, tribal members would correct the basemaps the university created using data supplied by the tribe and identify villages and communities that were uninhabited, excluded, or whose names were misspelled. This strategy was a great success. Through these and successive engagement activities, tribal members unknowingly became novice GIS technicians, securing their important role in the LUCIS process. LRP participants were also provided with a hard copy of the LUCIS criteria and a copy of the suitability criteria used in the 1979 land-use plan. Tribal members were asked to review the criteria, revise it as needed, and provide feedback on other considerations the project team should include in the model. The project team received no response. What was fruitful, however, was the use of more informal methods of information gathering to understand site planning needs. This approach included general conversations at council meetings, and even the development of a ring toss game that was used at the annual tribal rodeo to illustrate priorities among criteria.

Step 2: Tribal LUCIS model development

The tribal LUCIS model uses both a geographic information *system* and geographic information *science* perspective. Developing a modeling strategy requires a systematic approach to incorporate the interaction among political units, natural resources, and the conflict between land use and other tribal interests. The project team relied on the science of GIS to pinpoint where to focus their planning efforts. The science also helps set reasonable thresholds for change, and provides benchmarks to gauge when those thresholds are close to being met. Decision-making tools can help balance emotional and cultural values with scientific information.

One question in refining the model criteria was how the model would integrate respect for the tribal culture with practical land-use strategy. Based on personal experience, it is easy to oversimplify the nature of model inputs when describing the LUCIS process, or any scenario planning methodology. In this case, it was an even greater concern. The Aldrich creed tells a tribal member how to interact with the environment, what foods to eat, and how to work with the community, thus maintaining a balance with the world. As the project team developed the modeling criteria, it discussed the criteria's relationship to elements within the tribal creed. The idea was to reflect as many of the tribe's values as possible, either through the science or through LUCIS^{plus} tools appropriate for measuring community values.

In addition to using the A4 Community Values Calculator as a means to include and reflect local tribal values, the project team also translated feedback and comments from community meetings onto maps. The team used frequency or other spatial summary tools to summarize the maps. These final raster layers were included as a subobjective within the respective goal and objective for suitability. For example, during a community meeting, participants were asked to note on a paper map where they would like to see economic development and housing. These locations were summarized using frequency of selection as identified by the community, and inverse distance weighted (IDW) interpolation was applied to the final frequency layer. The IDW layer was included within the development stakeholder residential goal proximity objective as its own subobjective.

GIS used in the LUCIS strategy for this project, and in general, relies on a multitude of input data layers associated with the natural environment and existing structures and services. The natural environment data layers include elevation, slope, land cover, hydrologic information, and soils. Individually, these datasets provide a digital representation of what is on the ground. Collectively, with the use of analysis tools within GIS, these datasets can measure the physical suitability of certain land uses by establishing parameters appropriate for development. Additionally, these siting criteria allow for the evaluation of the economic impact of future development.

The presence of utilities such as electrical, sewer, and water are essential to the standard of living for future tribal development. Data from the Ellis University database included these features for many villages. Although the file provided coverage for the entire tribal nation, the data was community-specific. It did not include information on utility coverage between communities (figure 11.1). Additionally, the Ellis University data is generalized to the spatial extent of each community. It does not indicate which utility service is present on the line drawn. You are left to assume that the utilities identified in the attribute table (figure 11.2) are all on the same line, although several utility features had no associated data in the attribute table. Because of the data quality issues with this utility dataset, the modeling criteria initially excluded utilities, despite their value to the community.

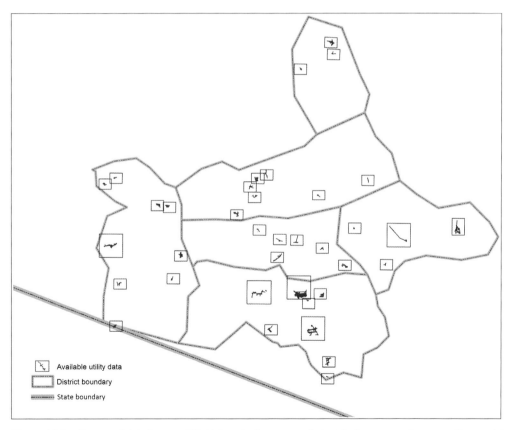

Figure 11.1. Data available from an Ellis University inventory of electrical, water, and sewer utility spatial data. Data is available within communities, but this dataset does not reflect the presence of infrastructure shared between communities.

Figure from the Arizona Board of Regents on behalf of the University of Arizona.

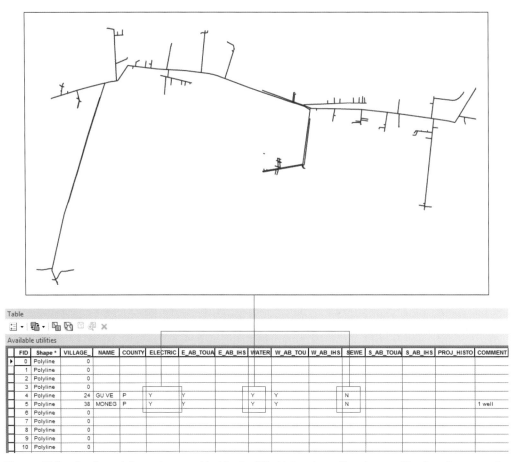

Figure 11.2. Attribute information provided in the Ellis University utility dataset. The line drawn on the map is generalized. There is no distinction between which utility service is on that line and the extent of each utility feature on the ground.

Figure from the Arizona Board of Regents on behalf of the University of Arizona.

Step 3: Suitability analysis and model refinement

Suitability modeling for the Aldrich Nation was complex. Because of individual district interests and an absence of regional or even coordinated tribal planning, the project team ran the LUCIS model twice: once for each individual district and once again for the Aldrich Nation as a whole. Also, because of the proximity of major metropolitan areas to the reservation and tribal reliance on regional transportation and infrastructure services, the analysis included a 10-mile buffer beyond Aldrich Nation boundaries (figure 11.3). The purpose of the double model was to illustrate the difference in opportunities when land-use policy is developed using only interjurisdictional resources versus intrajurisdictional resources.

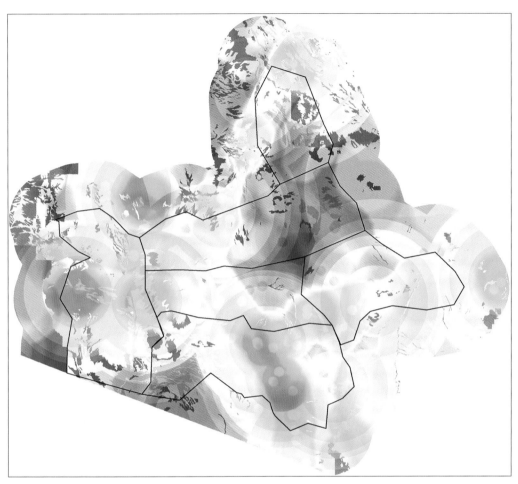

Figure 11.3. A physical suitability raster layer for the Aldrich Nation. Because of interactions within the region, the analysis extent exceeds the Aldrich Nation boundary.

Figure from the Arizona Board of Regents on behalf of the University of Arizona.

To evaluate model quality, particularly for the residential goal, the existing housing structures were overlaid on the final residential suitability raster layer. The result illustrates that most existing structures were placed in areas of high suitability. Typically, this condition would be seen as validating the quality of the model. In this instance, that was not the case. In step 2, utilities were excluded from the LUCIS criteria because of input data quality concerns. As a test, the project team was able to obtain specific utility data for a three-square-mile area in a district with unique land-use needs. The data was more comprehensive than the Ellis University utility inventory in that it provided a detailed spatial alignment of each infrastructure service (figure 11.4). A comparison of the final residential suitability excluding utilities with an illustration of the final residential suitability including detailed utility information reveals significant differences in suitability values. Without detailed

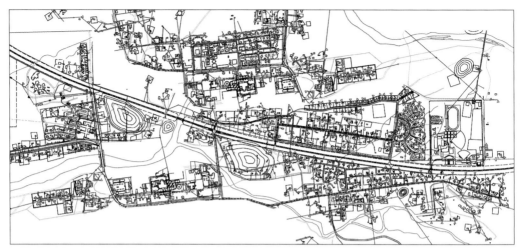

Figure 11.4. Detailed utility data provided by the Aldrich Nation Planning and Economic Development Department.

Figure from the Arizona Board of Regents on behalf of the University of Arizona.

utility information, the physical suitability raster layers provide little insight into the costs of expanding utilities to serve the new development. Thus, they provide an unclear picture of lands suitable for residential use. Therefore, the more specific and complete the information included in the analysis, the better the model can identify true opportunities for future growth and development. After a discussion with tribal leadership and the head of the utility authority, they were able to reach a broader understanding of the importance of having complete data to facilitate planning. Within a few weeks, the project team received complete utility information for the entire Aldrich Nation.

Provided with this example of the difference in decision-making power based on the availability of complete geospatial data, the Aldrich Nation began to understand the connection between quality geospatial data and sound land-use decision-making. Nonetheless, a theoretical disconnect remained between the initial LUCIS output, identifying existing locations of homes in highly suitable areas, and the lack of utility information. Tribal history indicates that tribal families would traditionally occupy different lands during the winter and summer, with each residence providing quality of life benefits depending on the season. As time evolved, families would often settle in one of the locations without regard to infrastructure access but primarily consider the location with the greatest cultural significance. Today, major complaints by tribal members to the tribe's utility office, housing association, and Planning and Economic Development Department reflect this conflict between location and suitability for services, which originated hundreds of years ago. Considering tribal tradition, the complex choice of where to locate is based less on access to services and more on tribal and cultural ties. Because the tribal LUCIS model includes both tribal and cultural considerations, future land-use decision-making can incorporate, address, and visually illustrate the impacts of this conflict.

Step 4: How conflict development informs geodesign

In keeping with the framework of maintaining the power of land-use decision-making at the district level, after the conflict layer was developed, the project team presented the results to tribal districts and communities. Participants were told a story that described what they were looking at and how the results were derived. Using the LUCIS conflict raster layer as a guide, participants were asked to rank the priority of different types of development and identify these priorities on the map. First, tribal members were asked to indicate where they would like to see economic development occur, then mixed-use development, and finally, locations they would like to see preserved or remain the same. After summarizing the feedback, the project team realized that although the land-use evaluation was executed in a rudimentary format, participants considered two things when making their decisions: places they lived or had a connection to and the assigned conflict value.

As in the other examples provided in this book, after conflict is assigned to each parcel, the "plus" in LUCIS^{plus} takes over. The process typically involves the allocation of land for particular uses depending on policy goals or population demands. In this tribal example, the "plus" scenarios are less about allocating future population, especially considering that the average annual population growth for the nation is a little over 1 percent per year. The allocation scenarios are more about using the conflict raster layer to inform design and future decision-making. The planning and land-use scenarios for the Aldrich Nation illustrate the range of opportunities for each individual district, and for the nation as a whole, to satisfy larger economic development, agriculture, heritage preservation, and conservation goals. The project team presented the tribe a description of impacts and future alternatives, considering various options of LUCIS opportunity.

Professor Ryan Perkl (2012 pers. comm.), of the University of Arizona, refers to geodesign as "data-driven design." The Aldrich Nation conflict map is an illustration of land-use opportunities that balance tribal values and scientific depth. But how can this example address pervasive development and preservation concerns? The Aldrich Nation remains a culture that values its language, traditions, and artifacts, so the project team had to address the question of how the conflict map lent itself to these values. The most logical response is that the conflict map facilitates a greater understanding of the interactions among stakeholders, and through collaborative design, tribal members can visualize the future Aldrich Nation.

The future land-use visualizations sparked a debate within the community. Nation elders were pessimistic of the future they saw because they were more concerned about daily challenges. They were worried about the lack of heating and cooling, hazardous conditions their children and grandchildren traveled through to school each day, and inadequate infrastructure. Younger tribal members were optimistic about the future and began to ask "what must we do to get there?" The tribal LUCIS^{plus} combined with geodesign provided tribal members with a blueprint for the future that could spark discussion and lead to effective policy change.

A public participation method employed during the early stages of the Aldrich LRP involved asking tribal members to close their eyes and imagine sitting on their back porch. They were asked to remember what they smelled, what they heard, but most important, what they saw. After a few moments, participants were asked to write down their ideas from the visualization exercise. The project team collected these documented thoughts and presented them again after the future land-use visualizations were presented. It was clear in many cases that the application of geodesign could create the future that tribal members envisioned. Further, the establishment of the land-based classification system helped allocate future land-use opportunities and informed more detailed design than the use of the general land-use designations alone.

Chapter summary

The following points are discussed in this chapter:

- The LUCIS^{plus} method was chosen for tribal scenario modeling because of its flexibility in incorporating community and cultural values.
- The LUCIS model proved to be important in helping the tribe understand that without quality geospatial data for decision-making, it could reach poor conclusions on future land use.
- Geodesign is data-driven design. Incorporating geodesign into the scenario planning process allows communities to understand the impacts of their decisions and visualize opportunities for the future.

Chapter 12

Analyzing and mapping affordable-housing alternatives

Abdulnaser Arafat

What this chapter covers

The housing suitability model (HSM) is an application of Land-Use Conflict Identification Strategy planning land-use scenarios (LUCIS[plus]). It is used to allocate suitable places for affordable housing. However, instead of conflict, the model uses *opportunity*. Opportunity occurs when high suitability for different goals is scored for a location. In HSM, the housing opportunity raster layer is based on four goals. The first goal is to evaluate land parcels' suitability for housing. This suitability is based on physical characteristics and neighborhood accessibility to services. It uses a model based on LUCIS residential suitability (Carr and Zwick 2007). The second goal is to evaluate travel cost in terms of the burden it places on low- and moderate-income populations. The third goal is to evaluate demand for affordable housing that is accessible to low- and moderate-income jobs. The fourth, and final, goal is to evaluate land parcels in terms of transit accessibility. The resultant opportunity layer represents a combination of these goals. It is generated using the same method as the conflict layer in LUCIS[plus].

In addition to the housing suitability components in HSM, the model also uses the LUCIS[plus] strategies to run scenarios that help planners allocate and preserve affordable-housing units. This chapter explains the hierarchal structure of the HSM model, the conflict/opportunity identification strategies, and how the allocation process uses the criteria evaluation matrix (CEM). CEM can also be used in the evaluation and scoring of existing affordable-housing sites. The identification of opportunity includes accessibility, driving cost, demand, and transit access.

The housing suitability model

Housing affordability can be evaluated by different methods. Traditionally, income in relation to housing cost has been a major factor in assessing affordability. The Center for Neighborhood Technology (CNT), a research center working on sustainability issues, added transportation cost to the affordability equation in 2007. It created the Housing + Transportation (H+T) Affordability Index, which evaluates the combined housing and transportation costs in relation to income. Other methods of affordability evaluation consider other variables in the process. HSM is one of these models. HSM is a sophisticated hierarchal suitability model that helps communities plan for attractive, equitable, and sustainable affordable housing. The model evaluates and identifies lands that are suitable for the development and preservation of affordable housing based on local preferences and planning expertise.

Generally, sustainability involves the interaction between the economy, social equity, and the environment. The model's goal is to develop affordable housing that (1) has less impact on the environment, (2) takes social justice into account, and (3) can be preserved under changing economic interests. HSM incorporates the three legs of sustainability in a sequential procedure that allocates suitable land for sustainable affordable-housing developments. To achieve the combined goals of affordability and sustainability, the model uses the LUCIS strategy and the population allocation procedure using combined raster layers (Carr and Zwick 2007). LUCIS and HSM also encourage community participation to find suitable lands for residential development. They use the analytic hierarchy process (AHP) to weight expert and local community preferences and evaluate suitability. HSM uses extra variables specific to affordable-housing locations to develop new affordable-housing sites and sustain existing development.

The main step in the process is conflict identification. LUCIS methodology identifies the conflict that the three major land uses (agriculture, conservation, and urban) pose to proposed land-use and allocation scenarios. HSM applies the LUCIS conflict methodology to the three major elements of sustainability. Thus, HSM uses environmental, social equity, and economic indicators to find places that are suitable for affordable housing. These broad indicators are then converted into more tangible indicators, such as employment accessibility, travel cost, and physical characteristics. Additional allocation steps use CEM to further refine these affordable-housing locations by adding extra socioeconomic conditions, policies, and incentives to the mix. Socioeconomic considerations include age distribution, access to education, poverty rate, distribution of ethnic groups, wealth, and income.

In addition to social-equity indicators, the affordable-housing allocation methodology generally focuses on objectives that lower the transportation burden, such as encouraging compact and transit-oriented development (TOD). This process uses multimodal transportation indicators such as transit accessibility and location-specific cost, which includes travel and housing costs.

HSM also uses CEM to evaluate existing "assisted housing" sites, or subsidized rental housing. CEM contains scores for each criterion to be evaluated, such as accessibility to employment using multimodal transportation systems. Existing affordability models have not yet been able to differentiate between low-, moderate-, and high-salary jobs in terms of accessibility, which this model investigates. HSM uses road network accessibility to estimate the opportunity of low- or moderate-salary employment for assisted-housing properties.

In technical terms, HSM is a hierarchal suitability model that uses LUCIS^{plus} methods to allocate suitable locations for affordable housing. The model consists of four main goals at the top level: (1) residential suitability, (2) travel or driving cost, (3) demand for affordable housing, and (4) transit accessibility to jobs. Using LUCIS, the model constructs suitability raster layers based on the four main goals and combines them into an opportunity layer. The opportunity layer is similar to the conflict layer, except that it identifies the opportunity for affordable housing rather than the conflict between land uses. Figure 12.1 shows the hierarchal structure of HSM.

The first goal of affordable-housing suitability, residential housing, is based on physical and neighborhood characteristics. If the suitability of the first goal is high, the preference for affordable housing is set as high. If the cost of travel (driving cost) is low, the travel cost preference, the second goal, is set as high. For the third goal, if the demand for affordable-housing units is high, the preference is also set as high. The fourth-goal preference is set to high when transit accessibility to jobs is high. The opportunity layer is thus a preference

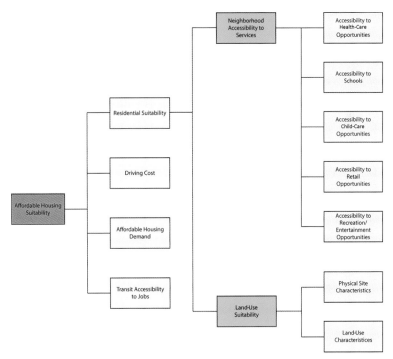

Figure 12.1. The housing suitability model (HSM) structure. HSM is an affordability index that considers residential suitability, driving cost, demand, and transit access to jobs in the determination of opportunities for housing affordability.

Figure from Abdulnaser Arafat and the Arizona Board of Regents on behalf of the University of Arizona.

matrix that uses LUCIS conflict strategies to combine the objectives for the result (Carr and Zwick 2007).

LUCIS^{plus} methods are used to evaluate the opportunity of a location for affordable housing. For example, a combination of low travel cost and high demand, combined with good physical and neighborhood characteristics, could be identified as affordable. At the same time, another location that has higher travel cost but high transit accessibility could also be regarded as affordable. The use of the conflict strategies adds flexibility to the choice of location for affordable housing without the trade-offs between goals that are usually the case in suitability models.

Physical and neighborhood characteristics preference layer

One of the goals of HSM is to generate suitability according to physical characteristics and neighborhood accessibility to services. The model then determines suitable locations for the allocation of residential housing. The structure of the suitability model contains many single-utility assignment (SUA) and multiple-utility assignment (MUA) layers that are weighted according to community preferences. The main suitability goals are land-use suitability (physical characteristics) and neighborhood accessibility to services. The MUA of the first goal is subdivided into two subobjectives, physical site characteristics and land-use characteristics. These subobjectives are built using a LUCIS residential suitability model. The subobjectives are reclassified based on zonal reclassification using the mean and standard deviation based on single-family and multifamily locations.

Neighborhood accessibility to services is subdivided into five subobjectives. Each subobjective has been constructed and evaluated using the A4 Network Distance and A4 Network Opportunity tools, discussed in more detail in chapter 13. The model weights and combines the network proximity and opportunity of services as suitability layers. The model also uses walking, biking, and driving network distance thresholds to evaluate the accessibility to services. During an HSM project, planners often hold meetings or webinars that include local planners from the different study areas or counties. These meetings and webinars have two purposes: to include local planners in the weighting process of the suitability model and to generate the weights. The weights are then used to update the model table of weights and generate the final HSM suitability layer. The tools used in the weighting process are explained in chapter 3. They include the A4 Community Values Calculator and A4 LUCIS Weights tool. Figure 12.2 shows an example of an equally weighted HSM goal preference layer for the Sun Rail region in east central Florida. The output layer is reclassified based on the mean and standard deviation in three preference values: low, moderate, and high. The high-preference areas are mainly areas that are preferred for residential development based on parcel and neighborhood characteristics as well as parcel accessibility to services. Many of these places, however, may not be affordable for all income categories (that is, low-income families).

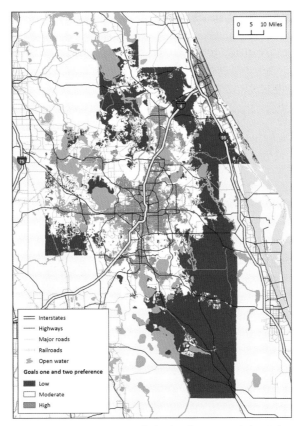

Figure 12.2. The combined preference layer of the land-use suitability (physical characteristics) and neighborhood accessibility to services goals. Housing affordability preference uses the same preference values as preference reclassification in other LUCIS^{plus} applications.

Figure from Abdulnaser Arafat and the Arizona Board of Regents on behalf of the University of Arizona. Water bodies from the US Geological Survey; state boundary from US Census Bureau TIGER/Line Files; interstates, highways, and major roads from the Florida Department of Transportation; railroads from the Federal Railroad Administration.

Travel cost preference layer

In recent affordability evaluation approaches, the spatial estimation of travel cost is used to evaluate location suitability. The Center of Neighborhood Technology was the first to estimate such cost spatially (CNT 2011). Other institutions, including the Shimberg Center for Housing Studies at the University of Florida, also conduct travel cost estimations. The US Department of Housing and Urban Development (HUD) has released an online Travel Cost Calculator (HUD 2013), which helps people calculate the location suitability of where they live. In this chapter, spatial interpolation of geocoded trip end points is used to estimate travel cost. The 2009 geocoded trip data from the National Household Travel Survey is used to estimate the travel cost for the base year 2009. Trips are classified into home-based trips and non-home-based trips. Home-based trips are further classified into work, shopping, social, recreation, and other. Thus, five trip categories are considered in the estimation of the travel cost for the base year. The process of estimation includes removing the trip outliers using the z-score and also using Moran's *I* and other GIS statistics tools.

After the outliers are removed, the trip data is used to estimate passenger miles traveled (PMT), which in turn is used to estimate the household vehicle miles traveled (VMT). The American Automobile Association (AAA) estimated value of dollars per mile ($/mile) is used to convert the estimated VMT into daily and monthly travel cost for the average-size household. The results are distributed spatially using a 10 × 10 meter raster and are also summarized to the parcel.

Regression models were established to estimate the travel cost in the future. The 2009 base year travel cost estimation is related to land-use and urban form variables, such as density, accessibility, road connectivity, and other characteristics of the built environment, for the year 2009. The methodology for estimated land-use and urban form variables includes using a 2.5 × 2.5 mile floating area to reduce the effect of the natural neighborhood boundaries. The floating area size and shape are taken from the results of research performed by the Shimberg Center to reduce the effect of modifiable areal unit problem (MAUP).

The independent variables used in the travel cost regression model include density indicators, such as population density and employment density; a 2009 land-use mix indicator captured by estimating the entropy; and connectivity of the 2009 road network. Road density, four-way intersections, and the gamma index, which relates the number of links to the number of intersections, help establish connectivity. The independent variables also include three disaggregate parcel-level variables: accessibility to major destinations using a gravity model based on the 2009 data, proximity to transit, and frequency of transit service.

Two regression models were used to relate the 2009 base-year travel cost to the 2009 land-use and urban form variables. The first regression model is an ordinary least squares (OLS) model. The model was checked for autocorrelation and the distribution of residuals where random. To reduce the error of prediction, geographically weighted regression (GWR) was used to relate travel cost to the urban form variables. The GWR model resulted in less prediction error, indicating that the GWR model is the best model. However, the global model might give a more tangible relationship between the variables than the GWR model. At the same time, the GWR model might capture local relationships between the variables, which the global OLS model cannot do. Figure 12.3 shows the observed travel cost generated by (a) spatial interpolation and the predicted travel cost using (b) GWR and (c) global regression.

To estimate the travel cost for the future, the independent variables should be estimated for the target year. For example, if the target estimation requires travel cost for the year 2015, the population, employment and land use, road network, major activity location, and transit stops and frequency should be used from that year. If future prediction of travel cost is needed, all the independent variables should be updated, using future land use and predictions on future population density and employment as well as suggested changes to the road network and transit service.

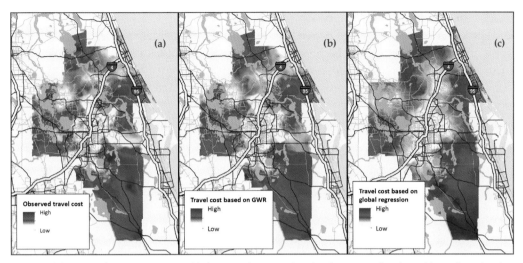

Figure 12.3. Travel cost estimation: (a) the travel cost estimation layer was developed based on density, entropy, connectivity, road density, and intersection links; (b) GWR was used to relate travel cost to urban form and reduce the error of prediction; and (c) using a global regression model, such as OLS, the input values could be checked for autocorrelation, resulting in a more tangible explanation of variable relationships.

Figure from Abdulnaser Arafat and the Arizona Board of Regents on behalf of the University of Arizona. Water bodies from the US Geological Survey; state boundary from US Census Bureau TIGER/Line Files; interstates, highways, and major roads from the Florida Department of Transportation; railroads from the Federal Railroad Administration.

To summarize, either spatial interpolation or regression can be used to create the travel cost layers. Urban form and land-use characteristics taken from the 5Ds of the built environment (density, diversity, design, distance to transit, and destination access), which correlate urban form to VMT (see Ewing and Cervero 2001; Ewing et al. 2008; Lee and Cervero 2007), are used. However, these layers are transformed to monthly values of travel cost. The travel cost layer is also reclassified according to transportation cost thresholds used for affordable-housing research. CNT (2011) used the combined percentage of 45 percent as the combined transportation and housing burden on monthly income. So if the cost of housing does not exceed 30 percent of income, the transportation cost should be no more than 15 percent of income for location affordability. Therefore, the travel cost layer is better reclassified to give a high preference to areas of transportation cost that do not exceed 15 percent of the target income group.

Demand preference layer

A report on housing impact analysis supported by HUD was drafted in 2006. Six fundamental determinants for housing demand were defined in the report and can be categorized into two aspects: household size and incomes, and the creation and location of jobs in an area (Dacquisto and Rodda 2006). To determine whether housing was affordable, HUD used estimates of area median income (AMI), which are calculated and published annually. For homeowners, other factors besides direct housing cost may have an impact on affordability (for example, indirect costs such as commuting distance [Bogdon and Can 1997]). M. Haffner and K. Heylen (2011) used an "opportunity cost" to conceptually integrate the concerns of demand-side affordability from the financial perspective of homeowners and renters.

In Florida, funding sources for affordable housing include HUD, the Department of Agriculture Rural Development (RD) program, and other local housing finance authorities (LHFAs), such as the Florida Housing Finance Corporation (FHFC). Other forms of assistance for affordable housing include public housing and rent vouchers. The Shimberg Center maintains affordable-housing data such as the assisted-housing inventory (AHI), which provides data on lower-income tenants and housing owners.

According to Shimberg Center research (Ray et al. 2009), most assisted-housing units (92 percent) serve tenants whose income is at or below 60 percent of AMI. Affordability ranges provided by HUD categorize low and moderate income at or below 80 percent of AMI. This chapter, however, identifies low- and moderate-income population as having incomes at or below 80 percent of AMI. Such an income level is in line with low- and moderate-wage categories used in the US Census Bureau's Longitudinal Employer-Household Dynamics (LEHD), which is the data source used in this study for estimating affordable-housing demand.

A. Arafat et al. (2012) studied the assisted-housing units in Orange County, Florida, in terms of their spatial distribution and spatial mismatch with employment. Their work concludes that studying the relationship between supply and demand spatially can assist decision-makers in the evaluation of existing AHI and in their mission of locating future assisted-housing units. However, the study did not include programs such as vouchers, public housing, and market rentals. The study also used a z-score, which is a descriptor of quantitative spatial distribution, to compare supply and demand. The study flags the importance of studying the relationship between supply and demand in the process of allocating new locations for affordable housing or evaluating location suitability of existing properties for affordable-housing funding programs.

Technically speaking, the first step in creating the preference for affordable-housing demand is to estimate the demand for affordable housing generated by low- and moderate-wage employment. In this study, a spatial comparison is performed at different scales. The opportunity for low- and moderate-wage employment (that is, the demand for affordable housing) was determined within three differently sized areas surrounding a representative set of random points. The three scales of analysis relate to the average distances people travel from home using different modes of transportation: walking, biking, and driving. In this way, the demand investigates the opportunity for employment that can be reached using different modes of transportation relative to the location of affordable housing. For example, within a walking distance of each random point, the demand layer shows the level of opportunity for low- and moderate-wage employment for people who might work in those jobs and walk to work.

The methodology for determining parcel-level employment plays an important role in the study, especially in the incorporation of socioeconomic characteristics from the census data to a parcel-level scale. Quarterly Workforce Indicators (QWI) data is summarized according to North American Industry Classification System (NAICS) sectors by census block and land-use type. The density of employees by industrial sector is then calculated for each census block. Employee densities by industrial sector are assigned to individual property parcels using multipliers from a cross-tabulation methodology.

In some census blocks, the QWI data indicates employees of an industrial sector that is not consistent with the land uses reported by the property parcel data. Such discrepancies can arise, for example, when the QWI data has been incorrectly geocoded to a nearby census block in which a corresponding land use does not exist. In these instances, it is not possible to cross-tabulate the two datasets, which results in some employees remaining unassigned to blocks and therefore parcels. For this reason, a second round of allocation is performed to assign the remaining employees to property parcels. For this process, areas and employees are summarized to, and densities calculated at, the larger census block group level.

The A4 Network Opportunity tool uses the ArcGIS Network Analyst extension to provide the spatial estimation of demand for the three transportation modes. The estimation creates a random set of 50,000 points that have access to the transportation network. These points are the origins in the origin-destination (OD) cost matrix and the focal points in later steps of the analysis. Three network distance buffers are applied using the A4 LUCIS Tools, based on three different travel modes. In this study, a distance of 0.5 mile is considered a walking distance, 1.8 miles is the transit distance, and eight miles is the driving distance. These distance thresholds are obtained from analyzing the trip data from the National Household Travel Survey (NHTS) 2009 data. Figure 12.4 shows the demand based on the driving network buffers. The demand layers are transformed into suitability and then weighted and combined to produce the demand preference layer.

Transit access preference layer

Literature on location choice shows that the choice of residential location depends on attributes derived from location theories. Attributes such as housing cost, transportation cost, distance to the central business district (CBD), accessibility, space, leisure, and environmental quality are frequently used in location choice models (Yamada 1972). Historically, people have left the inner city to live in suburban locations that provide more spacious and luxurious homes at an affordable price. This trend is often referred to as "drive until you qualify." Housing policies and incentives make suburban residential locations available. These suburban locations are then promoted by low pricing and low interest rates on mortgages.

As a result of people living far away from jobs, the use of cars has increased. The use of cars is also encouraged by low gas prices and the increased mobility resulting from the construction of highways and freeways. This car-dependent suburban housing has caused urban sprawl, longer commuting distances, and higher transportation costs as a result of

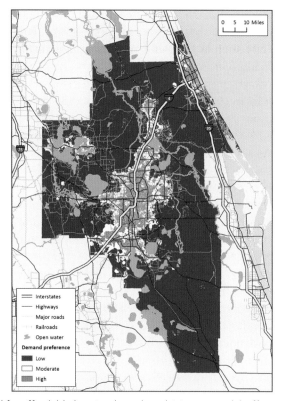

Figure 12.4. Demand for affordable housing based on driving network buffers.

Figure from Abdulnaser Arafat and the Arizona Board of Regents on behalf of the University of Arizona. Water bodies from the US Geological Survey; state boundary from US Census Bureau TIGER/Line Files; interstates, highways, and major roads from the Florida Department of Transportation; railroads from the Federal Railroad Administration.

rising gas prices. In terms of housing affordability, suburban housing has become a heavier burden for householders and less affordable as a result of increasing transportation costs (Lipman 2006).

CNT (2011) evaluates the location affordability of housing by estimating the housing and transportation cost in regard to income. In CNT's work, the transportation cost includes transit accessibility using a regression model. Housing is considered affordable if the combined housing and transportation cost does not exceed 45 percent of income. However, the actual spending on combined transportation and housing nationally exceeds 50 percent of household income. CNT shows that increasing the commuting distance by living away from work reduces the housing cost, but it also increases the total housing and transportation cost. Thus, the "drive until you qualify" trend has placed an extra burden on people and reduced the affordability of suburban residential locations.

Transit research on the national level shows an increase in transit ridership in metropolitan areas and CBDs (TCRP 1998). Transit studies also show that transit ridership trends increase for low-income and older populations (TCRP 1998). However, ridership also depends on the urban form and level of service of the transit system. Commuting to work by car or transit is a controversial and complex issue for low-income populations. For example, J. Grengs (2009) shows that transit in a city such as Detroit is not advantageous for low-income populations when using cars may increase the employment rate and accessibility to jobs.

The research on coordinating land-use transportation suggests the importance of accessibility and urban form characteristics in reducing the cost of driving associated with residential location and increasing the use of other modes of transportation such as walking and transit (Cervero and Kockelman 1997). Thus, locating residential land uses in places that have high accessibility to services, transit, and employment reduces the travel cost for people driving to work. At the same time, it encourages the use of other transportation modes such as walking, biking, and public transit.

HSM values the importance of transit accessibility and recognizes the trade-offs between using cars and transit. For this reason, transit access is included in the conflict/opportunity layer and not in the suitability structure. However, this section of the chapter focuses on creating the transit accessibility layers and using these layers in the allocation of affordable housing. The methodology for creating the transit scores can be divided into four categories:

1. Downstream assignment of opportunity stops based on the relationship between these stops and employment opportunities
2. Assigning the cost field by weighting the travel distance by the frequency of the service
3. Estimating the network cost, including transfers
4. Estimating a walkability score raster, which includes the walking distance from residential parcel to transit stops

The identification of the destination or downstream stop starts by identifying the activities within the walking distance between each stop and assigning the employment to the transit stop. The walking distance buffer is a quarter-mile network distance from the transit stop at the center of the buffer. The employment opportunity is joined to the downstream stops and identifies stops that have access to employers. The user can set a threshold for selecting stops depending on the number of employees. However, the destination stops can be considered as the whole set of bus stops. The downstream transit stops could also be upstream transit stops for different transit trips. The transit accessibility measurement depends on creating a stop-to-stop, OD cost matrix, which may have size limitations depending on the computer and the size of the study area. However, if the downstream stops are not significant to serving employment, the analysis could ignore those stops.

The final transit access score combines the walkability score, which represents the trip from residential location to transit stop, and the bus transit accessibility. Transit accessibility is determined according to the transit access methodology explained in chapter 14. This transit access methodology is used to create transit access layers for the study area based on the actual transit stop and route frequency combination, plus the parcel data from which the employment opportunities were derived. The estimated walkability scores and transit

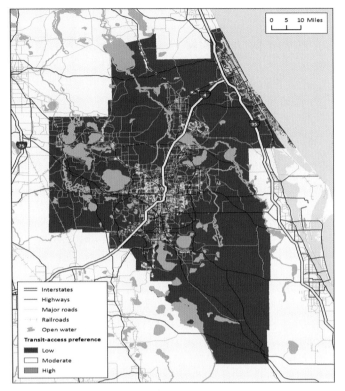

Figure 12.5. Transit access preference raster layer.

Figure from Abdulnaser Arafat and the Arizona Board of Regents on behalf of the University of Arizona. Water bodies from the US Geological Survey; state boundary from US Census Bureau TIGER/Line Files; interstates, highways, and major roads from the Florida Department of Transportation; railroads from the Federal Railroad Administration.

access scores for residential parcels are transformed into suitability to create the transit access suitability raster layer. The suitability layer for transit accessibility is then reclassified into three preferences: low, moderate, and high (figure 12.5).

Access-driving-demand-transit opportunity layer

In the access-driving-demand-transit (ADDT) opportunity layer, the four preference layers are combined into a conflict/opportunity layer using LUCIS conflict strategies (Carr and Zwick 2007). The first layer is the preference based on physical characteristics and neighborhood accessibility, and relates to the A in the ADDT opportunity layer. The second layer is the driving cost preference, and corresponds to the first D. The third layer is the demand for affordable-housing preference, the second D, and the final layer is the transit access, or the T in the opportunity layer. The LUCIS methodology is applied in the following section, "Identifying affordable-housing locations," which describes how the conflict between these four

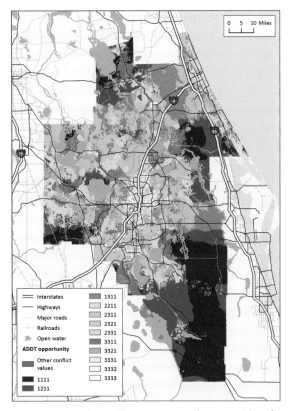

Figure 12.6. ADDT opportunity raster layer. The opportunity layer can identify areas that have high preferences in location suitability, which includes transportation cost. It also identifies places that may not have low driving cost but have alternative transportation options such as high transit accessibility.

Figure from Abdulnaser Arafat and the Arizona Board of Regents on behalf of the University of Arizona. Water bodies from the US Geological Survey; state boundary from US Census Bureau TIGER/Line Files; interstates, highways, and major roads from the Florida Department of Transportation; railroads from the Federal Railroad Administration.

components can be expanded into a discussion of opportunity for affordable housing. The generated ADDT opportunity layer is shown in figure 12.6.

Identifying affordable-housing locations

As mentioned earlier in this chapter, in the case of affordable housing, the opportunity layer is easier to understand than the conflict layer. The opportunity components A, D, D, and T have the same direction in preference for affordable-housing opportunity. For example, for the opportunity value of 3222, the location has a high preference for residential housing and a moderate preference for travel cost, demand of affordable housing, and transit access. An ideal conflict value is 3333, which has a high preference for all four components. Usually, affordable housing is not allocated in a low-preference area or in areas that have a

1 preference value as an opportunity value. However, if transit is not available, the planner may compensate a value of 1 in transit accessibility with a value of 3 in travel cost. For example, the value 2321 has a moderate preference for physical characteristics, high preference for driving cost, moderate preference in terms of demand, and low preference for transit access. Such a place might be chosen for affordable-housing allocation because it has a low driving cost, which means that people may use the car mode in their trip to work.

Having the value of 2222 in the opportunity layer means that the location is moderately preferred for affordable housing based on the evaluation of the four components. A value of 3221 indicates a high preference for accessibility, moderate preference based on the driving cost and demand value, and low preference for transit accessibility. However, using a refined opportunity for affordable housing based on reclassification, values of 2 and 3 are considered places of opportunity, whereas a value of 1 in any category negates the opportunity for affordable housing. High or moderate transit accessibility does not cancel the need for good neighborhood characteristics. Therefore, for high to moderate opportunity affordable-housing scenarios, masking locations that have low preference in one of the four components can refine the opportunity layer. Figure 12.7 shows the refined opportunity layers for affordable housing for the region. Both layers have moderate or high preferences for each of the four affordable-housing variables and do not contain the low-preference categories.

Using CEM to allocate affordable housing

Because accessibility and travel cost impact location suitability (Arafat 2011), this chapter uses the ADDT opportunity layer for the affordable-housing allocation process. The refined ADDT opportunity layer includes the land that is moderately or highly preferred for affordable housing based on the four goals that generate the opportunity layer. However, other factors also influence the selection of land for affordable housing. These extra conditions may address a certain policy or priority based on a certain scenario.

You can assign an affordable-housing final preference score using the allocation tools for land and people in land-use models such as LUCIS. Acres for affordable housing are used as the allocation field, using the A4 Allocation toolset presented in chapter 3. This allocation process is based on iterations. The first iteration targets locations with the highest preference, repeating iterations until the last one targets the lowest preference. Parcels and locations that do not meet the criteria set in the conditions are left unranked, which means that their preference is lower than the lowest preference in the ranking ladder. However, these unranked parcels may still be suitable for affordable housing, even if they do not satisfy the ranking conditions used in the specified scenario.

The allocation tools work on CEM, which is a complex raster layer (see chapter 3). This combined layer is composed of several layers that represent the opportunity layer, as well as other layers that can help refine the places for affordable housing or add restrictions or constraints on the process. The layers can also represent a change in policy that must be tested, such as new transit lines. For example, the scenario can investigate the opportunity

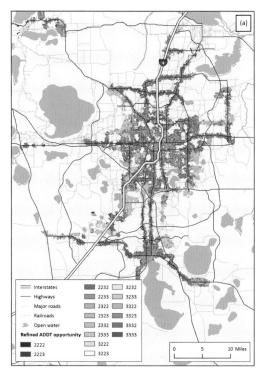 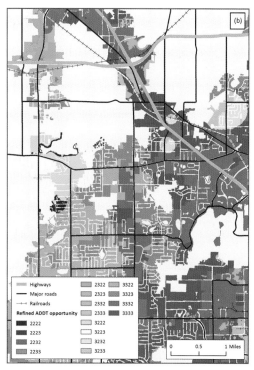

Figure 12.7. The refined ADDT opportunity layer includes only land identified as moderate or high preference for each of the affordable-housing variables. Most of the moderate to high ADDT opportunity exists in the middle of the east central Florida region, with other pockets in the northeast region. At a large scale (a), it does not seem as if there is much variation in accessibility, driving cost, demand, or transit access. At a small scale (b), these accessibility variables can change from high opportunity for affordable housing for all four variables to moderate opportunity within a short distance.

Figure from Abdulnaser Arafat and the Arizona Board of Regents on behalf of the University of Arizona. Water bodies from the US Geological Survey; state boundary from US Census Bureau TIGER/Line Files; interstates, highways, and major roads from the Florida Department of Transportation; railroads from the Federal Railroad Administration.

for allocating affordable housing based on ADDT, with the addition of constraints based on compact development. The scenario can look at livability indicators such as walkability and also refine these locations according to their density and other variables that are important for compact development. A compact-development scenario can also target areas that are underused when compared with their surroundings. Other variables such as enterprise zones, Qualified Census Tracts, and locations that qualify for Community Reinvestment Act (CRA) funding can be used in the allocation process.

The compact-development scenario may differ from one county to another, depending on the availability of data. For example, the Livability Index includes walkability and crime for Orange County, Florida, although no data for walkability exists for the other counties. The following section, "Creating the CEM raster layer for the allocation," explains the raster layers that might be used for allocation in addition to the ADDT opportunity layer.

Creating the CEM raster layer for the allocation

The ArcGIS Combine tool is used to create the base CEM (figure 12.8 shows the value attribute table [VAT]) for each county. If a region is considered, a raster layer for those different counties should be added to the combined raster layer. The combined raster layer is a layer enumeration tool that can hold multiple layer values for each cell. The raster layers that are used to create the combined raster layer may be different from one scenario to another and will also depend on the political region and boundary. The following list of layers is typically included in an HSM CEM raster layer:

- *Parcel layer*, which shows the parcel boundary. This raster layer is generated by rasterizing the property appraiser shapefile, using the ArcGIS Polygon to Raster tool. For large datasets, the Zonal Statistics tool uses this file to summarize the model output layers to parcels. Zonal Statistics using parcel geometry is used on all the layers in the combined raster layer to limit the size of CEM. However, there are cases when the parcel-level CEM is not desirable. These situations include the following:
  - When the parcels are large, because the suitability variation will be considerably high within the boundary of the parcels
  - When the required accuracy suggests the use of a smaller unit than the parcel
- *Suitability layer.* These layers are the output of HSM, which is generated using LUCIS. The two goals of HSM, land-use suitability (physical characteristics) and neighborhood access to services, are used in the combined raster layer for single-family and multifamily residential.
- *Transit access score*, which represents the estimated transit accessibility before reclassification into suitability or preference. However, it can also be a score of 1 to 100 that shows the relative value of transit access.
- *Travel cost,* as explained earlier in the chapter. CEM can include a dollar amount of the travel cost or a score of 1 to 100, indicating the relative values of travel cost. If a score for locations is required, a travel cost score of 1 to 100 can be used in CEM.
- *Zoning layer.* Census tracts, blocks, block groups, and traffic analysis zones (TAZ) are layers used for summary purposes for the allocated lands. These layers can also be used to set conditions or prioritize selected zones.
- *Demand layer.* These layers are generated and transformed into preference layers to be included in the opportunity/conflict raster layer. However, it is useful to have the actual demand values as counted by the A4 Network Opportunity tool and use them in the combined raster layer. CEM might also include a demand score from 1 to 100 for evaluation purposes.
- *Supply layer.* These layers represent a count for affordable-housing units counted using the A4 Network Opportunity tool. The network distance thresholds for walking, biking, and driving are typically used to estimate the supply of affordable housing. The same methodology for generating the demand layers is used for the supply layers.

The opportunity points used here are the affordable-housing units such as AHI units and other units administered or funded by affordable-housing programs. The AHI units can also be affordable rental units or mortgages from the housing market.

- *Underutilized density layer.* The "underutilized" value for land is a number that compares the cell density of a location with the surrounding area density in residential units. The surrounding neighborhood is a quarter mile (Manhattan distance) that surrounds each cell. The underutilized density is defined as the number of residential units that could be added to the cell to match its average surrounding density value. A cell that has a density value greater than its surrounding neighborhood is assigned zero (0) for underutilized density. The importance of this layer is to capture the parcels that might be preferable for redevelopment.
- *Allocation mask.* These masks can be used to identify infill parcels, redevelopment parcels, or greenfield parcels. The masks can also be used to prioritize areas such as developments of regional impact (DRIs), transportation corridors, and so forth.
- *Proximity layer.* These layers can help in the allocation process. Examples of these layers are the distance to CBD, distance to major employment centers, and so forth.
- *Policy layer.* The planner creates these layers to prioritize the allocation. An example of a policy layer is multiple rings around an employment center, in which the allocation progresses from the inner buffer to the outer buffer.
- *Livability and HUD regional Fair Housing Equity Assessment (FHEA) layer.* The term *livability* here is a general term for a category of layers that represents walkability and safety. This layer, however, can be different from one county to another, depending on the availability of data. The layer is a utility assignment layer. Therefore, it could contain one layer as an SUA, such as a crime avoidance suitability layer, and another layer to estimate a walkability or a "bikability" MUA. The FHEA indicators can also be used to generate layers for use in the CEM allocation. These indicators include dissimilarity, poverty concentration, and so forth (HUD 2012).

Table

presun1

RowId	VALUE	COUNT	REGION_NEWPAR	COUNTY	SF_G1	SF_G2	MF_G1	MF_G2	TRANS_TR	TRANS_WR	TRVCOSTT	BLKGRP	DISSINDEX	EPA	TRACTIND	ASSTDMDB	ASSTDMDD	ASSTDMDW	ASSTSUPB
0	1	399	227847	1	67	17	65	38	0	0	57	831	56	851	374	1	35	0	0
1	2	386	204536	1	74	22	63	42	0	0	70	1062	139	990	425	1	44	0	0
2	3	88	1136924	1	68	17	59	38	0	0	60	1062	139	990	425	1	35	0	0
3	4	284	1136954	1	63	17	66	40	0	0	57	831	56	851	374	1	35	0	0
4	5	420	1136955	1	70	17	82	40	0	0	58	831	56	851	374	1	36	0	0
5	6	162	1136956	1	67	18	81	42	0	0	59	831	56	851	374	1	37	0	0
6	7	373	204537	1	73	23	62	42	0	0	72	1062	139	990	425	1	44	0	0
7	8	1	204537	1	73	23	62	42	0	0	70	1062	139	990	425	1	44	0	0
8	9	3	204538	1	73	23	62	42	0	0	72	1062	139	990	425	1	44	0	0
9	10	370	204538	1	73	23	62	42	0	0	70	1062	139	990	425	1	44	0	0
10	11	393	204539	1	74	22	63	42	0	0	68	1062	139	990	425	1	44	0	0
11	12	1	204539	1	74	22	63	42	0	0	60	1062	139	990	425	1	44	0	0
12	13	54	1136925	1	68	17	59	38	0	0	59	1062	139	990	425	1	35	0	0
13	14	398	204540	1	74	22	63	42	0	0	68	1062	139	990	425	1	44	0	0
14	15	399	204541	1	73	22	62	42	0	0	69	1062	139	990	425	1	44	0	0
15	16	396	204561	1	73	22	63	43	0	0	68	1062	139	990	425	1	44	0	0
16	17	3	204561	1	73	22	63	43	0	0	69	1062	139	990	425	1	44	0	0
17	18	93	1136922	1	68	17	59	38	0	0	60	1062	139	990	425	1	35	0	0
18	19	173	1136923	1	67	17	59	38	0	0	59	1062	139	990	425	1	35	0	0
19	20	241	1136950	1	66	16	65	40	0	0	58	831	56	851	374	1	38	0	0
20	21	400	1136951	1	70	17	77	41	0	0	59	831	56	851	374	1	39	0	0
21	22	324	1136952	1	70	18	84	43	0	0	60	831	56	851	374	1	39	0	0
22	23	393	207624	1	74	22	64	43	0	0	67	1062	139	990	425	1	44	0	0
23	24	440	202792	1	74	23	64	43	0	0	67	1062	139	990	425	1	45	0	0
24	25	444	202793	1	73	23	63	43	0	0	67	1062	139	990	425	1	45	0	0

1 (0 out of 1450241 Selected)

presun1

Figure 12.8. An example CEM for the Sun Rail region generated using the Combine tool. Each column is a criterion used in the CEM allocation process.

Figure from Abdulnaser Arafat and the Arizona Board of Regents on behalf of the University of Arizona.

Simple scoring procedures using CEM

Selected components from CEM (see figure 12.8) can be used to generate a total score representing affordability. The procedure can be simplified by transforming all the suitability and cost fields and the evaluation criteria into a score of 1 to 100. To get the total score, add a total score field to CEM and calculate that field using the Calculate Field tool. For example, equation (12.1) combines scores, as follows:

$$W1 * A + W2 * DC + W3 * DE + W4 * T \quad (12.1)$$

where:
A = the combined goal1 and goal2 scores,
DC = the driving cost score,
DE = the demand for affordable housing score,
T = the combined transit access score, and
$W1$–$W4$ = weights as decided by the planner or the group of experts.

Figure 12.9 shows a sample affordable-housing score map using equal weights. The user can also include other components from CEM in the scoring, including supply and other FHEA indicators.

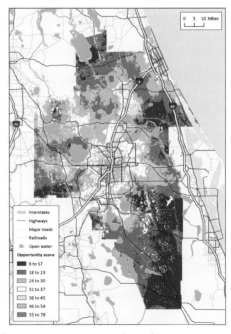

Figure 12.9. Affordable housing score using equal weights.

Figure from Abdulnaser Arafat and the Arizona Board of Regents on behalf of the University of Arizona. Water bodies from the US Geological Survey; state boundary from US Census Bureau TIGER/Line Files; interstates, highways, and major roads from the Florida Department of Transportation; railroads from the Federal Railroad Administration.

Advanced allocation and scoring procedures using CEM and scenario tables

The first step in using the A4 Allocation toolset to allocate or rank land for affordable housing is to understand the differences between allocating land and allocating population using CEM. CEM is a container to hold the raster layers that are an output of suitability or other evaluation techniques. Therefore, CEM is similar in both cases. The first difference is that acreage instead of population is used to allocate affordable housing. Therefore, the population field in the A4 Allocation toolset actually represents acres in this type of allocation. Another difference is that you need a score instead of an allocation year, so you can change the allocation year field to an affordable housing score using an A4 Allocation tool.

The A4 Allocation by Table tool is the best option for this situation. The scenario and policy table (SPT) (see chapter 3) includes the set of conditions, which are technically SQL equations used to select the lands that fit these criteria. Each iteration is a line in the table, and the user decides the score of the first iteration, second iteration, and so forth. Each score should represent one line of iteration. However, the score can be the same for different iterations, which depends on your score assignment. For example, you may want the first two iterations to have a score of 9 and the next two iterations to have a score of 8, the next two to have a score of 7, and so forth. Therefore, if you use CEM from figure 12.8, the region field in the scenario table can simply be the name of the county that you are working on (for example, County = 1, where "1" is Orange County). The region field can also be more complex by defining a certain selection threshold as a smaller study area, as in query 12.1.

> **Query 12.1.**
> `County = 1 and (SFG1 >= 70 OR SFG2 >= 70),`

where:
County = the county number,
SFG1 = the single-family physical characteristics suitability, and
SFG2 = the single-family neighborhood access suitability.

The same method can also be used to construct a masking expression for Mask1 through Mask5 in the scenario table, as explained in chapter 3. However, the user can choose to use the four opportunity components instead of a conflict/opportunity value in CEM.

Chapter summary

The following points are discussed in this chapter:

- Both LUCIS and LUCIS^{plus} methods can be used to build HSM.
- Statistical methods, as well as deterministic models, can be used to generate suitability raster layers.
- For HSM, the opportunity involves combining nonconflicting goals so that they have a positive effect on each other. However, conflict combines conflicting goals so that the same area may be needed, for example, for both agriculture and urban use.
- You can build a CEM to allocate new affordable-housing units and then use that same CEM to evaluate existing affordable-housing units.
- You can use CEM in direct scoring, which has some trade-offs in the scoring of affordable-housing locations. However, the planner can at any time identify the different scores used to generate the final score and also change the weights used in combining different components.
- CEM and SPT can be used to prioritize the allocation of affordable housing and transform the allocation ranking into a final score.
- The A4 Allocation by Table tool can be used to set the conditions of the allocation using SPT.

ArcGIS tools referenced in this chapter

Tool Name	Version 10.2 Toolbox/Toolset
A4 Allocation by Table	A4 LUCIS Tools/A4 Allocation
Combine	Spatial Analyst Tools/Local
Polygon to Raster	Conversion Tools/To Raster
Weighted Sum	Spatial Analyst Tools/Overlay
Zonal Statistics	Spatial Analyst Tools/Zonal

References

Arafat, A. 2011. "Evaluating Accessibility and Travel Cost as Suitability Components in the Allocation of Land Use." A Case Study of Identifying Land for Affordable Housing in Three Counties in Florida. http://ufdc.ufl.edu/UFE0043123/00001.

Arafat, A., A. Blanco, W. O'Dell, Y. Zou, and R. Wang. 2012. "Allocation and Preservation of Affordable Housing: A Spatially Discriminated Supply Demand Analysis Based on Parcel Level Employment Assignment." Paper presented at the Urban Affairs Association UAA 2012 conference, Pittsburgh, PA, April 18–21.

Bogdon, A. S., and A. Can. 1997. "Indicators of Local Housing Affordability: Comparative and Spatial Approaches." *Real Estate Economics* 25 (1): 43–80.

Carr, Margaret H., and Paul D. Zwick. 2007. *Smart Land-Use Analysis: The LUCIS Model*. Redlands, CA: Esri Press.

Cervero, R., and K. Kockelman. 1997. "Travel Demand and the 3Ds: Density, Diversity, and NT Design." *Transportation Research Part D: Transport and Environment* 2 (3): 199–219.

CNT (Center for Neighborhood Technology). 2007. "Uncovering the Hidden Assets of Established Communities." *Housing + Transportation Affordability Report*. Retrieved January 20, 2010. http://www.oak-forest.org/UserFiles/File/Planning_&_Zoning/Research_&_Studies/H-T-Oak-Forest.pdf.

———. 2011. *True Affordability and Location Efficiency*. Retrieved January 30, 2011. http://htaindex.cnt.org.

Dacquisto, D. J., and D. T. Rodda. 2006. *Housing Impact Analysis*. Last accessed November 15, 2011. http://www.huduser.org/Publications/pdf/hsgimpact.pdf.

Ewing, R., K. Bartholomew, S. Winkelman, J. Walters, and D. Chen. 2008. *Growing Cooler: The Evidence on Urban Development and Climate Change*. Washington, DC: Urban Land Institute.

Ewing, R., and R. Cervero. 2001. "Travel and the Built Environment: A Synthesis." *Transportation Research Record: Journal of the Transportation Research Board* 1780:87–114.

Grengs, J. 2009. "Job Accessibility and Modal Mismatch in Detroit." *Journal of Transport Geography* 18 (1): 42–54. doi:10.1016/j.jtrangeo.2009.0.01.012.

Haffner, M., and K. Heylen. 2011. "User Costs and Housing Expenses: Towards a More Comprehensive Approach to Affordability." *Housing Studies* 26 (4): 593–614.

HUD (US Department of Housing and Urban Development). 2012. Regional Fair Housing Equity Assessment Portal. http://Portal.hud.gov.

———. 2013. "HUD Secretary Donovan and DOT Secretary Fox Unveil Tool to Provide US Renters and Homeowners with New Housing and Transportation Calculator." HUD No. 13–168. http://portal.hud.gov/hudportal/HUD?src=/press/press_releases_media_advisories/2013/HUDNo.13–168.

Lee, R. W., and R. Cervero. 2007. "The Effect of Housing near Transit Stations on Vehicle Trip Rates and Transit Trip Generation: A Summary Review of Available Evidence." Retrieved April 15, 2009. http://www.reconnectingamerica.org/public/show/hcd_tod_resource_paper_2008_09_20.

Lipman, B. 2006. *A Heavy Load: The Combined Housing and Transportation Burdens of Working Families*. Washington, DC: Center for Housing Policy.

Ray, A., D. Nguyen, W. O'Dell, P. Roset-Zuppa, and D. White. 2009. "The State of Florida's Assisted Rental Housing." Accessed February 8, 2012. http://www.policyarchive.org/handle/10207/bitstreams/20727.pdf.

TCRP (Transit Cooperative Research Program). 1998. *Transit Markets of the Future: The Challenge of Change*. TCRP Report 28. Washington, DC: National Academy Press.

Yamada, H. 1972. "On the Theory of Residential Location: Accessibility, Space, Leisure, and Environmental Quality." *Papers in Regional Science* 29:125–35. doi:10.1111/j.1435–5597.1972.tb01537.x.

Chapter 13
Advanced A4 tools for accessibility

Abdulnaser Arafat

What this chapter covers

Since 1991, federal legislation has encouraged research related to the coordination and impacts of land use and transportation. This legislation includes the Intermodal Surface Transportation Efficiency Act (ISTEA); the Safe, Accountable, Flexible, Efficient Transportation Equity Act: A Legacy for Users (SAFETEA-LU); and the Moving Ahead for Progress in the 21st Century Act (MAP-21). The sprawling development patterns of many of today's first-ring suburbs are attributed to government subsidies that encouraged leapfrog development. The advent of the interstate highway system unlocked open land to tract housing and new subdivisions. A growing population and continued urban expansion continue to generate demand for more freeways. Researchers have identified this problem and now recommend planning for accessibility instead of mobility. This is where Land-Use Conflict Identification Strategy planning land-use scenarios (LUCIS^{plus}) comes in.

LUCIS^{plus} is an automated GIS-based land-use modeling strategy that uses suitability and criteria evaluation matrices (CEMs) to envision future growth. The original LUCIS framework used Euclidean proximity and point density layers to model accessibility to services. However, the new land-use/transportation approach in LUCIS^{plus} uses accessibility to facilities based on proximity to create suitability layers. The LUCIS^{plus} tools in A4 LUCIS Tools estimate accessibility based on network analysis using streets and crosswalks for walking and biking and road networks for driving. The LUCIS^{plus} tools use bus routes, stops, and service frequency for transit modes. This chapter explains the accessibility estimation methods and how to incorporate them in land-use modeling. The chapter also introduces four advanced A4 LUCIS Tools, the A4 Accessibility toolset used to estimate accessibility to services and amenities.

Distance estimation methods

As described in chapter 2, land-use suitability layers are organized in hierarchal structures to construct single-utility and multiple-utility layers. Preferences and community values are used to construct these utility layers (Carr and Zwick 2007). Accessibility is reflected using proximity to facilities such as highways and transit stations. This proximity estimation discriminates between destinations based only on distance. Other important variables such as the attraction of the destinations and method of transportation also play a pivotal role in travel behavior (Hopper 1989; Krumm 1980; Levinson 1995). Different mathematical models are also used to estimate accessibility.

Accessibility is the potential to interact, and mobility is the potential to move. Thus, the term *accessibility* is connected to destinations, and the term *mobility* is connected to the routing of networks and vehicles. Accessibility, for example, measures the number of jobs in a certain area, the number of destinations in a specified area, or the availability of choices between modes of transportation. Mobility deals with traffic delays and level of service (Handy 2004). Although accessibility could be different between modes, traffic congestion is not considered because it is a mobility measure, according to the definition. According to S. Hanson (2004), *accessibility* is the number of opportunities within a distance or travel time, whereas *mobility* refers to the ability to move between different sites. Hanson (2004) explains that because the distance between activities increases as density decreases, accessibility becomes dependent on mobility. This connection means interdependency among density, accessibility, and mobility. The relationship between accessibility and mobility is stronger for regional destinations compared with neighborhood destinations, particularly when different modes of transportation are considered (Hanson 2004; Salomon and Mokhtarian 1998). For regional destinations, using highways and freeways increases accessibility's dependency on mobility. So when transit is involved, accessibility may depend on the level of service and thus on mobility.

Several other definitions of accessibility are also relevant. Accessibility is also defined as the ease with which a destination can be reached, which is an important factor in location decision choices. This definition connects accessibility as a function of land-use and transportation patterns. Accessibility is also defined as the ease with which people can participate in activities. This definition acknowledges that destination activities and location properties are important factors in accessibility (Primerano and Taylor 2004).

The accessibility can be divided into personal or place accessibility, depending on whether the personal characteristics of the traveler are considered. The measurement includes the magnitude of the distance for each location in a gravity-cumulative approach. The accessibility of a place considers the number of activities at a certain distance from a place. These are simple methods for calculating accessibility. More advanced methods of time-space analysis are needed to address the effect of time on accessibility (Hanson 2004). However, performing time-space analysis on a disaggregated level of data at a dependable level of accuracy is not always possible in places with poor records of travel activity.

To simplify, choosing whether to use personal accessibility or place accessibility for analysis depends on whether personal characteristics are included in the estimation of accessibility. For example, it is possible to use personal values to estimate accessibility for existing urban development, but it is more complicated to predict personal values for new development. Generally, the estimation of accessibility is a topological or opportunity measurement, or both. Topological estimations are an estimation of physical proximity from origin-destination (OD), which includes the measurement of distance, such as the distance to the nearest location. Opportunity models measure a density or attraction of accessible places. Incorporating both aspects of measurement gives the relative accessibility, which can be clearly shown in gravity models. This relative accessibility, if accumulated on a large scale, will result in a measure of absolute accessibility (Levinson and Krizek 2008).

Accessibility measurement

The traditional LUCIS framework uses proximity to determine access to services by estimating the Euclidean OD distance. The ArcGIS Euclidean Distance tool estimates the distance between any cell in the raster layer to the nearest facility or destination feature (Esri 2011). Facilities such as schools and hospitals are considered favorable locations in terms of proximity, whereas facilities such as prisons and noise-generating sources are regarded as a disadvantage in terms of proximity. For example, increasing distances to amenities will generally lead to decreasing suitability, meaning that as the distance away from the amenity increases, the suitability of a location decreases. Conversely, for facilities that are less desirable to locate nearby, suitability will increase as the distance away from the facility increases.

The reclassification into decreasing and increasing suitability usually follows a sequential process that includes descriptive statistics. To obtain descriptive statistics, calculate the zonal statistics of the Euclidean distances for all the residential parcels. From the zonal statistics result, the mean and standard deviation are calculated, and their values are used to assign the suitability values. However, proximity does not represent the actual OD travel distance. Nor does it discriminate between destinations according to their size or attractiveness.

Proximity can also be projected by estimating the network distance from any cell in a raster to the nearest facility or destination feature. However, the estimation process is not a direct output of ArcGIS tools. Network Analyst Tools can be used to locate the nearest network points for each origin and its nearest destination. Using the shortest-path method, the network distance is also calculated. Then the values for the distance of the whole trip are summed and assigned a score for the origin, allowing the user to create a network distance raster layer. The output network distance raster layer is shown in figure 13.1. The network distance raster layer may be a better estimation of travel distance to services, but it does not discriminate between destinations according to their size or attraction. The raster layer estimates the distance only to the nearest facility. Furthermore, the ArcGIS Network Analyst extension is used for destinations that can be reached by walking or driving using street networks. It is not used to estimate the proximity to noise generators, sources of pollution, or other point sources for which the Euclidean distance might be a better estimation than the network distance. Proximity does not discriminate between services that a person might choose as a destination. Rather, it estimates only the distance to the nearest facility. However, if a destination has more shopping options, for example, a person might choose to travel farther for shopping than to merely the closest destination.

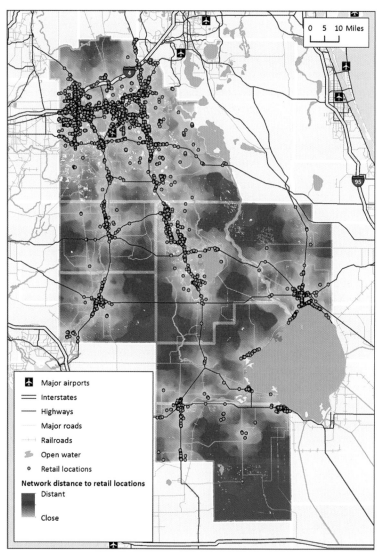

Figure 13.1. After Network Analyst locates the nearest network point using OD, a network distance raster layer is created. In the Heartland region of central Florida, a suitability raster layer is generated using the results of the network distance raster layer for proximity to retail locations.

Figure from Abdulnaser Arafat and the Arizona Board of Regents on behalf of the University of Arizona. Water bodies from the US Geological Survey; state boundary from US Census Bureau TIGER/Line Files; interstates and highways from the Florida Department of Transportation; parcels from the Florida Department of Revenue and the Hillsborough County property appraiser.

How accessibility is measured varies, ranging from linear distance to network distance, travel time, and the number of activities within a distance from an attraction or a certain residential location. Land-use change and prediction models use accessibility estimations to model the change of land use over time. Suitability models of location typically use topological accessibility based on proximity (Carr and Zwick 2007).

Many mathematical forms are used to estimate accessibility. C. R. Bhat et al. (2002) summarize accessibility measurements into different equations for cumulative opportunity and gravity. However, the equations are applied to either a small random data sample or to aggregated and zonal-level data such as traffic analysis zones (TAZs). These estimations are classified as Gaussian, composite impedance, activity distance, and in-vehicle travel time. Table 13.1 compares the variables used for accessibility estimation.

In the absence of measured travel times, the accuracy of topological accessibility estimations depends on the method used to estimate distance. Typically, distance measurements in land-use research use one of three methods: Euclidean distance, rectilinear distance (Manhattan), and network distance. For network distance, measure the length of street segments as a percentage of the whole street network or measure the actual distance traveled (Zhao et al. 2003). The use of travel time may be a more sophisticated method and take more variables into consideration. However, barriers can be included in network distance to give a more accurate indication of travel distance. A. Arafat et al. (2008) compare network distance to Euclidean and Manhattan distance in research on school siting. Their research found that the use of network distance gives a better estimation of walking distance than Euclidean or Manhattan distance. They also found that the area used for estimating population, which is an accessibility indicator, is exaggerated when using a Euclidean buffer. Transportation research in Texas uses the network distance to build accessibility indicators (Bhat et al. 2002). The travel distance was obtained from travel surveys, which may not be available on a disaggregated level. An alternative methodology can be used to generate the network distance at a parcel level using the ArcGIS Network Analyst extension, which can calculate OD distance following road networks. In this methodology, the shortest network distance can be measured from each origin to each destination (Arafat et al. 2008).

Traditionally, proximity as a straight-line distance has been used as an accessibility measurement in deterministic land-use models (see Carr and Zwick 2007), while gravity models have been used to measure accessibility in statistical and stochastic models (see Waddell 2002). Other measurements of accessibility such as opportunity access are also used in statistical models (see Handy 2004; Hanson 2004). However, using gravity models on a parcel-level scale requires generating huge OD matrices that contain billions of records for a large county study area. These OD matrices can exceed the capacity of the hardware and software used in the analysis, even for high-performance computers. Therefore, a methodology to create smaller representative datasets is used in the research in this chapter. Network analysis can be used to estimate the distance. This distance estimation method can also generate the capture area for opportunity estimation, which has historically used Euclidean distances. Arafat et al. (2008) discuss the differences between capture areas based on network and Euclidean distances in terms of the number of students within a walking distance to schools. They recommend the use of network distance to generate the capture area.

The measurement differences between network, Manhattan, and Euclidean distance (figure 13.2) are fundamental to understanding the basis of LUCIS^{plus} tool calculations. The network distance is the distance traveled between two locations using the road network. The Manhattan distance is the distance traveled between two locations following a grid network based on a strictly horizontal or vertical path. The Euclidean distance is the straight-line distance between two locations. Distance estimation is used to evaluate suitability in the proximity component and to create the capture area for the opportunity suitability component.

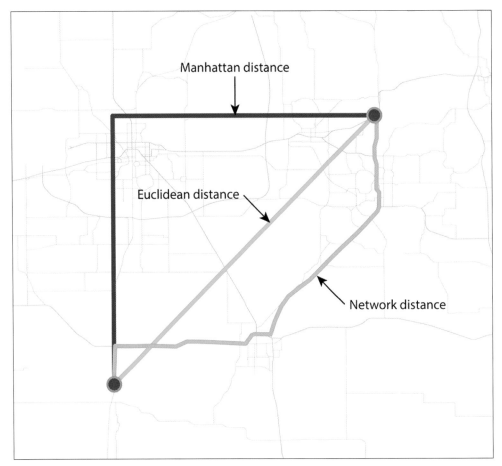

Figure 13.2. Network distance, Manhattan distance, and Euclidean distance are often used in distance estimation to measure proximity and in calculating a capture area in suitability analysis. The network distance is the distance traveled between two locations using the road network. The Manhattan distance is the distance traveled between two locations following a grid network, while the Euclidean distance is the straight-line distance between two locations.

Figure from the Arizona Board of Regents on behalf of the University of Arizona.

Table 13.1. Variables used in accessibility equations

Article	Distance	Network Distance	Opportunity	Gravity General	Gravity Hanson
Carr and Zwick (2007)	D				
Levinson and Krizek (2008)	D or TM		O	O, TM, α	
Arafat et al. (2008)		ND			
Bhat et al. (2002)		TM, TT	O	O, TM, α	O, TT, α
Handy (2004)			O		
Hanson (2004)			O		
Waddell et al. (2007)				O, TM, α	
Grengs (2009)					O, TT, α
Srour et al. (2002)	D		O (Driving time buffer)		
Bhat and Guo (2004)	TT			O, TM, α	

Note: The variables are *D*: distance, *ND*: network distance, *TM*: travel miles, *TT*: travel time, *α*: decay factor, and *O*: opportunity of attraction.

LUCIS^{plus} tools for estimating accessibility

Creating a distance raster layer is one of the simplest methods to measure accessibility. Creating Manhattan or network distance in a distance raster layer is not a functionality included in ArcGIS software. LUCIS^{plus} adds tools for use in ArcGIS to model accessibility. Four Python-based tools are created for that purpose: the A4 Manhattan Distance tool, A4 Network Distance tool, A4 Network Opportunity tool, and A4 Network Accessibility tool. These tools make up the A4 Accessibility toolset in A4 LUCIS Tools.

The A4 Manhattan Distance tool (figure 13.3) calculates the x,y components of the proximity distance using the distance value and the direction from the output Euclidean raster layer. The distance raster layer and direction raster layer are both outputs of the Euclidean distance proximity tools included in ArcGIS. The Python script uses the distance and direction generated by the Euclidean Distance tool to estimate the Manhattan distance for any cell in the output raster layer.

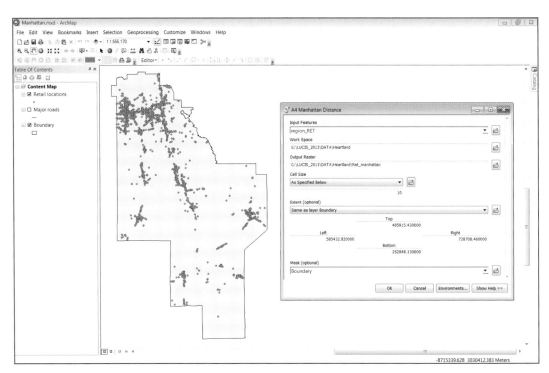

Figure 13.3. The A4 Manhattan Distance tool calculates the x,y components of the proximity distance.

Figure from the Arizona Board of Regents on behalf of the University of Arizona. Regional boundary from US Census Bureau TIGER/Line Files; parcels from the Florida Department of Revenue.

The A4 Network Distance tool (figure 13.4) provides the shortest network distance to the nearest facility. The A4 Network Distance tool estimates the distance from each cell to the nearest destination and generates a network distance raster layer instead of a Euclidean distance raster layer. To reduce the memory used by the tool and increase the processing speed, the tool uses random origins and then estimates the OD distance using Network Analyst.

The procedure the A4 Network Distance tool uses includes generating a set of origins out of the random points and a set of destinations from the point feature class of facilities. A network database of all the streets is also prepared, which the tool calls for analysis. The tool uses the Network Analyst tools to find the closest facility for each origin using the set of destination points. This closest distance is then joined to the origin.

The user generates the random origins before using the tool. The use of the random sample is optional, and you can replace it with the entire population. The output raster layer is created by interpolating the results of the distance estimated for each origin point to the closest destination. For validation of the interpolation results, you can also generate training and testing samples and compare the output raster layers for both datasets.

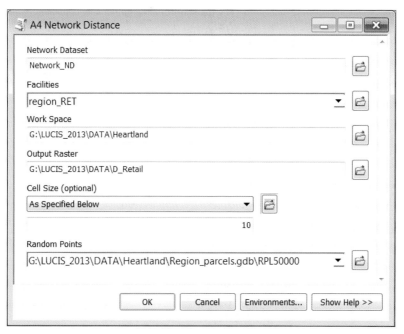

Figure 13.4. The A4 Network Distance tool provides the shortest network distance to the nearest facility.

Figure from the Arizona Board of Regents on behalf of the University of Arizona.

Opportunity access is also a common method for estimating accessibility and can be defined as the number of opportunities within a defined area. Weighting can be adjusted according to the type of use or size of the facility. For example, the floor area can be used to weight retail. The opportunity accessibility is estimated according to equation (13.1).

$$A_i = \sum_i O_i C_{i'} \quad (13.1)$$

where:
O_i = the weight or the attraction of a facility, and
C_i = 1 for distance < buffer distance; otherwise, C_i = 0.

The opportunity is calculated by accumulating the attraction value of each service, which depends on the type of service. This attraction could be the square feet for retail services, the number of beds for hospitals, or any other criteria that can be used to discriminate between the preference of one service to another. ArcGIS includes many tools to estimate the opportunities around a location, including the Point Density and Focal Statistics tools. These tools typically work on Euclidean-based distance. However, a Manhattan distance can be defined as a capture area for opportunity estimated by focal statistics.

The A4 Network Opportunity tool (figure 13.5) estimates opportunity within a certain travel time or travel distance based on street networks. The tool result is a raster layer. The A4 Network Opportunity tool requires the user to suggest random points that represent the origin for each trip estimated. The set of destinations for each origin is the set of destination points that lie within a certain network distance from the origin. The A4 Network Opportunity tool uses Network Analyst to generate an OD matrix using the neighborhood size as a break point. The weight for the destination is then joined to each trip end point, and all the trip weights associated with each origin are then summarized to the origins.

You can use the entire population as the origin or enter representative random points. It is more accurate to use the whole population. However, the greater the number of origin points, the slower the processing speed. The tool uses the values of attraction as weights associated with the random points in an interpolation function to generate the final raster layer.

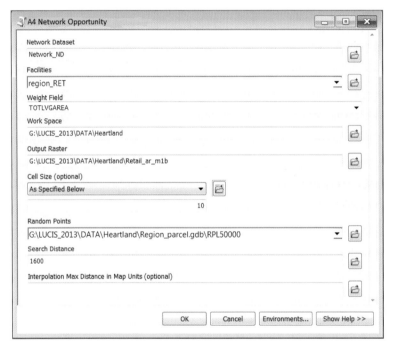

Figure 13.5. The A4 Network Opportunity tool estimates opportunity within a certain travel time or travel distance based on street networks.

Figure from the Arizona Board of Regents on behalf of the University of Arizona.

Gravity accessibility can also be used to estimate accessibility using the A4 Network Accessibility tool. According to Bhat et al. (2002), the accessibility measure is summarized into different equations for cumulative opportunity and gravity measures. Gravity measures are classified as Gaussian, composite impedance, activity-distance, and in-vehicle travel time measures. Equation (13.2) uses the gravity model to calculate accessibility.

$$A_i = \sum_J \frac{O_j}{d_{ij}^\alpha},\qquad(13.2)$$

where:
O_j = the weight or attraction of the facility,
d_{ij} = the distance from each origin to each destination,
J = the number of destinations connected to each origin, and
α = the distance decay factor.

The distance as impedance estimation is a gravity estimation. However, the gravity model is a simple model for use on a zonal level such as TAZs. On a parcel level, the model will generate an OD matrix that contains hundreds of millions of trip combinations if applied to a large study area such as an urban county, region, or metropolitan statistical area (MSA). Applying the gravity mathematical equation at that level is impractical because of hardware and software limitations.

The A4 Network Accessibility tool (figure 13.6) uses a gravity model to estimate network access. The tool solves the size problem of the OD matrix by establishing a representative random sample of origins to capture the accessibility to destinations and interpolate the results to predict the accessibility from any point. To provide flexibility, the use of random points is optional. Points are added to the input dataset for the new tool. The methodology of the A4 Network Accessibility tool is similar to the methodology of using the OD cost matrix to estimate opportunity. The only difference in the process is associating the distance and the decay factor in a mathematical equation, where the number of people willing to make a trip decreases as the distance increases, instead of summarizing the opportunity.

Gravity models can be built on available opportunities and the travel distances to these locations. Gravity models using statistical methods are used for complex modeling of location, such as modeling the employment opportunities for residential locations (Waddell et al. 2007). Parcel-level opportunity and gravity accessibility indicators using either Euclidean or network distance are not yet used in land-use suitability analysis. The simple accessibility estimation, defined as proximity using the Euclidean distance measurement, is typically used in LUCIS models (Carr and Zwick 2005). Gravity, opportunity, and distance models are all used to model accessibility (see Waddell 2002; Handy 2004; Hanson 2004). However, they are applied on either a small random data sample, on aggregated and zonal level data such as TAZs, or are based on Euclidean distance estimations. The LUCIS[plus] tools illustrate how network or Manhattan distance can be used as alternatives to Euclidean distance.

Determining accessibility is important in coordinating land use and transportation. Building on the methods used in *Smart Land-Use Analysis: The LUCIS Model* (Carr and Zwick

2007), LUCIS models have been modified to account for network distance instead of Euclidean distance in accessibility estimation. This reliance on network distance also leads to better coordination between land use and transportation. The automation tools in LUCIS^{plus} affect the overall LUCIS framework in two ways: by replacing the Euclidean proximity and density grids in the suitability models and allocating future population during postprocessing of a new scenario in CEM.

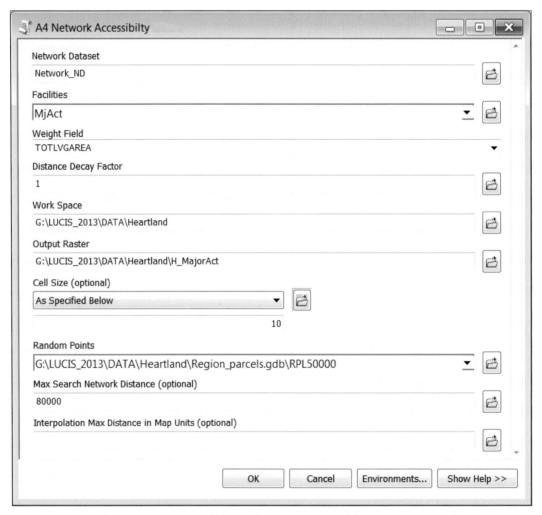

Figure 13.6. The A4 Network Accessibility tool uses a gravity model to estimate network access.

Figure from the Arizona Board of Regents on behalf of the University of Arizona.

In general, the opportunity accessibility estimation captures how many services are within a specified distance. The scoring for services can be different, depending on the type of service such as the square footage of a retail store or the number of beds in a hospital. The neighborhood or surrounding distance can be Euclidean distance, Manhattan distance, or network distance. The ArcGIS Point Density, Point Statistics, and Focal Statistics tools can be used to calculate the Euclidean- and Manhattan-based distance opportunities. The cumulative opportunity score within the specified distance from a parcel is assigned an opportunity score for that parcel (or cell in the raster layer). However, the A4 Network Opportunity tool estimates the opportunity accessibility within a certain network distance or drive time threshold. Figure 13.7 shows retail density within a one-mile driving distance.

The parcel-level accessibility estimation used in this chapter is based on multimodal transportation accessibility. It is applied at the local level to estimate the access to neighborhood services by walking or biking. And it is also applied at the local and regional levels using the driving mode to estimate parcel-level access based on parcels, jobs, and driving distance. Chapter 14 describes a different application of the accessibility tools applied to transit stops, routes, frequency, and transfers for modeling suitability in Hillsborough County. Transit data, parcel data, and amenity data are also used to create transit accessibility layers.

The A4 Network Accessibility tool has a limited application in the suitability model. It is often used as a variable in CEM for the allocation of population in future scenarios. The planning of future-population allocation incorporates running multiple scenarios using LUCIS^{plus} and the CEM method. These scenarios evaluate multiple criteria, including transportation, economic development, and sea level rise.

Network accessibility is important in scenario building. For example, accessibility to major activity centers, logistic terminals, and ports is central to the economic development scenario. As the travel distance increases, accessibility becomes more dependent on mobility and transportation networks. An individual's choice of destination depends on combined attraction, distance, and friction in the effort to travel between two points, factors that are more evident at longer distances. Therefore, the A4 Network Accessibility tool is one of the main tools used to prioritize the allocation of future population in CEM.

Generating an accessibility raster layer for major retail and commercial activities serves as an example of this type of accessibility estimation. In estimating this variable, the major activities in retail and commercial are identified as the destination points. This list can be generated by assigning a size threshold. The resultant activities are integrated within walking distances to reduce the number of major activities. However, the values of attractions are accumulated at the integrated points. To reduce the size of the OD matrix, the random points are used as origin points, and the major activities are considered destination points.

The resultant OD matrix that contains the network distances generated by Network Analyst is joined to major activities to calculate the attraction for each trip. The trips from each parcel to each activity (destination) are used to estimate the accessibility of the parcels to major activities (figure 13.8). The gravity access equation (13.2) is used. In gravity or network access estimation, decay in travel distance is also important and can be estimated from a trip diary or activity model. The A4 Network Accessibility tool also accepts decay factors as an input to the tool.

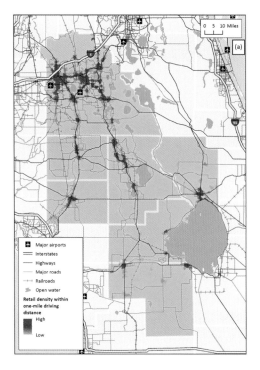
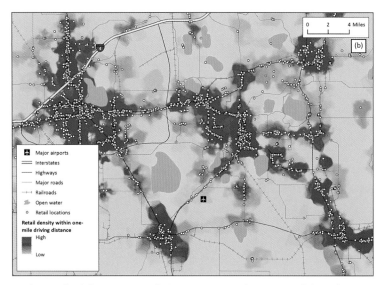

Figure 13.7. The result of the A4 Network Opportunity tool is a network-based opportunity raster layer. Retail density at a regional scale, as in (a) the Heartland region in central Florida, and at a local scale, as in (b) Polk County, are calculated based on an opportunity accessibility score.

Figure from Abdulnaser Arafat and the Arizona Board of Regents on behalf of the University of Arizona. Water bodies from the US Geological Survey; state boundary from US Census Bureau TIGER/Line Files; interstates and highways from the Florida Department of Transportation.

Chapter 13 Advanced A4 tools for accessibility

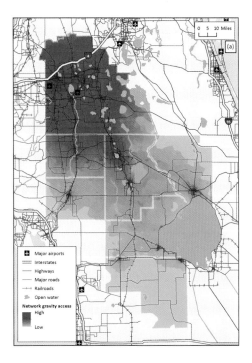

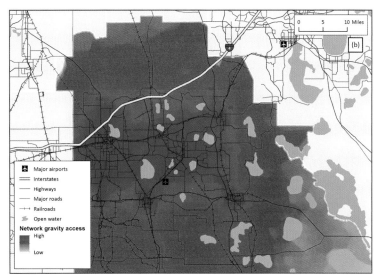

Figure 13.8. The result of the A4 Network Accessibility tool is a distance decay-based raster layer. Accessibility for retail and commercial locations at a regional scale, as in (a) the Heartland region in central Florida, and at a local scale, as in (b) Polk County, Florida, are calculated using the attraction to the locations and the gravity access equation (13.2).

Figure from Abdulnaser Arafat and the Arizona Board of Regents on behalf of the University of Arizona. Water bodies from the US Geological Survey; state boundary from US Census Bureau TIGER/Line Files; interstates and highways from the Florida Department of Transportation.

Tool validation and estimation accuracy

The A4 Accessibility tools, except for the Manhattan Distance tool, can be validated for accuracy. Validation of the tools depends on two main estimation steps. For the A4 Network Distance tool, step 1 is to use the tool to estimate the accessibility score for each point in the population. The use of random points helps speed up processing. Step 2 is to use ArcGIS Network Analyst to calculate the network accessibility score as the network distance to the nearest destination for any point. The results of step 1 are validated by getting the identical results in step 2.

For the A4 Network Opportunity tool, step 1 is to use the tool to estimate the accessibility score based on the density inside a network driving shed, or network service area. Step 2 is to generate the service area in Network Analyst, and the service areas are used to capture the density of services inside the network buffer. The results of step 1 are validated by getting the identical results in step 2.

For the A4 Network Accessibility tool, the accessibility score in step 1 reflects the gravity equation (13.2) applied to network distance, as well as the attraction and decay factor. These steps are estimation steps and so should be free of error. To make sure that the estimation is error free, a sample from the random points is estimated in step 2 using Network Analyst. The sample is compared with the corresponding list generated as an intermediate step by the tools. For the study area of the Heartland region, the two estimations are identical.

However, the A4 Accessibility tools use a prediction step in the final stage. This step is the inverse distance weighted (IDW) interpolation. IDW is an exact interpolator, which means that it does not change the values of the input points. Theoretically, the more points used for the interpolation, the fewer interpolation errors will be generated.

The tools used in the study area generate an access raster layer for many locations. The number of random points used varies between 5,000 and 10,000 points, depending on the area of study. The prediction errors in the study area in all cases were small. However, determining the number of random points and the distribution of these points to establish a standard prediction error may not be an easy task. It is recommended that you create random point samples as a subset for training and testing and compare the results for verification purposes.

These new A4 Accessibility tools are used in chapter 14 to create suitability raster layers for allocating future population. The transportation variables used include transit access and access by driving or walking, combined with other population allocation prioritization raster layers used in a LUCIS^{plus} CEM.

Tool limitations

The LUCIS^{plus} tools introduced in this chapter demonstrate three main types of accessibility estimations: proximity, opportunity, and gravity. The A4 Accessibility tools use a road network to estimate accessibility scores. Therefore, distance or time can be used as a cost field in the analysis. The tools can also be updated to include traffic congestion and road network barriers. The choice of the suitability tool for estimating accessibility depends mostly on the study area and the rationale behind the accessibility score estimation. For example, in investigating the Florida National Household Travel Survey dataset NHTS 2009, it is not difficult to conclude that the frequency of shopping trips decreases as the travel distance increases. This hypothesis is true for shopping trips that are more than two miles long. So if the shopping trip distance is less than two miles, people may choose to shop in a farther shopping center within the two-mile network service area. Thus, opportunity estimation is more suitable for determining neighborhood access.

The comparison among the three methods of estimation is complex, and it may not always be practical to incorporate all three in a suitability model. In the Heartland region, however, all three tools have been used in suitability models. For the neighborhood and local accessibility level, a combination of network-based opportunity and network-based proximity are used to model suitability. In suitability modeling, the closer the shopping center, the more suitable the location is for residential, which is a direct application of the network proximity tool. In terms of a commercial location, the higher the density of commercial square footage, the more suitable the location is for commercial use. The latter is a direct application of the opportunity tool. For regional destinations, the choice of where to shop depends on the attraction of the destination as well as the travel time or distance. Applying these factors is an application of the gravity model, which is a direct application of the A4 Network Accessibility tool. However, the shopper may not travel 50 miles to shop, even if the attraction of the destination is high. The element of choice necessitates including decay and friction factors in the gravity equation. Also for regional destinations, accessibility is more dependent on mobility (Hanson 2004), which increases the need to include such factors as road capacity, barriers, and congestion in estimating accessibility.

The tools introduced in this chapter are parcel-level population tools that should be free of estimation error. The parcels in the study area are used as origins, and the tools create an accessibility score for each OD. The tools create an accessibility raster layer from the values of the access score for each parcel. This process is time consuming, and its success depends on processor speed, computer memory, and software limitations. Further, the use of a random sample can reduce estimation time. Yet it will also lead to some estimation error because of the interpolation of values using the origin location. To make the tool flexible, the use of population or sample data is optional so that you can decide on the trade-off between time and accuracy. You may also decide to run the tool on training and testing samples for verification purposes.

The LUCIS^{plus} tools are automated and simple to use. Because these tools account for the network distance and travel time, they are recommended for use in land-use modeling instead of the proximity tools that are based on Euclidean distance. However, proximity to such factors as noise and pollution, which is not dependent on road networks, is thus not dependent on network distance. Therefore, Euclidean distance should continue to be used in these types of estimations. The LUCIS^{plus} tools can also be applied to distance and travel time. The inclusion of barriers and congestion will be studied in future research.

Chapter summary

The following points are discussed in this chapter:

- LUCIS^{plus} uses accessibility and network distance to create suitability raster layers.
- Accessibility estimation is based on network distance and includes proximity, opportunity, and gravity models.
- The A4 Accessibility tools are GIS customized tools that can be used in the same way as any geoprocessing tool. They can also be used within the ArcGIS ModelBuilder application or be called by other Python programs.
- Euclidean proximity estimation is still used in LUCIS to estimate the proximity to locations that is not dependent on streets, such as proximity to noise-generating locations.
- The A4 Accessibility tools can also be used to estimate accessibility to destinations by walking, biking, driving, and transit. The estimation includes sidewalks and crosswalks for walking and biking and streets for driving. For transit, the estimation is based on transit routes, frequencies, and stop locations.
- The A4 Accessibility tools can be integrated in suitability models to create suitability raster layers. The A4 Accessibility tool output can be used as extra criteria for the allocation of population in CEM.

ArcGIS tools referenced in this chapter

Tool Name	Version 10.2 Toolbox/Toolset
A4 Manhattan Distance	A4 LUCIS Tools/A4 Accessibility
A4 Network Distance	A4 LUCIS Tools/A4 Accessibility
A4 Network Opportunity	A4 LUCIS Tools/A4 Accessibility
A4 Network Accessibility	A4 LUCIS Tools/A4 Accessibility
Euclidean Distance	Spatial Analyst Tools/Distance
Focal Statistics	Spatial Analyst Tools/Neighborhood
Make Service Area Layer	Network Analyst Tools/Analysis
Make Closest Facility Layer	Network Analyst Tools/Analysis
Make OD Cost Matrix Layer	Network Analyst Tools/Analysis
Point Density	Spatial Analyst Tools/Density
Point Statistics	Spatial Analyst Tools/Neighborhood

References

Arafat, A., R. L. Steiner, and I. Bejleri. 2008. "A Method for Measuring Network Distance Using Network Shortest Distance and Spatial Interpolation." Paper presented at ACSP-AESOP Fourth Joint Congress, Chicago, IL, July 6–11.

Bhat, C. R., S. Handy, K. Kockelman, H. Mahmassani, A. Gopal, I. Srour, and L. Weston. 2002. *Development of an Urban Accessibility Index: Formulations, Aggregation, and Application.* Report No. FHWA/TX-02-4938-4. Austin, TX: University of Texas, Austin.

Carr, Margaret H., and Paul D. Zwick. 2005. "Using GIS Suitability Analysis to Identify Future Land-Use Conflicts in North Central Florida." *Journal of Conservation Planning* 1 (1). Retrieved March 15, 2009. http://www.journalconsplanning.org/2005/index.html.

Carr, Margaret H., and Paul D. Zwick. 2007. *Smart Land-Use Analysis: The LUCIS Model.* Redlands, CA: Esri Press.

Esri. 2011. *ArcGIS 10 Help Documentation.* Redlands, CA: Esri.

Grengs, J. 2009. "Job Accessibility and Modal Mismatch in Detroit." *Journal of Transport Geography* 18 (1): 42–54. doi:10.1016/j.jtrangeo.2009.0.01.012.

Handy, S. 2004. "Planning for Accessibility: In Theory and Practice." In *Proceedings from the Access to Destination Conference.* Minneapolis, MN: University of Minnesota.

Hanson, S. 2004. "The Context of Urban Travel: Concepts and Recent Trends." In *The Geography of Urban Transportation,* edited by S. Hanson and G. Giuliano, 3rd ed., 3–29. New York: Guilford Press.

Hopper, Kevin G. 1989. *Travel Characteristics at Large-Scale Suburban Activity Centers.* Transportation Research Board, NCHRP #323. Washington, DC: National Research Council.

Krumm, Ronald J. 1980. "Neighborhood Amenities: An Economic Analysis." *Journal of Urban Economics* 7:208–24.

Levinson, David. 1995. "Location, Relocation, and the Journey to Work." Paper presented at the Western Regional Science Conference, San Diego, CA, November.

Levinson, D., and K. Krizek. 2008. *Planning for Place and Plexus: Metropolitan Land Use and Transport.* New York: Routledge.

Primerano, F., and M. Taylor. 2004. "An Accessibility Framework for Evaluating Transport Policies." In *Proceedings from the Access to Destination Conference*, Nov. 8–9. Minneapolis, MN: University of Minnesota.

Salomon, I., and P. L. Mokhtarian. 1998. "What Happens When Mobility-Inclined Market Segments Face Accessibility-Enhancing Policies?" *Transportation Research Part D: Transport and Environment* 3 (3): 129–40.

Srour, I., K. Kockelman, and T. Dunn. 2002. "Accessibility Indices: A Connection to Residential Land Prices and Location Choices." Paper presented at the 81st Annual Meeting of the Transportation Research Board (CD-ROM), Washington, DC, January 13–17.

Waddell, P. 2002. "UrbanSim: Modeling Urban Development for Land Use, Transportation, and Environmental Planning. *Journal of the American Planning Association* 68 (3): 297–314.

Waddell, P., C. R. Bhat, N. Eluru, L. Wang, and R. M. Pendyala. 2007. "Modeling the Interdependence in Residential and Workplace Choices." *Journal of the Transportation Research Board,* 84–92.

Zhao, F., L. Chow, A. Gan, and I. Ubaka. 2003. "Forecasting Transit Walk Accessibility: Regression Model Alternative to Buffer Method." *Transportation Research Record* 1835 (8): 34–41.

Chapter 14

Analyzing and mapping transportation accessibility and land use

Abdulnaser Arafat

What this chapter covers

Literature on coordination of land use and transportation generally studies travel behavior and transportation system characteristics and how they relate to land use. The land-use–transportation relationship is bidirectional: land use impacts transportation, and transportation impacts land use. Simple accessibility estimations are traditionally used to capture the impact of land use on transportation (Hanson 2004, 3–29; Giuliano 2004, 237–73). Meanwhile, sophisticated and detailed activity-based models can be used to model the impact of travelers, transportation facilities, and networks on land use. The urban form variables that affect trip generation and vehicle miles traveled (VMT), such as density, land-use mix, and connectivity, can be used to study the impact of land use on transportation. The availability of transit as a travel mode also reduces VMT. The land-use and transportation literature (see Ewing and Cervero 2001) amply covers the impact of urban form on transportation, yet planners and decision-makers concerned about location choice and land-use modeling seem to make this connection only indirectly. However, studying the impacts of land use on transportation could be a major factor in helping planners solve transportation network problems, make improvements on transportation networks, and influence the calculation of impact fees. Land-use planning can also reduce travel cost and contribute to a cleaner environment by reducing travel miles and trips by car.

One traditional method of evaluating the suitability of land parcels is to determine the proximity of parcels to nearby services. This proximity uses a straight-line (Euclidean) distance in deterministic land-use modeling (see Carr and Zwick 2007). However, people may choose to travel farther than the shortest distance to access better services. Proximity measurement does not discriminate between services by their type, quality, attraction, or other variables deemed important in travel behavior research. The Land-Use Conflict Identification Strategy planning land-use scenarios (LUCIS^{plus}) models integrate accessibility and proximity to evaluate the physical suitability of land parcels for different land uses. The notion of accessibility in LUCIS^{plus} goes beyond the proximity measurement and adds opportunities within defined neighborhoods, taking account of their relative attractiveness.

The latest development in accessibility research provides different methods for estimating accessibility to destinations. Chapter 13 compares the mathematical models used to estimate accessibility in transportation and land-use planning. Gravity, opportunity, and distance (proximity) models are all traditionally used in statistical and stochastic models for this purpose. However, some of these metrics are used on an aggregate scale and can be modified for use in a spatially aware environment such as LUCIS^{plus}. LUCIS^{plus} analysis and models use high-resolution raster layers that may have a 10 × 10 meter cell size. This level of analysis is often more detailed than standard parcel-level analysis. This chapter discusses two applications of tools in the A4 Accessibility toolset in A4 LUCIS Tools: the A4 Network Opportunity tool and A4 Network Accessibility tool. These applications create a quantitative measurement of accessibility.

Using network distance to replace Euclidean distance

Land-use models often use proximity to amenities and services as criteria to decide the suitability of a land parcel. The Euclidean distance raster layer represented this proximity in many of the original LUCIS models. The new trend in LUCIS^{plus} is to replace the estimation of Euclidian distance with the estimation of network distance to represent the proximity to different land-use types. The network distance simulates proximity using any of the transportation modes and networks. For example, if the transportation mode is the automobile, the network used is the road network. If the transportation mode is transit, the network used might be a bus route or train route.

In many cases, the Euclidian distance is still the best proximity estimation method, especially in evaluating subobjectives such as distance to noise generators. The general idea in deciding whether to use Euclidean distance or network distance proximity is to model the relationship between the origin and destination and whether the origin generates transportation trips. More simply, how is the destination reached? If the road or street network can reach the destination, network proximity is used. Otherwise, Euclidian distance may be a good choice. Figure 14.1 shows the submodel for determining suitability using the A4 Network Distance tool to capture the proximity to retail.

The network dataset in the model in the figure is the 2011 TIGER/Line Shapefile for streets. The network dataset can be either a line feature class or line shapefile with appropriate

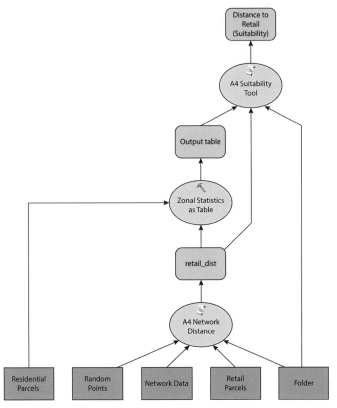

Figure 14.1. The ArcGIS ModelBuilder model for the distance to retail suitability subobjective. The model uses the A4 Network Distance tool in place of the Euclidean Distance tool because the destination can be reached by the road network. Thus, the A4 Network Distance tool provides a better proximity estimation.

Figure from the Arizona Board of Regents on behalf of the University of Arizona.

attributes for conversion to a network dataset. The street network can be generated using the ArcCatalog application in ArcGIS.

Trip calculation requires an origin and a destination. For proximity estimation, only the nearest destination is used as the destination of the trip. For the study area of Hillsborough County, Florida, a set of random locations represent the origins of all generated trips. These locations consist of 10,000 random points that are located near existing streets. The destination points will differ between models and the land-use relationship you are attempting to measure. The model in figure 14.1 calculates the network distance between residential and retail land uses.

The A4 Network Distance tool is a population-scale tool. That is, the tool works on the entire population. However, running the entire population in the tool can be slow and may cause computer memory errors. Therefore, it can be beneficial to use a random sample of

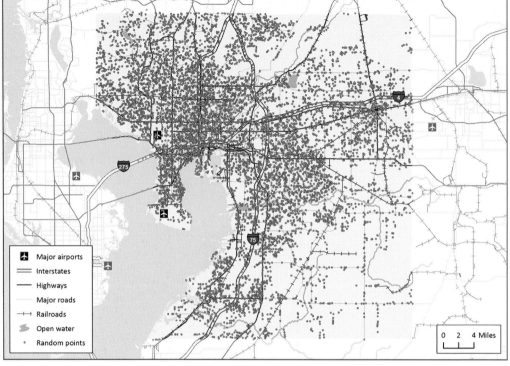

Figure 14.2. Set of 10,000 origin random points generated for Hillsborough County, Florida, at a regional level. These random points represent a sample of origins used to better approximate road network accessibility.

Figure from Abdulnaser Arafat and the Arizona Board of Regents on behalf of the University of Arizona. Water bodies from the US Geological Survey; state boundary and street network from US Census Bureau TIGER/Line Files.

representative origin points. As to the entire population, the origin points would be the set of points generated from transforming the parcel polygon into a point feature class using the ArcGIS Feature to Point tool. Or you can use the Create Random Points tool to generate a random set of points. Figure 14.2 shows the result of generating a random set of 10,000 points. The random points can be generated near streets to ensure easier access to the road network.

As explained in chapter 13, the A4 Network Distance tool generates a raster layer that represents the predicted network distance from each cell to the nearest point in the destination set. The tool uses an inverse distance weighted (IDW) interpolation function. One of the main characteristics of IDW is that it is an exact interpolator. So the interpolated raster layer has zero error at the random point locations used as an input for IDW values. Figure 14.3 shows the output distance raster layer that represents the distance to the nearest entertainment location. The entertainment point feature class is generated from the property

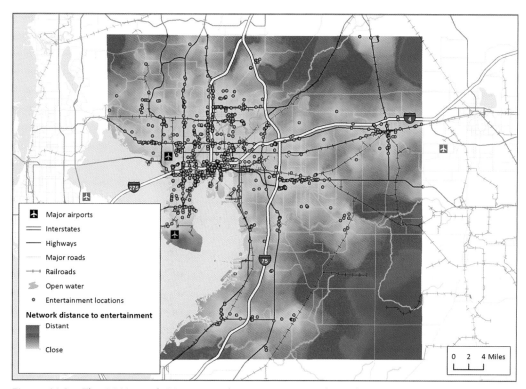

Figure 14.3. The A4 Network Distance tool generates a raster layer that represents the predicted network distance from each cell to the nearest point in the destination set. The network distance raster layer illustrates the network proximity to the nearest entertainment locations.

Figure from Abdulnaser Arafat and the Arizona Board of Regents on behalf of the University of Arizona. Water bodies from the US Geological Survey; state boundary from US Census Bureau TIGER/Line Files; interstates and highways from the Florida Department of Transportation.

appraiser data and parcel data for Hillsborough County. The figure also shows the TIGER/Line street network for Hillsborough County. It is always wise to use training and testing samples to compare results and verify the model to ensure the desired accuracy.

The ArcGIS Zonal Statistics as Table tool is used to transform proximity to suitability (see figure 14.1) and establish the thresholds for suitability transformation. The basis for the reclassification is the average distance from residential parcels. The scoring is increased or decreased according to one-quarter standard deviations. The A4 LUCIS Suitability tool, explained in chapter 3, is also used to assign the suitability values using a decreasing suitability function. The resultant suitability values range from 9 to 1.

Network proximity analysis assumes that the mode of transportation is driving. It considers the road network, direction of travel, and distance to estimate travel cost. The same network proximity analysis can also be performed using driving time instead of driving distance. For network proximity analysis, the road feature class should be converted to a network dataset using ArcCatalog. Creating the network for analysis involves the following steps:

1. Choose the line feature class to be used in creating the network.
2. Define the connectivity between different links.
3. Model the elevation. This modeling is beneficial for modeling the intersection topology, especially if there are bridges or tunnels.
4. Define one-way roads and the constraints for other road types.
5. Define the cost field in the line feature dataset. By default, ArcGIS Network Analyst will take the length field as distance cost. However, you can define the cost field using another cost variable in the street data attribute table.
6. Model directions, if needed.
7. Build the network dataset and resolve any dataset errors that may arise.

Geodatabases or shapefiles can be used to create the network dataset in ArcCatalog in a step-by-step, interactive, user-friendly environment. The quality of the network greatly depends on the line network data. The default flow direction comes from the data structure unless you indicate otherwise. For example, if the line is created by digitizing from point 1 to point 20 sequentially, the driving direction will, by default, be from 1 to 20 for one-way streets. Flow direction is especially important when modeling highway exits.

Using the A4 Network Distance tool to estimate accessibility by transit system

The current approach used by planners for surface transportation planning is a multimodal transportation system approach. Multimodal transportation modes include walking, biking, transit, and driving. Many computer software programs can model and simulate these modes. Transportation systems are analyzed in the context of the larger network. An alternative to driving may be a combination of walking, biking, and transit. Many types of networks can be created for multimodal transportation modeling. These networks include the following:

- *The driving network*. Generated by defining the roads, intersections, and rights-of-way, assuming the trip between two points is performed by driving.
- *The pedestrian network*. Considers sidewalks, crosswalks, and roads as the pedestrian path. The pedestrian network can also include "social roads," which can be a shortcut paved or unpaved/unimproved road.
- *The biking network*. Considers bike lanes, routes, sidewalks, and roads in modeling the trips that use bikes.
- *The transit network*. Can include bus routes, light rail, heavy rail, or a combination among them. Can use the route line feature class, stop point feature class, and the bus schedule in modeling this network.
- *Combined networks*. Combines two or more transportation networks into one network.

An example of the combined network is the combination between walking and transit modes to model accessibility to different land uses. The traveler can walk from home to the transit stop, ride transit, and then walk again to get to the destination. The four main components of this accessibility estimation are

- the network proximity to the closest transit stop to the origin of the trip,
- the trip length,
- the frequency of transit service, and
- the network proximity of the destination to the nearest transit stop.

Figuring the proximity to and from transit stops is one type of accessibility estimation. Other types of accessibility estimation, such as opportunity access or gravity access, can replace network proximity. Opportunity access is important in cases in which there are many transit stops going to different destinations within a specified walking distance. The gravity model can include a transit stop score and weight different stops by distance. Suitability modeling estimates and combines these accessibility elements and weights them. The following sections of this chapter explain the process of transit accessibility estimation applied to Hillsborough County, Florida.

Proximity of origins to transit stops

This section focuses on network proximity and the use of the A4 Network Distance tool to estimate the proximity of origins to transit stops. The proximity of origins to transit stops is important for initiating trips by transit. Some people walk to transit stops while other people ride their bikes or drive. However, the access score assumes that people will walk to transit stops, and therefore the proximity to the nearest transit stop is important. This proximity can be estimated using the A4 Network Distance tool. The network is a combination of the road network, sidewalks, and crosswalks. Social roads can also be included in this network. However, the walkability to a transit stop does not depend only on the network distance. Other parameters such as the completeness of the network, road connectivity, design, street view, and safety are also considered in estimating walkability. These parameters are used to generate suitability layers that are incorporated in LUCIS^{plus}. To estimate the relationship between the origin and transit stops, Network Analyst generates a walking service area around the transit stops. The resulting network service area is dissolved to create one polygon and used to generate 10,000 random points of origin.

The random locations are chosen near sidewalks or streets to ensure easy access to the road network. These random locations form the origin from which the walking distance is estimated using the A4 Network Distance tool. The random locations are also used to generate the output for network proximity to transit stops. Figure 14.4 illustrates the output raster layer for proximity to transit stops generated by the A4 Network Distance tool. This estimation can be improved to consider the number of transit stops that can be reached by walking. The A4 Network Opportunity tool performs this estimate. The final result is a weighted overlay of the suitability obtained by the A4 Network Distance tool with the suitability obtained by the A4 Network Opportunity tool.

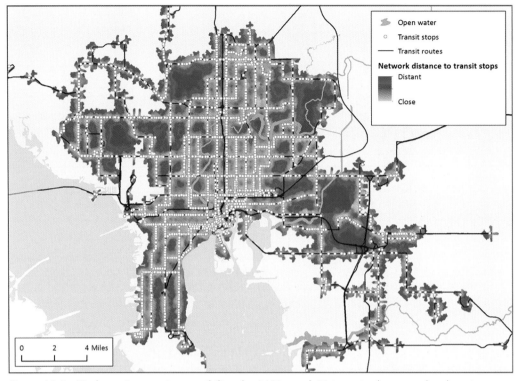

Figure 14.4. To determine transit accessibility, the A4 Network Distance tool uses random locations near sidewalks or streets to measure the accessibility to transit stops. The proximity layer illustrates the network proximity to stops generated by the A4 Network Distance tool.

Figure from Abdulnaser Arafat and the Arizona Board of Regents on behalf of the University of Arizona. Water bodies from the US Geological Survey; state boundary from US Census Bureau TIGER/Line Files; interstates and highways from the Florida Department of Transportation; transit routes and stops from the Florida Transit Information System.

The Zonal Statistics as Table tool is used to transform the accessibility layer into a suitability layer. The zonal statistics are based on the statistics used in the A4 LUCIS Suitability tool. For example, if you are working on multifamily suitability, the A4 LUCIS Suitability tool uses the multifamily parcels as zones for the reclassification procedure. The model in figure 14.1 shows the complete process, from creating the proximity raster layer to creating the suitability map from the distance raster layer. The reclassification method is decreasing suitability, which means that as the transit stop gets farther away from the location, suitability decreases. Figure 14.5 shows the suitability for multifamily residential based on distance to transit stops.

Estimating transit access to destinations by frequency/trip length

Traditional accessibility models use the proximity to transit stops as an accessibility measurement. However, some transit accessibility estimation models also use transit frequency

Chapter 14 Analyzing and mapping transportation accessibility and land use

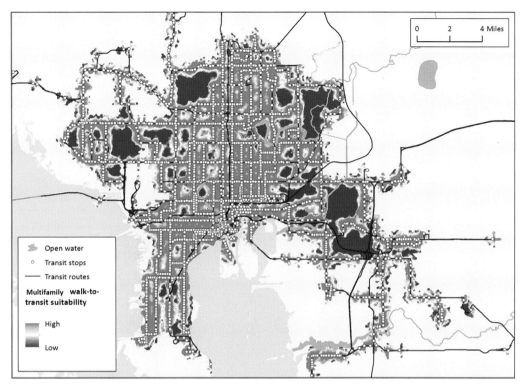

Figure 14.5. The final suitability raster layer for multifamily residential based on the walking distance to transit stops shows that suitability decreases as the transit access points become farther away from multifamily residential locations.

Figure from Abdulnaser Arafat and the Arizona Board of Regents on behalf of the University of Arizona. Water bodies from the US Geological Survey; state boundary from US Census Bureau TIGER/Line Files; interstates and highways from the Florida Department of Transportation; transit routes and stops from the Florida Transit Information System.

to create a score for the transit stop. The A4 Network Accessibility tool creates an accessibility score based on the trip from the transit stop to the destination. This score includes the travel time/distance, service frequency, and transfer points in the trip. The frequency is weighted by the travel distance, which assumes that the longer the trip length, the longer the traveler is willing to wait for transit. This frequency weighted by distance is applied to the network dataset and creates a cost field, which is the result of dividing each route distance by the frequency of service. The resultant cost field is used in linear referencing to the network dataset. The transit stop is the origin in the access score, and the score created from the trips to destinations are joined back to the network dataset as a score for each transit stop. IDW is used to generate the raster layer from the transit stop scores. The interpolation is done inside the dissolved walking distance service area. IDW weights the stops by distance from the origin, similar to the gravity estimation of accessibility.

The model in figure 14.1 measures the suitability of multifamily residential to existing commercial. The A4 Network Accessibility tool uses the multifamily suitability layer to convert the transit stop accessibility score to suitability parameters. Figure 14.6 shows the resultant suitability raster layer based on accessibility scores.

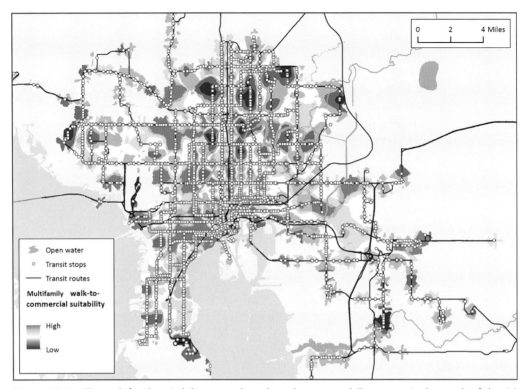

Figure 14.6. The multifamily suitability raster layer based on accessibility scores is the result of the A4 Network Accessibility tool reclassifying the suitability raster layer for multifamily to existing commercial. The result illustrates the transit accessibility relationship between multifamily residential and commercial developments.

Figure from Abdulnaser Arafat and the Arizona Board of Regents on behalf of the University of Arizona. Water bodies from the US Geological Survey; state boundary from US Census Bureau TIGER/Line Files; interstates and highways from the Florida Department of Transportation; transit routes and stops from the Florida Transit Information System.

The final suitability is generated by combining the suitability raster layers shown in figures 14.5 and 14.6 using the ArcGIS Raster Calculator tool. The model uses the same steps to generate accessibility raster layers using various accessibility estimation procedures such as opportunity. The only difference is that the model contains the A4 Network Opportunity tool or the A4 Network Accessibility tool instead of using the A4 Network Distance tool.

Using network opportunity to replace point density

Using a similar approach to the network proximity estimation, you can estimate the opportunity within a specific-size neighborhood by creating a network service area around a point of interest. For example, if you want to estimate the opportunity of retail within a one-mile driving distance from point A, create a one-mile service area. Then calculate the square footage of retail within the network service area. However, creating a service area around each point in the map is not practical, and creating a raster layer from this method is not easy. An alternative approach is to generate an origin-destination (OD) cost matrix and set a breakpoint according to the value of the distance (in this case, one mile). The origins in the matrix are the random points. Then you can determine the destination dataset (in this case, retail points). After creating the OD cost matrix, sum the total square footage of retail per origin and interpolate it to create the accessibility raster layer.

LUCIS models have traditionally used point density to estimate the density of specific land uses within a user-set neighborhood size. Instead of using a circular buffer as used in the ArcGIS Point Density tool, the A4 Network Opportunity tool uses a network service area in the OD cost matrix. The breakpoint is equal to the radius of the Euclidean buffer. The weighted A4 Network Opportunity tool estimates the sum of values of the input point feature classes within the service area. When the tool was applied to Hillsborough County, 10,000 random points were used to estimate the opportunity and generate the accessibility raster layer. Figure 14.7 shows a typical submodel for estimating network opportunity.

The first raster layer generated from the submodel illustrates the opportunity. The opportunity is the summation of the point values within a network service area. The Zonal Statistics as Table tool uses the evaluated parcels to establish the threshold for assigning suitability values. The A4 LUCIS Suitability tool uses an increasing suitability method to assign suitability values. Figure 14.8 shows an example of the opportunity for retail land use using a network service area of one mile.

Using network gravity access as a suitability layer

Gravity access uses the attraction of the destination and the weight of the attraction to estimate OD accessibility. The distance traveled from each origin to each destination determines the weight of the attraction. Chapter 13 explains the gravity access equation (13.2) and the parameters used. The traditional gravity access equation was limited to small random data samples or aggregated data, such as zonal-level data on the order of traffic analysis zones (TAZs). LUCIS^{plus} uses a parcel-level suitability evaluation that captures a combination of access by opportunity and by distance in a multiple-utility assignment (MUA). The distance estimation methods introduced in chapter 13—network, Manhattan, and Euclidean distance—evaluate the proximity component. Distance estimation is used to create the neighborhood for network opportunity. LUCIS^{plus} uses the automated A4 Network Accessibility tool to estimate the gravity accessibility at a parcel level. The methodology for estimating the accessibility score applies gravity access equation (13.2) to an OD matrix. The origins are the set of random points, and the destinations are the set of destinations chosen for the model. Figure 14.9 shows a model for using the gravity accessibility estimation explained in chapter 13 to create the suitability.

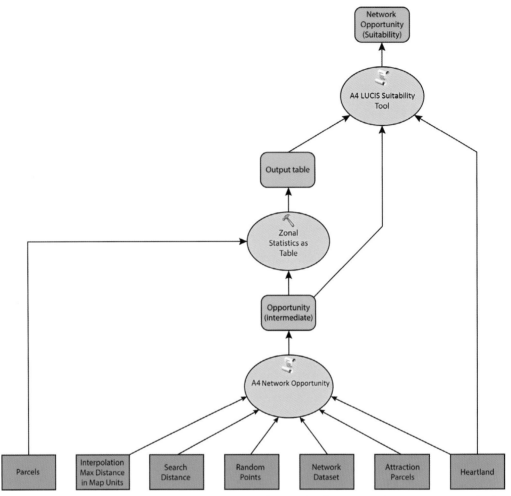

Figure 14.7. The suitability estimation model using the weighted A4 Network Opportunity tool. The weighted A4 Network Opportunity tool is used in place of the Point Density tool to estimate network opportunity and generate the accessibility raster layer. The Zonal Statistics as Table tool is used to establish ranges for suitability reclassification.

Figure from the Arizona Board of Regents on behalf of the University of Arizona.

The typical LUCIS model does not contain gravity access as a submodel in the hierarchical suitability structure. This type of layer can be used in the criteria evaluation matrix (CEM) to prioritize an allocation based on accessibility. One example is the accessibility to shopping, which considers the total square footage in addition to the network distance. The raster layer in figure 14.10 shows the accessibility to shopping using the A4 Network Accessibility tool.

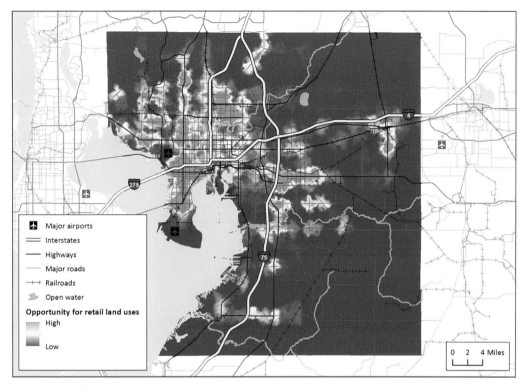

Figure 14.8. The final suitability raster layer identifies network opportunity using the retail square footage estimation and a network service area of one mile. Suitability is assigned as a function of decreasing opportunity in which the distance from the retail network service area increases.

Figure from Abdulnaser Arafat and the Arizona Board of Regents on behalf of the University of Arizona. Water bodies from the US Geological Survey; state boundary from US Census Bureau TIGER/Line Files; interstates and highways from the Florida Department of Transportation.

The impacts of accessibility on transportation and land-use planning

This chapter explains how to incorporate accessibility in LUCIS^{plus} models. The main objective is to achieve better coordination between land use and transportation. Historically, transportation planning was regarded as planning for better mobility. This focus led to the construction of highways and freeways to increase the capacity of the road network and decrease travel time. But the expansion of the road network is a case of transportation impacting land use and leads to low-density development. This cycle of bidirectional impact between land use and transportation, and transportation and land use, can be better coordinated by studying how land use impacts transportation and how transportation

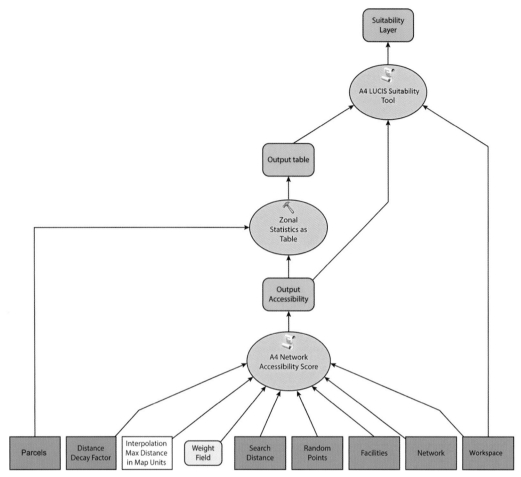

Figure 14.9. The A4 Network Accessibility tool is included in the model to create the gravity accessibility suitability at a parcel level.

Figure from the Arizona Board of Regents on behalf of the University of Arizona.

impacts land use. The land-use impact on transportation can be measured by urban form variables, such as density, diversity, network design, destination, and distance. Reid Ewing and Robert Cervero (2001) called the first four variables the *4Ds*, and added the fifth to make it the *5Ds* in a later work. The 4Ds research provides a method for decreasing VMT by increasing density, improving road connectivity and accessibility, increasing land-use diversity, and developing nearby transit. The LUCIS[plus] approach incorporates these variables within the suitability model or CEM. This chapter discusses some of these variables, such as the network proximity to transit stops and accessibility. Simple ArcGIS tools such as Line Density and Point Density, as well as Focal Statistics, can be used to estimate variables such as density, diversity, and connectivity or network design. Accessibility estimation can capture the overall impact of transportation on land use.

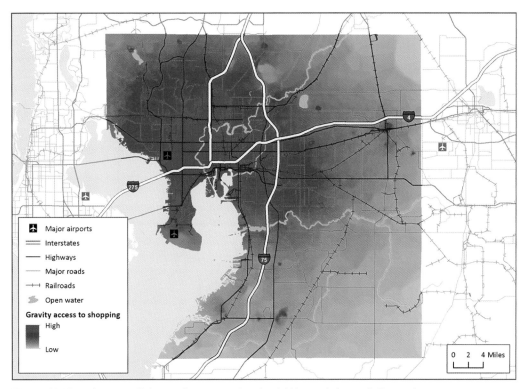

Figure 14.10. The accessibility to shopping using the A4 Network Accessibility tool.

Figure from Abdulnaser Arafat and the Arizona Board of Regents on behalf of the University of Arizona. Water bodies from the US Geological Survey; state boundary from US Census Bureau TIGER/Line Files; interstates and highways from the Florida Department of Transportation.

Although this chapter details how network distance can replace Euclidean distance in modeling proximity, LUCIS modeling still uses Euclidean distance when appropriate. The chapter explains some of the cases in which Euclidean distance is used. The chapter also applies the A4 Network Distance tool to capture the network proximity to transit stops by walking. The chapter also explains the A4 Network Opportunity tool and uses it to replace the Point Density tool. The A4 Accessibility tools are used to capture the accessibility among different land uses, considering the distance traveled, transit frequency, and availability of transfer stations. The chapter also shows an example of creating a gravity access suitability raster layer that can be used to prioritize the population allocation procedure.

Chapter summary

The following points are discussed in this chapter:

- Various accessibility estimation tools are used in the land-use modeling of Hillsborough County.
- The road accessibility estimation is based on the Hillsborough TIGER/Line street network, while the transit accessibility estimation considers the road network and transit routes, stops, and frequency.
- The LUCIS^{plus} model incorporates network proximity and network opportunity estimation into its suitability raster layers.
- The LUCIS^{plus} CEM uses the A4 Network Accessibility tool to help prioritize population allocation.
- The use of accessibility and the 5Ds—density, diversity, network design, destination, and distance—leads to better coordination between land use and transportation. This coordinated planning encourages multimodal transportation, reduces travel costs, and leads to a cleaner environment.

ArcGIS tools referenced in this chapter

Tool Name	Version 10.2 Toolbox/Toolset
A4 Network Accessibility	A4 LUCIS Tools/A4 Accessibility
A4 Network Distance	A4 LUCIS Tools/A4 Accessibility
A4 Network Opportunity	A4 LUCIS Tools/A4 Accessibility
Euclidean Distance	Spatial Analyst Tools/Distance
Focal Statistics	Spatial Analyst Tools/Neighborhood
Make Service Area Layer	Network Analyst Tools/Analysis
Make Closest Facility Layer	Network Analyst Tools/Analysis
Make OD Cost Matrix Layer	Network Analyst Tools/Analysis
Point Density	Spatial Analyst Tools/Density
Point Statistics	Spatial Analyst Tools/Neighborhood

References

Carr, Margaret H., and Paul D. Zwick. 2007. *Smart Land-Use Analysis: The LUCIS Model*. Redlands, CA: Esri Press.

Ewing, Reid, and Robert Cervero. 2001. "Travel and the Built Environment: A Synthesis." *Transportation Research Record: Journal of the Transportation Research Board* 1780:87–114.

Giuliano, G. 2004. "Land-Use Impacts of Transportation Investments: Highway and Transit." In *The Geography of Urban Transportation*, edited by Susan Hanson and George Giuliano, 3rd ed., 237–73. New York: Guilford Press.

Hanson, Susan. "The Context of Urban Travel: Concepts and Recent Trends." In *The Geography of Urban Transportation*, edited by Susan Hanson and George Giuliano, 3rd ed., 3–29. New York: Guilford Press.

Index

4Ds, 348
5Ds, 348
100-year base storm analysis, 241–47; assessing land-use implications, 251–56; property parcel-based analysis, 248–50

A

A4 Accessibility toolset, 321–29; Manhattan Distance tool, 321–22; Network Accessibility tool, 325–30; Network Distance tool, 322–23, 330; Network Opportunity tool, 324–25, 327–28, 330; tool limitations, 330–31; validation and estimation accuracy, 330
A4 Allocation by Table tool, 62–65, 69, 311
A4 Allocation tools, 60–63, 68–69
A4 Community Values Calculator, 296
A4 Detailed Allocation tool, 62
A4 LUCIS Community Values Calculator, 24, 54–57
A4 LUCIS Suitability tool, 49–53, 339, 342, 345
A4 LUCIS Weights tool, 56–60, 296
A4 Manhattan Distance tool, 321–22
A4 Network Accessibility tool, 325–30; estimating gravity accessibility at parcel level, 345–49
A4 Network Distance tool, 322–23, 330, 338–39; estimating access by frequency/trip length, 342–44; estimating accessibility by transit system, 340–41; estimating proximity of origins to transit stops, 341–43
A4 Network Opportunity tool, 301, 324, 325, 327–28, 330; replacing point density using network opportunity, 344–45, 347

A4 Overlay tools, 53–60; Community Values Calculator, 547; Weights tool, 57–60
A4 Standard Trend Allocation tool, 61–62
A4 Suitability tools, 48–53
access-driving-demand-transit (ADDT) opportunity layer, housing suitability model (HSM), 304–5
accessibility, 315–16. *See also* A4 Accessibility toolset; transportation accessibility and land use
  5Ds, 348; distance estimation, 316–17; vs. mobility, 316; measuring accessibility, 317–21
ACOND, 64
acquisition of land for conservation or agriculture preservation, 183–84
Acres field, raster layer VAT, 40
Add Field tool (ArcGIS), 36, 40–41
ADDT opportunity layer, housing suitability model (HSM), 304–5
adjusted growth rate (AGR), 121
Advanced Land-Use Analysis for Regional Geodesign: Using LUCIS^{plus}, 2, 7–9
affordable-housing alternatives, 293–96; access-driving-demand-transit (ADDT) opportunity layer, 304–5; CEM scoring procedures, 310–11; creating CEM raster layer, 308–9; demand preference layer, 299–301; identifying affordable-housing locations, 305–6; physical and neighborhood characteristics preference layer, 296–97; transit access preference layer, 301–4; travel cost preference layer, 297–99; using CEM to allocate affordable housing, 306–7
age cohort population, 110–11

aggregated urban preference layer, 29
agriculture preservation and protection, 181–82; acquisition and land-use options, 183–84; agriculture preservation and land-use planning, 193–96; final agriculture and conservation conclusions, 203–6; green infrastructure, 184–87; high-hazard flood areas, 201–4; identifying conservation opportunities, 189–92; identifying lands for preservation or protection, 198–201; LUCIS agriculture goals, 196–98; LUCIS ecological significance goals, 187–88
Aldrich Nation, 12. *See also* tribal lands
allocating affordable housing, 306–7; CEM scoring procedures, 310–11; creating CEM raster layer, 308–9
Allocation by Table, 62–65, 69, 311
allocation scenario 1, land-use and population allocation, 36–43
allocation scenario 2, TOD, mixed use, and redevelopment, 63–69
allocation scenario 3, increase in residential density for infill and greenfield development, 153–57
allocation scenario 4, increased density and proximity restriction, 158–64
allocation scenario 5, increased density using mixed-use development, 164–78; employment density adjustment, 169–71; redevelopment employment mixed-use allocations, 172–77
Allocation tools. *See* A4 Allocation tools
ALLYRIND variable, industrial employment allocation queries, 130, 132
ALLYRPOP variable: alternative land-use allocation queries, 156; industrial employment allocation queries, 130, 141; redevelopment residential population queries, 96–97, 105
alternative urban mixed-use opportunity, 151–52; allocation scenario 3, increase in residential density for infill and greenfield development, 153–57; allocation scenario 4, increased density and proximity restriction, 158–64; allocation scenario 5, increased density using mixed-use development, 164–78; increased density as an alternative land-use allocation concept, 152–53; reducing trend sprawl, 152
analytic hierarchy process (AHP), 2, 24, 47; scale of importance, 55
analyzing. *See* affordable-housing alternatives; alternative urban mixed-use opportunity; conservation and agriculture preservation and protection; employment land-use futures; natural disasters; residential land-use futures; transportation accessibility and land use
ArcGIS: Add Field, 36, 40–41; Allocation by Table, 63–64, 69; Combine, 36–43; data classification methods, 26; Detailed Allocation, 69; Euclidean Distance, 49, 317; Extract by Attributes, 215–16; Extract by Mask, 212–13; Focal Statistics, 327, 348; Geostatistical Analyst extension, 26; Line Density, 348; ModelBuilder, 21–22, 337; Network Analyst, 301, 317–18; Point Density, 327, 348; Point Statistics, 327; Polygon to Raster, 308; Raster Calculator, 344; Reclassify, 49; Select By Attributes, 33–34; Spatial Analyst, 48–49; Spatial Analyst toolbox, 57; Standard Trend Allocation, 69; Summary Statistics, 216–19, 221, 224; Weighted Overlay, 81; Weighted Sum, 57; Zonal Statistics, 308; Zonal Statistics as Table, 49, 53, 81, 254, 339, 342, 345–46
ArcGIS for Server, 260–64
ArcGIS public accounts, 260
ArcGIS queries. *See* queries
ArcGIS Viewer for Flex, 259–60; creating applications, 264–74; Design page, 273; Layout page, 269–72; Maps page,

265–67; Preview page, 274; using viewer applications, 275–80; Widgets page, 267–69
ArcMap. *See also* A4 LUCIS Community Values Calculator; creating ArcGIS for Server services, 260–64
assisted-housing inventory (AHI), 300
automation tools, 48; A4 Allocation, 60–69; suitability assignment (A4 Suitability tools), 48–53; suitability overlay (A4 Overlay tools), 53–60

B

basemaps, Flex Viewer, 278–79
basic exponential projection, 75–76
basic linear projection, 74–75
bikability, 309. *See also* walkability; estimating accessibility by transit system, 340–41
Boolean operators, 46
Bureau of Economic and Business Research (BEBR), 73; population projections, 84
Bureau of Labor Statistics website, 114

C

Carr, Margaret H., 1, 18, 46, 62
carrying capacity, 76–77
CEMs (criteria evaluation matrices), 7. *See also* alternative urban mixed-use opportunity; conservation and agriculture preservation and protection; affordable housing allocation, 306–11; allocating population, 68; Conflict values, 41; creating raster layers with Combine tool, 36–43; industrial land-use allocation, 127–28; linking to census blocks, 110; Plandev values, 41; raster layer output, 65, 69; raster layer VAT, 39; residential allocation. *See* residential allocation using CEM; summarizing land-use results. *See also* trend allocation, 61–62; Urbsuit values, 41
CENBLKID, 110
census blocks, linking to CEMs, 110

Center of Neighborhood Technology (CNT), 297, 299, 302
Cervero, Robert, 348
city boundaries, 222
City field, raster layer VAT, 39
classes, 26
CLIP (Critical Lands and Waters Identification Project), 187
CNT (Center of Neighborhood Technology), 297, 299, 302
CNTERSLCE19, 158, 172
coastal flooding. *See* storm surge
COG LUCIS visioning exercise, 38
collaborative GIS, 2
collapsed preference, 25–31
Combine tool (ArcGIS), 36–43
commercial: land-use suitability, 136–47; mixed-use allocations, 172–77
Community Par, 57
community values. *See also* A4 LUCIS Community Values Calculator; equal-interval reclassification, 25, 27, 29; geometric interval reclassification, 25; reclassification of, 25–31; standard-deviation reclassification, 25, 27, 29; integrating, 21–24
complex MUAs, 19
CONF, 64
conflict, 19, 31. *See also* summarizing land-use results; conservation and agriculture preservation and protection, 188–201; defined, 19; greenfield preference/conflict matrix, 87–88; major, 30–31; minor, 31; moderate, 27, 30; residential suitability, 87–88; tribal lands, 291–92
Conflict field, raster layer VAT, 39, 41
conservation and agriculture preservation and protection, 181–82; acquisition and land-use options, 183–84; agriculture preservation, 193–96; final agriculture and conservation conclusions, 203–6; green infrastructure, 184–87; high-hazard flood areas, 201–4;

identifying conservation opportunities, 189–92; identifying lands for preservation or protection, 198–201; LUCIS agriculture goals, 196–98; LUCIS ecological significance goals, 187–88
conservation easements, 183
contribution, 46
contributory rule, 46
Count field, raster layer VAT, 39
County field, raster layer VAT, 39
creating applications in Flex Viewer, 264–74
criteria evaluation matrices. *See* CEMs
Critical Lands and Waters Identification Project (CLIP), 187

D

data classification methods in ArcGIS, 26
data extraction/summarization, 211–16; Extract by Mask tool, 212–13
data gathering for tribal lands, 284–86
data inventory and preparation (LUCIS step 2), 19
data sharing. *See* sharing maps and data
data summaries, 109–11
defined interval method, 26
defining and mapping land-use suitability (LUCIS step 3), 19–21
defining goals and objectives (LUCIS step 1), 18
demand preference layer, housing suitability model (HSM), 299–301
density increases: allocation scenario 3, infill and greenfield development, 153–57; allocation scenario 4, increased density and proximity restriction, 158–64; allocation scenario 5, increased density using mixed-use development, 164–78
Design page, Flex Viewer, 273
Detailed Allocation tool, 62, 69
development rights, 183
developments of regional impact (DRIs), 211–14, 219
digital elevation model (DEM) shorelines, 236, 238–39
distance: accessibility measurement, 317–21; estimating access by frequency/trip length, 342–44; estimating accessibility by transit system, 340–41; estimating proximity of origins to transit stops, 341–43; estimation methods, 316–17; Euclidean Distance tool, 49, 317; replacing Euclidean distance with network distance, 336–40
dominance, 46
dominance rule, 46
driving cost, 295, 304–7, 310
Dunn Jr., Edgar S., 118

E

ecological significance models, 187–88
ecological significance suitability index, 188
economic base analysis, 115–18
economic base location quotients, 118–19
economic base theory, 114
Ellis University, 12. *See also* tribal lands
employment: density adjustment, 169–71; mixed-use redevelopment, 168, 172–77
employment base factor (EBF), 169
employment land-use allocation, 123–24; commercial, retail, service, and institutional land-use suitability, 136–47; employment density, 125–27; industrial employment allocation, 127–36; NAICS categories and land-use classifications, 125
employment land-use futures, 113–14; calculating LUCIS new employment change (LNEC), 123; calculating population change proportion (PCP), 123; economic base analysis, 115–18; economic base theory, 114; employment statistics, 114; location quotient (LQ), 115, 117–19; modeling local employment

changes using shift-share, 118–19; North American Industry Classification System (NAICS) codes, 114; Quarterly Census of Employment and Wages website, 115; shift-share projections, 122; shift-share projections by employment sector, 119–21, 123; US Department of Labor Bureau of Labor Statistics website, 114
enumeration, 46
EPA (Environmental Protection Agency), 184
equal-interval reclassification, 25, 27, 29
Euclidean distance, 349; replacing with network distance, 336–40; measurement, 317–21, 325–27
Euclidean Distance tool (ArcGIS), 49, 317
Ewing, Reid, 348
Excel: basic exponential projections, 75–76; basic linear projections, 74–75; modified exponential projections, 76–77
exponential projection, 75–77
Extract by Attributes (ArcGIS), 215–16
Extract by Mask (ArcGIS), 212–13
eye of the hurricane, 230
eye wall, 230

F

farming. *See* agriculture preservation and protection
Federal Emergency Management Administration (FEMA), 229
fee simple acquisition, 183
five-step LUCIS process, 18–25
Flex Viewer (ArcGIS Viewer for Flex), 259–60; creating applications, 264–74; Design page, 273; Layout page, 269–72; Maps page, 265–67; Preview page, 274; using viewer applications, 275–80; Widgets page, 267–69
Flood Insurance Rate Maps (FIRM), 240
flooding. *See also* storm surge; flood areas, 201–4

Florida Bureau of Economic and Business Research (BEBR), 73; population projections, 84
Florida green infrastructure program, 184–85
Focal Statistics (ArcGIS), 327, 348

G

geodesign, 7–8, 18; tribal lands conflict, 291–92
geographically weighted regression (GWR), 298
geometric interval reclassification, 25–27, 29
Geostatistical Analyst extension (ArcGIS), 26
GFCONFLICT, 190, 200–201
GIS layer overlay, 46–48
goals, 18
gravity accessibility, 324–25, 327, 329, 345–46, 348–49
green infrastructure, 184–87
green infrastructure maps, 185
greenfield, 3, 60. *See also* alternative urban mixed-use opportunity; CEMs; balancing with infill and redevelopment, 6–7; increase in residential density for infill and greenfield development, 153–57
greenfield CEM for residential population allocation, 91, 106–9
greenfield conflict matrix, 30–36, 87–88
greenways, 185
greyfields, 3

H

Hazus-MH (Multi-Hazard Loss Estimation Methodology), 229; 100-year base storm analysis, 241–47; 100-year base storm analysis, property parcel-based, 248–50; coastal flood definitions, 236–40; Coastal Flood Model, 235–40; Flood Model, 229, 240; Wave Height Model, 238
Heartland region, Florida, 11–12
high-hazard flood areas, 201–4
Hillsborough County, Florida, 9–10; basic residential analysis, 80–84; developments of regional impact

(DRIs), 211; digital elevation model (DEM), 239; greenfield conflict raster layer, 32; population projections, 84; residential suitability, 84–88; Select By Attributes query results, 34; Tampa Hillsborough Expressway Authority project, 175–78; traffic analysis zones (TAZs), 220; transit-oriented development (TOD), 175, 178

histograms: aggregated urban preference layer, 29; class specifications, 26

Housing and Urban Development (HUD); affordable-housing study, 299–300; Travel Cost Calculator, 297

housing suitability model (HSM), 293–96; access-driving-demand-transit (ADDT) opportunity layer, 304–5; CEM scoring procedures, 310–11; creating CEM raster layer, 308–9; demand preference layer, 299–301; identifying affordable-housing locations, 305–6; physical and neighborhood characteristics preference layer, 296–97; transit access preference layer, 301–4; travel cost preference layer, 297–99; using CEM to allocate affordable housing, 306–7

HTML: Flex Viewer applications, 270–72

Hurricane Sandy, 234–35

hurricanes, 229–33; 100-year base storm analysis, 241–50; assessing land-use implications associated with 100-year base storm surge, 251–56; Hazus-MH coastal flood definitions, 236–40; Hazus-MH Coastal Flood Model, 235–37, 240; Hazus-MH Flood Model, 229, 240; Hazus-MH Wave Height Model, 238; Saffir-Simpson Hurricane Wind Scale, 231, 235; storm surge, 232–35

I

identifying potential land-use conflict (LUCIS step 5), 25

increased density. *See* density increases

industrial employment allocation for Hillsborough County, 127–36

infill, 3, 60. *See also* alternative urban mixed-use opportunity; CEMs; balancing with redevelopment and greenfield, 6–7; increase in residential density for infill and greenfield development, 153–57

infill CEM for residential population allocation, 91, 100–105

institutional land-use suitability, 136–47

integrating community values to determine land-use preference (LUCIS step 4), 21–24

interaction, 46

interaction rule, 46

inverse distance weighted (IDW), 338

J

Jankowski, Piotr, 47

Jenks, 26

L

land acquisition for conservation or agriculture preservation, 183–84

land-use complex multivariable raster layer, 36–43

land-use conflict. *See* LUCIS land-use conflict methodology

land-use prediction vs. land-use visioning, 23

layer overlay, 46–48

Layout page, Flex Viewer, 269–72

lease, 183

Line Density (ArcGIS), 348

linear projection, 74–75

linking CEMs to census blocks, 110

livability, 307, 309

local constant term (LCT) calculation, 120

local growth rate (LGR) calculation, 121

location quotient (LQ), 115, 117–19

Longitudinal Employer-Household Dynamics (LEHD), 300

LUCIS (Land-Use Conflict Identification Strategy), 1; agriculture goals, 196–98; assessing land-use implications associated with 100-year base storm surge, 251–56; ecological significance models, 187–88; greenfield conflict matrix, 30–36, 87–88; identifying conservation opportunities, 189–92; strategy, 6; tribal lands model development, 286–87, 288–90; value of, 47
LUCIS five-step process, 18–25
LUCIS land-use conflict methodology, 25–31
LUCIS Model, The, 1–2, 18
LUCIS new employment change (LNEC), 120, 122–23
LUCIS Suitability tool, 49–53, 342, 345
LUCIS urban mixed-use opportunity matrix. *See* mixed-use opportunity matrix
LUCIS web application, 259–60; ArcGIS for Server, 260–64; ArcGIS Viewer for Flex, 259–60, 267, 276–80; creating Flex Viewer applications, 264–74; using Flex Viewer applications, 275–80
LUCIS Weights tool, 56–60
LUCIS^{plus}, 2, 7–8

M

major conflict, 30–31
Malczewski, John, 47
Manhattan (rectilinear) distance measurement, 319–23, 325, 327
Manhattan Distance tool, 321–22
manual class breaks, 26
mapping. *See also* affordable-housing alternatives; alternative urban mixed-use opportunity; conservation and agriculture preservation and protection; employment land-use futures; land-use suitability (LUCIS step 3), 19; natural disasters; residential land-use futures; sharing maps and data; transportation accessibility and land usage cohort population, 110–11
Maryland green infrastructure program, 185–87
mean normalized adjusted employment factor (MNAEF), 169–70
median center of activity/employment, 158–59, 161
Microsoft Excel: basic exponential projections, 75–76; basic linear projections, 74–75; modified exponential projections, 76–77
minor conflict, 31
mixed-use development, 88; residential allocation, 91–92
mixed-use opportunity, 32–36, 89–90. *See also* alternative urban mixed-use opportunity; urban mixed-use opportunity matrix, 34–36
mobility, 316
ModelBuilder, 21–22; Allocation by Table tool, 63; distance to retail suitability, 337; weights table, 60
moderate conflict, 27, 30
modified exponential projection, 76–77
Mrs. Smith, 2
multiattribute method, 46–47
multicriteria decision-making (MCDM) method, 46–48
multifamily land-use suitability, 88–90
multifamily mixed-use allocations, 172–77
multifamily residential suitability. *See* residential suitability; residential allocation using CEM
Multi-Hazard Loss Estimation Methodology. *See* Hazus-MH
multiobjective method, 46–47
multiple-utility assignments (MUAs), 19, 46; housing suitability model (HSM), 296
multivariable raster layers, creating with Combine tool, 36–43
multivariate regression, 79

N

NAICS codes. *See* North American Industry Classification System (NAICS) employment codes
national growth rate (NGR) calculation, 120–21
Native American lands. *See* tribal lands
natural breaks (Jenks), 26
natural disasters, 229; 100-year base storm analysis, 241–47; 100-year base storm analysis, property parcel-based, 248–50; assessing land-use implications, 251–56; Hazus-MH coastal flood definitions, 236–40; Hazus-MH Coastal Flood Model, 235–40; Hazus-MH Flood Model, 229, 240; Hazus-MH Wave Height Model, 238; hurricanes, 230–33; Saffir-Simpson Hurricane Wind Scale, 235; sea level rise, 241; storm surge, 232–35
neighborhood characteristics preference layer, housing suitability model (HSM), 296–97
neighborhood choice factors, 3–4
Nelson, Arthur C., 3
Network Accessibility tool, 325–30; estimating gravity accessibility at parcel level, 345–49
Network Analyst (ArcGIS), 301, 317–18; walking service area, 341
network distance, 319–21, 325; replacing Euclidean distance, 336–40
Network Distance tool, 322–23, 330, 338–39; estimating access by frequency/trip length, 342–44; estimating accessibility by transit system, 340–41; estimating proximity of origins to transit stops, 341–43
Network Opportunity tool, 301, 324–25, 327–28, 330; replacing point density using network opportunity, 344–45, 347
NEWPOP variable, 97
Norquist, John, 6
North American Industry Classification System (NAICS) employment codes, 114–20; linking to land-use classifications, 125
Nyerges, Timothy L., 47

O

objectives, 18, 20–21, 23; defining, 18
OID field, raster layer VAT, 38
opportunity: A4 Network Opportunity tool, 324–30; housing suitability model (HSM), 293; tribal lands, 285
ordered weighted averaging (OWA), 47
ordinary least squares (OLS) model, 298
origin-destination (OD), 317, 345; estimating Euclidean OD distance, 317–21
output tables: A4 Community Values Calculator, 57; A4 LUCIS Suitability tool, 49, 53; Zonal Statistics as Table tool, 53
overlay, 46–48
overlay tools, 53–60; Community Values Calculator, 54–57; Weights tool, 57–60

P

pairwise comparison, 47; analytic hierarchy process (AHP), 24
passenger miles traveled (PMT), 298
People field, raster layer VAT, 40
physical suitability raster layers, 289–90
place-making, 5–6
Plandev field, raster layer VAT, 40–41
planned unit development (PUD), 40
"plus" in LUCIS^{plus}, 2
Point Density (ArcGIS), 327, 348
Point Statistics (ArcGIS), 327
Polygon to Raster (ArcGIS), 308
population allocation: allocating population into CEM, 68; greenfield CEM, 91, 106–9; infill CEM, 91, 100–105; redevelopment CEM, 91–100; summarizing data, 109–11
population allocation tools, 60–63, 69

population change proportion (PCP), 120, 122; calculating, 123
population density increase scenarios. *See* density increases
population growth, 2
population projection, 73–79, 84; basic exponential, 75–76; basic linear, 74–75; Hillsborough County, 84; modified exponential, 76–77; multivariate regression, 79; proportional, 78–79
preference, 19
Preview page, Flex Viewer, 274
projecting population. *See* population projection
proportional population projection, 78–79
proximity restrictions, 158–64
publishing maps and data. *See* sharing maps and data
purchase with lease-back, 183

Q
quantiles, 26
Quarterly Census of Employment and Wages, 115
Quarterly Workforce Indicators (QWI), 301
queries: affordable housing allocation, 311; allocating agriculture and conservation lands, 181–200; employment land-use allocation, 128–34, 137–46; greenfield residential allocation, 106, 109; increased residential density allocation, 155–56, 164–67; industrial employment allocation, 170–71; infill residential development, 104–6; land-use selections, 210, 213, 220; redevelopment employment mixed-use allocations, 172; redevelopment residential population, 95–100; service and institutional employment allocation, 173–74; TOD, mixed use, and redevelopment, 64

R
Randolph, J., 183
Raster Calculator (ArcGIS), 344
raster data extraction/summarization, 211–16; Summary Statistics tool, 216–21, 224
raster layer VAT. *See* value attribute table (VAT)
raster layers: creating with Combine tool, 36–43; mixed-use opportunity, 88–90; output, 65, 69; single-family residential, 85–88
recession of 2008, 2–3
reclassification histograms, 29
reclassification of community values, 25, 27, 29; equal interval, 25, 27, 29; geometric interval, 25; standard deviation, 25, 27, 29
Reclassify tool (ArcGIS), 49
rectilinear distance (Manhattan) measurement, 319–23
REDEVCONFLICT variable, 97, 165, 168, 172; industrial employment allocation queries, 141, 143–46
redevelopment, 3, 60. *See also* alternative urban mixed-use opportunity; CEMs; balancing with infill and greenfield, 6–7; employment mixed-use allocations, 172–77; mixed-use redevelopment employment, 168; summarizing redevelopment data, 219–23; trend redevelopment calculation, 92
redevelopment CEM for residential population allocation, 91–100
Regional Economic Model Inc. (REMI), 113
Res_year field, raster layer VAT, 40
RESFSUIT100I variable, 156
residential allocation using CEM, 88–91; greenfield CEM, 91, 106–9; infill CEM, 91, 100–105; mixed-use allocation, 91–92; redevelopment CEM, 91–100; summarizing data, 109–11; trend redevelopment calculation, 92

residential density increase scenarios. *See* density increases
residential land-use futures; basic residential analysis, 80–84; population projection, 73–79; residential suitability, 84–88
residential preference, 86
retail: distance to retail suitability, 337; mixed-use allocations, 172–77; retail land-use suitability, 136–47
rectilinear (Manhattan) distance measurement, 319–23, 325, 327

S

Saaty, T. L., 47
Saffir-Simpson Hurricane Wind Scale, 231, 235
scenario and policy tables (SPTs), 48, 62–64, 66; affordable housing allocation, 311
scenario planning, 2
scenarios: allocation scenario 1, land-use and population allocation, 36–43; allocation scenario 2, TOD, mixed use, and redevelopment, 63–69; allocation scenario 3, increase in residential density for infill and greenfield development, 153–57; allocation scenario 4, increased density and proximity restriction, 158–64; allocation scenario 5, increased density using mixed-use development, 164–78; transit-oriented, 68
schema: infill CEM, 101–3; redevelopment CEM, 93–95
scoring housing affordability using CEM, 310–11
sea level rise, 229, 241; 100-year base storm analysis, 241–47; 100-year base storm analysis, property parcel-based, 248–50; assessing land-use implications associated, 251–56
Select By Attributes (ArcGIS), 33–34
selection queries, 210, 213, 220
service land-use suitability, 136–47

Sfresden field, raster layer VAT, 40
sharing maps and data, 259–60; ArcGIS for Server, 260–64; ArcGIS Viewer for Flex, 259–60, 267, 276–80; creating Flex Viewer applications, 264–74; using Flex Viewer applications, 275; WMS (web map service), 261–64
shift term calculation, employment growth, 122
shift-share technique: modeling local employment changes, 118–19; shift-share projections, 119, 121–23; using shift-share projections, 120
single-family residential suitability, 84–88. *See also* residential allocation using CEM; residential suitability
single-utility assignments (SUAs), 19, 46; housing suitability model (HSM), 296
smart growth, 3, 5
Smart Land-Use Analysis: The LUCIS Model, 1–2, 18, 325
Smart Quantiles, 26
Spatial Analyst tool (ArcGIS), 48–49
Spreadsheet Models for Urban and Regional Analysis, 118
spreadsheets: basic exponential projections, 75–76; basic linear projections, 74–75; modified exponential projections, 76–77
SQL queries. *See* queries
standard deviation reclassification, 25–27, 29
Standard Trend Allocation tool, 61–62, 69
statement of intent, 18
statistical software: multivariate regression, 79
stillwater elevations (SWEL), 237–38
storm surge, 229–30, 232–35; 100-year base storm analysis, 241–47; 100-year base storm analysis, property parcel-based, 248–50; assessing land-use implications, 251–56; Hazus-MH coastal flood definitions, 236–40; Hazus-MH Coastal Flood Model, 235–37, 240; Hazus-MH Flood Model, 229, 240; Hazus-MH Wave

Height Model, 238; Saffir-Simpson Hurricane Wind Scale, 235
subobjectives, 18, 20–21, 23
suitability, 19. *See also* summarizing land-use results; residential, 84–88
suitability analysis; tribal lands model, 288–90; workflow, 54
suitability assignment tools, 48–53
suitability objectives, 18
suitability overlay tools, 53–60; Community Values Calculator, 54–57; Weights tool, 57–60
summarizing data, 109–11
summarizing land-use results, 209; city boundaries, 222; developments of regional impact (DRIs), 211–14; Extract by Attributes tool, 215–16; extracting/summarizing raster data, 211–16; query criteria, 210; Summary Statistics tool, 216–19, 221, 224; traffic analysis zones (TAZs), 218–23
Summary Statistics (ArcGIS), 216–19, 221, 224
sustainability, 3

T

tables; A4 Allocation by Table tool, 62–65, 69, 311; output tables, 49, 53, 57; updating, 57
Tampa Hillsborough Expressway Authority project, 175, 178
traffic analysis zones (TAZs), 218–23
transaction costs, 6
transects, Hazus-MH Coastal Flood Model, 240
transit access preference layer, housing suitability model (HSM), 301–4
transit accessibility. *See* accessibility
transit-oriented development (TOD), 5–6; allocation scenario 5, increased density using mixed-use development, 175–78
transit-oriented scenarios, 68
transportation accessibility and land use, 335–36; estimating access by frequency/trip length, 342–44; estimating accessibility by transit system, 340–41; estimating proximity of origins to transit stops, 341–43; impacts of accessibility on planning, 347–49; replacing Euclidean distance with network distance, 336–40; replacing point density, 344–45, 347; using network gravity access as a suitability layer, 345–46, 348–49
travel cost preference layer, housing suitability model (HSM), 297–99
trend allocation, 61–62
trend infill residential development, 105
trend redevelopment calculation, 92
trend scenarios, 152
trend sprawl, 152
tribal lands, 283–92; conflict, 291–92; data gathering, 284–86; focus on development, 285; long-range plan (LRP) planning process, 284; LUCIS model development, 286–87; LUCIS model refinement, 288–90; suitability analysis, 288–90

U

undivided interest, 183
unique numeric identifier, 88
updating tables, 57
urban growth boundary (UGB), 158
urban mixed-use opportunity matrix, 34–36; Combine tool, 36–43
urban services boundary (USB), 158
urban sprawl. *See* alternative urban mixed-use opportunity
Urbsuit field, raster layer VAT, 39, 41
US Census Bureau Longitudinal Employer-Household Dynamics (LEHD), 300
US Department of Housing and Urban Development (HUD) Travel Cost Calculator, 297

US Department of Labor Bureau of Labor Statistics, 114
US Environmental Protection Agency (EPA) green infrastructure definition, 184
utilities, 46; utility service data, tribal lands, 287–90

V

vacant lands, 91; greenfield CEM, 107–8; infill CEM for residential population allocation, 103–4; new allocation of vacant residential lands, 108–9
validation of A4 Accessibility tools, 330
value attribute table (VAT); affordable-housing allocation, 309; fields, 36, 38–40; LUCIS greenfield conflict raster layer, 27, 30; summarizing land-use results, 219, 224
Value field, raster layer VAT, 38
variables, queries. *See* queries
vehicle miles traveled (VMT), 298, 335
visioning, 2

W

walkability, 307, 309; estimating access by frequency/trip length, 342–44; estimating accessibility by transit system, 340–41; estimating proximity of origins to transit stops, 341–43; proximity of origins to transit stops, 342–43; walking service area, 341
Wave Height Model (Hazus-MH), 238
web application, 259–60; ArcGIS for Server, 260–64; creating Flex Viewer applications, 264–74; using Flex Viewer applications, 275–80
web map service (WMS), 261–64
Web Maps page, Flex Viewer, 265–67
weighted linear combination (WLC), 46–47
Weighted Overlay tool (ArcGIS), 81
Weighted Sum tool (ArcGIS), 57
weights, 47; A4 Community Values Calculator, 55–56; A4 LUCIS Weights tool, 56–60; scoring housing affordability using CEM, 310
widgets, Flex Viewer, 267–69

Z

Zonal Statistics (ArcGIS), 308
Zonal Statistics as Table (ArcGIS), 49, 53, 81, 254, 339, 342, 345, 346
Zwick, Paul D., 1, 18, 46, 62